Forum for Interdisciplinary Mathematics

The *Forum for Interdisciplinary Mathematics* is a Scopus-indexed book series. It publishes high-quality textbooks, monographs, contributed volumes and lecture notes in mathematics and interdisciplinary areas where mathematics plays a fundamental role, such as statistics, operations research, computer science, financial mathematics, industrial mathematics, and bio-mathematics. It reflects the increasing demand of researchers working at the interface between mathematics and other scientific disciplines.

Tanmoy Som · Oscar Castillo ·
Anoop Kumar Tiwari · Shivam Shreevastava
Editors

Fuzzy, Rough and Intuitionistic Fuzzy Set Approaches for Data Handling

Theory and Applications

Editors
Tanmoy Som
Department of Mathematical Sciences
Indian Institute of Technology BHU
Varanasi, India

Oscar Castillo
Computer Science at the Graduate Division
Tijuana Institute of Technology
Tijuana, Mexico

Anoop Kumar Tiwari
Department of Computer Science
and Information Technology
Central University of Haryana
Mahendergarh, Haryana, India

Shivam Shreevastava
Department of Mathematics, School
of Basic and Applied Sciences
Harcourt Butler Technical University
Kanpur, Uttar Pradesh, India

ISSN 2364-6748 ISSN 2364-6756 (electronic)
Forum for Interdisciplinary Mathematics
ISBN 978-981-19-8568-3 ISBN 978-981-19-8566-9 (eBook)
https://doi.org/10.1007/978-981-19-8566-9

Mathematics Subject Classification: 03B52, 03E72, 03B20, 26A33, 35R09, 94D05, 90C70, 11D04, 91D30, 90C26, 68T07

This Springer imprint is published by the registered company Springer Nature Singapore Pte Ltd.
The registered company address is: 152 Beach Road, #21-01/04 Gateway East, Singapore 189721, Singapore

Dedicated to my parents
and
beloved spouse Sarbani and daughters
Tannishtha & Tejaswita

Preface

The main objective of this book is to facilitate both the theoretical background and applications of fuzzy, intuitionistic fuzzy and rough sets in the area of data science. This book provides various individual and hybridization techniques of fuzzy and intuitionistic fuzzy sets with rough sets and their applications for data handling along with some theoretical base. Some concepts of fuzzy sets, intuitionistic fuzzy sets and rough sets and different algebraic and logical operations performed over these sets are discussed for better understanding of their applications. This book focuses on various learning techniques for data handling with emphasizing the problems of uncertainty, vagueness, imprecision, inconsistency and ambiguity available in data sets.

Machine learning techniques are effectively implemented to solve a diversity of problems in pattern recognition, data mining, bioinformatics, etc. Due to the advancement of high-throughput assay systems in modern laboratories and the development of modern Internet-based technology, large volume of data sets is created every day. Data size is continuously enlarging in the form of data instances (tuples) as well as the data attributes (features). Imbalanced and high-dimensional data sets affect the learning of classifiers. This may reduce the average performance of most of the machine learning algorithms, especially in the presence of redundant and/or irrelevant features. So, the overall accuracy decreases and imbalanced data sets result in classifier bias towards the majority class and is the cause for lower sensitivity. There are basically two approaches to handle such problems: firstly, internal approaches, which are included within existing algorithms for handling class distribution and secondly, external approaches, which involve resampling and feature selection approaches. External approaches have an advantage over internal approaches as they are independent of machine learning algorithms and can be easily used with any machine learning algorithm.

This book investigates both imbalanced and high-dimensional data sets. In the present book, dimensionality reduction, rule extraction, decision-making, classification, etc., techniques and time series forecasting approaches are developed, implemented and discussed for handling both rows (instances) and columns (features) of an information system (data set) and future prediction of patterns. Various studies are carried out in different aspects to take research to the next higher level by focusing more on improving the performances of various classifiers.

A time series is a sequence of statistical observations of the recorded data over a period at uniform intervals yearly, monthly, weekly or daily. Time series analysis is an important tool for forecasting the future values on the basis of past observations. In the last few decades, several researchers have developed many methodologies, and tools have emerged to deal with the forecasting processes. A forecast is an estimate for observations of uncertain future events. Time series forecasting has been emerging as one of the useful tools to predict the future behaviour in many practical fields such as financial trading, economics, marketing, tourism demand and many other branches of science and engineering. Several chapters provide a very carefully crafted introduction to the basic and logical concepts and techniques of fuzzy set theory, intuitionistic fuzzy set theory, rough set theory and their hybridization for data handling, time series forecasting with modelling and optimization aspects. It also provides the basic back ground of intuitionistic fuzzy sets (IFS) and intuitionistic fuzzy numbers (IFN). Further, rough set theory and its applications for data handling are also discussed. Rough set theory is a notable tool for data science in different domains because of its character of analysis friendly. From the literature, it can be observed that rough set theory is very useful for both practitioners and researchers as it does not require any external information. Moreover, combination of fuzzy and intuitionistic fuzzy sets with rough sets is proposed, discussed and implemented, and a brief discussion of deep learning concept is presented, which can be extended with the hybridization of rough and fuzzy sets to handle uncertainty in a much better way.

Finally, this book concludes with supplementary studies to the related fields of data science, including rule induction, decision-making, pattern recognition, dimensionality reduction, time series forecasting, etc., by using fuzzy, intuitionistic fuzzy, rough sets theories and their hybridizations along with modelling and optimization aspects also. Entire book offers various innovative frontiers for the continuous advancement of the core technologies presented in the area of computational intelligence. This book is predominantly envisioned for senior undergraduates, postgraduates, medical scientists, professional engineers and researchers handling data science, However, it advances a candid demonstration of the underlying notions that any fellow with a non-specialist scope would be capable to understand and implement.

Varanasi, India — Tanmoy Som
Tijuana, Mexico — Oscar Castillo
Mahendergarh, India — Anoop Kumar Tiwari
Kanpur, India — Shivam Shreevastava

Contents

About the Editors

Tanmoy Som is Professor and former Head at the Department of Mathematical Sciences, Indian Institute of Technology (Banaras Hindu University), Varanasi, India. He completed his Ph.D. from Institute of Technology, Banaras Hindu University, India, in 1986. Earlier, he had been the Head of the Department of Mathematics (twice) at Assam Central University, Silchar, India. His research interests are in functional analysis, optimization and mathematical modelling, fuzzy set theory, soft computing and image processing. He has successfully guided 18 Ph.D. students and published more than 100 papers and several book chapters in reputed international journals, proceedings and edited books. He has published several papers on metric fixed point theory, optimization modelling and recently on applied analysis, soft computing and fuzzy geometry.

He has completed a BRNS-BARC funded project titled "Fractional calculus approached solutions for two dimensional ground water contaminations in unsaturated media" during 2014–18 jointly with Prof. S. Das of the Indian Institute of Technology (IIT-BHU) Varanasi. Professor Som is Guest/Handling Editor of *International Journal of Fuzzy Computation and Modelling* since 2018, and an editorial board member and reviewer of several reputed journals including *IEEE Transactions*, American Mathematical Society. He has jointly edited along with D. Ghosh, D. Giri, Ram N. Mohapatra, K. Sakurai and E. Savas the proceedings, *Mathematics and Computing: ICMC 2018, Varanasi, India, January 9–11, Selected Contributions*. He has delivered several invited talks at national and international conferences, seminars, and refresher courses and has organized four international events as the main convener. He holds the Vice-President position of Calcutta Mathematical Society. He has made short academic visits to Texas A&M University, Kingsville (2012) and the University of California, Berkeley (2016).

Oscar Castillo is Professor of Computer Science at the Graduate Division, Tijuana Institute of Technology, Tijuana, Mexico. In addition, he is serving as Research Director of Computer Science and Head of the research group on Hybrid Fuzzy Intelligent Systems. Currently, he is President of Hispanic American Fuzzy Systems Association (HAFSA) and Past President of International Fuzzy Systems Association (IFSA). He holds the Doctor of Science degree (Doctor Habilitatus) in Computer Science from the Polish Academy of Sciences (with the dissertation "Soft Computing and Fractal Theory for Intelligent Manufacturing"). Professor Castillo is also Chair of the Mexican Chapter of the Computational Intelligence Society (IEEE). He also belongs to the Technical Committee on Fuzzy Systems of IEEE and to the Task Force on "Extensions to Type-1 Fuzzy Systems". He is Member of NAFIPS, IFSA and IEEE. He belongs to the Mexican Research System (SNI Level 3).

His research interests are in type-2 fuzzy logic, fuzzy control, neuro-fuzzy and genetic fuzzy hybrid approaches. He has published over 300 papers in several journal, authored 10 books, edited 50 books, more than 300 papers in conference proceedings and more than 300 chapters in edited books; in total more than 1000 publications according to Scopus (H index = 66) and more than 1200 publications according to Google Scholar (H index = 80). He has been Guest Editor of several successful special issues of the following journals: *Applied Soft Computing*, *Intelligent Systems*, *Information Sciences*, *Nonlinear Studies*, *Fuzzy Sets and Systems*, *JAMRIS*, and *Engineering Letters*. He is currently Associate Editor of *Information Sciences*, *Engineering Applications of Artificial Intelligence*, *Complex and Intelligent Systems*, *Granular Computing*, and *International Journal on Fuzzy Systems*. Finally, he has been elected IFSA Fellow in 2015 and MICAI Fellow member in 2017. He has been recognized as Highly Cited Researcher in 2017 and 2018 by Clarivate Analytics because of having multiple highly cited papers in the Web of Science.

Anoop Kumar Tiwari is Assistant Professor at the Department of Computer Science and Information Technology, Central University of Haryana, Mahendergarh, Haryana, India. He holds the Ph.D. degree in computer science from the Department of Computer Science, Institute of Science, Banaras Hindu University, India. His research interests are in computational intelligence, machine learning and bioinformatics. He has published more than 18 publications with h index of five and more than 131 citations according to Google Scholar. He has qualified reputed UGC-NET and GATE exams. He is also Reviewer of some reputed journals: *Journal of Intelligent and Fuzzy Systems*, *International Journal of Fuzzy System Applications*, and *Expert Systems with Applications*.

Shivam Shreevastava is Assistant Professor at the Department of Mathematics, School of Basic and Applied Sciences, Harcourt Butler Technical University, Kanpur, India. He has completed his B.Sc. from Banaras Hindu University, Varanasi; M.Sc. from the Indian Institute of Technology Kanpur; and Ph.D. from the Indian Institute of Technology (BHU) Varanasi. He has qualified several competitive examinations like JAM, CSIR/UGC JRF, GATE, NBHM MA/M.Sc. fellowship. His main research interest includes soft computing, attribute reduction via fuzzy rough set-based techniques. More than 11 research publications are there to his credit in reputed journals/edited book chapters of national and international repute. He has delivered several invited talks at national and international conferences/workshops/refresher courses. He is also Reviewer of some journals of international repute like *International Journal of Fuzzy Systems*, *Soft Computing*, *International Journal of Fuzzy System*, and *The International Journal of Fuzzy System Applications*.

Chapter 1
Fuzzy Sets and Rough Sets: A Mathematical Narrative

Mihir Kumar Chakraborty and Pulak Samanta

1.1 Introduction

Fuzzy set theory was introduced by L. A. Zadeh in the year 1965 [53] and rough set theory was introduced by Z. Pawlak in 1982 [38]. Inventors of both the theories were, formally speaking, computer scientists. Their motivations originated from the problems/issues belonging to the domain of computer science. This will be apparent from the following quotations:

> Whether the particular concept defined in this paper *will prove to be of value in system design or analysis remains to be seen.* It is clear, though, that in one form or another, the notion of fuzziness will come to play an important role in pattern classification, control, system optimization and other fields since fuzziness is a basic and pervasive part of life that cannot be avoided. ... (Zadeh, 1965, First paper [53])

> The rough set concept *can be of some importance*, primarily in some branches of artificial intelligence such as inductive reasoning, automatic classification, pattern recognition, learning algorithm, etc. (Pawlak, 1982, First paper [38])

In subsequent years, world has witnessed enormous usefulness of the two theories in computer science and its applications, in the above two quotes, though, both the scientists were a bit skeptic about the future of their theories as evidenced from the *italic* portions in the above paragraphs.

However, the insight of the two great minds while proposing their theories had been so deep that the theories surpassed the boundary of computer science and extended to the domains of mathematics, philosophy, linguistics and even natural

---

Both authors contributed equally to this work.

M. K. Chakraborty (✉)
School of Cognitive Science, Jadavpur University, 132, Raja Subodh Chandra Mallick Road, Jadavpur, Kolkata 700032, West Bengal, India
e-mail: mihirc4@gmail.com

P. Samanta
Department of Mathematics, Ramakrishna Mission Residential College, Ramkrishna Mission Ashram, Main Road, Narendrapur, Kolkata 700103, West Bengal, India
e-mail: pulak.samanta@rkmrc.in

T. Som et al. (eds.), *Fuzzy, Rough and Intuitionistic Fuzzy Set Approaches for Data Handling*, Forum for Interdisciplinary Mathematics,
https://doi.org/10.1007/978-981-19-8566-9_1

sciences. In our opinion, both Zadeh and Pawlak were basically philosophers in the broad sense of the term as well as visioneries.

In this article, we shall present some theoretical aspects of both the theories in which Chakraborty along with the group of researchers around him has been engaged for over past forty years. Areas touched upon will be primarily the algebras arising out of the theories (Sects. 1.2 and 1.3) interrelationship between fuzzy sets and rough sets (Sect. 1.4) and glimpses of the logical aspect (Sect. 1.5).

1.2 Algebraic Aspects

1.2.1 Initial Definitions

Although these are well known now a days, in order to link the current article to the source, we would like to present the initial definitions of fuzzy sets and rough sets.

Definition 1 Fuzzy set: A fuzzy set in the universe of discourse X is a mapping $\tilde{A}$ from X to the unit interval [0, 1]. It is, in fact, a generalization of the characteristic function of a set within the universe. In this sense, fuzzy sets generalize crisp sets. The notions of intersection, union and complementation of crisp sets (or their characteristic functions) are generalized to fuzzy sets $\tilde{A} \cap \tilde{B}$, $\tilde{A} \cup \tilde{B}$ and $\tilde{A}^c$ by the following functions respectively:

$$\tilde{A} \cap \tilde{B}(x) = \min(\tilde{A}(x), \tilde{B}(x)) \equiv (\tilde{A}(x) \wedge \tilde{B}(x)),$$
$$\tilde{A} \cup \tilde{B}(x) = \max(\tilde{A}(x), \tilde{B}(x)) \equiv (\tilde{A}(x) \vee \tilde{B}(x))$$
and
$$\tilde{A}^c(x) = 1 - \tilde{A}(x)$$
for all $x \in X$.

The value $\tilde{A}(x)$ is said to be the membership degree of x in the fuzzy set $\tilde{A}$. Hence, here the notion of membership is 'gradual' unlike the crisp case when it abruptly changes from 1 to 0 (or otherwise). However, crisp sets represented by their characteristic functions are special cases of fuzzy sets.

A fuzzy n-ary relation $\tilde{R}$ is naturally defined by a function

$$\tilde{R} : X^n \longrightarrow [0, 1].$$

In case of binary $\tilde{R}$, one reads as 'x is related to y to the degree $\tilde{R}(x, y)$.' These are fuzzy relations on the crisp universe X. A more general notion of fuzzy relation over a fuzzy set was introduced and developed [13]. The natural question of generalization of notions reflexivity, symmetry, anti-symmetry, transitivity, etc. arises for which we refer to [14, 21]. The set of fuzzy sets forms a quasi-Boolean algebra with respect to the operations $\cap, \cup, ^c$. A qBa is short of Boolean algebra in that the law of contradiction (equivalently, the law of excluded middle) does not hold [46].

For the original definition of rough sets, one may see Pawlak's paper of 1982 [38]. Here, the universe X is endowed with a partition in it generated by an equivalence relation. Formally, it begins with a pair (X, R) called the approximation space, X being a non-empty set and R an equivalence relation (reflexive, symmetric and transitive) in it. Any subset $A \subseteq X$ is then approximated by two sets $\underline{A}$ (the lower approximation) and $\overline{A}$ (the upper approximation) given by the following definitions:

$$\underline{A} = \{x \in X \ : \ [x]_R \subseteq A\} \ \text{ and}$$
$$\overline{A} = \{x \in X \ : \ [x]_R \cap A \neq \phi\}$$

where $[x]_R$ is the equivalence class (or block) of the element x formed by the relation R. Clearly, $\underline{A} \subseteq A \subseteq \overline{A}$.

The set $Bd(A) = \overline{A} \setminus \underline{A}$ is called the boundary of A. In [42], a set A is called a 'rough set' if $Bd(A) \neq \phi$. According to this definition, 'rough' is an adjective applicable to an ordinary set A, as in case of 'finite/infinite sets' or 'open/closed sets.' On the other hand, by 'fuzzy set' is meant a different mathematical object, namely the function from X to [0, 1], the word 'fuzzy' is not a qualifier. However in his first paper [38] as well as his book [41], Pawlak defined a rough set as an equivalence class $[A]_\approx$ of the power set $P(X)$ of X determined by the equivalence relation $\approx$ given by $A \approx B$ if and only if $\underline{A} = \underline{B}$ and $\overline{A} = \overline{B}$, i.e., the sets are 'roughly equal.' This definition is equivalent to defining a rough set as a pair $< \underline{A}, \overline{A} >$. Clearly, there may be sets $A \neq B$ such that $< \underline{A}, \overline{A} > = < \underline{B}, \overline{B} >$ both A, B belonging to the same equivalence class $[\cdot]_\approx$. Several other definitions were also introduced for which we refer to [4, 12]. The lower–upper definition has, however, been the most popular one. It is to be noted that the pair $< \underline{A}, \overline{A} >$ will be called a rough set even though $Bd(A) = \overline{A} \setminus \underline{A} = \phi$. In such a situation, $\underline{A} = \overline{A} = A$. Because of this property, a subset A of X which is the union of some equivalence classes is considered to be 'crisp' relative to the approximation space $< X, R >$. In the case when R completely discretises X, that is, all the equivalence classes are singletons, $\overline{A} = \underline{A} = A$ for all $A \subseteq X$. In this sense, 'rough set' may be taken as a generalization of the notion of set.

Union, intersection and complementation of rough sets are defined by

$$< \underline{A}, \overline{A} > \sqcap < \underline{B}, \overline{B} > = < \underline{A} \cap \underline{B}, \overline{A} \cap \overline{B} >,$$
$$< \underline{A}, \overline{A} > \sqcup < \underline{B}, \overline{B} > = < \underline{A} \cup \underline{B}, \overline{A} \cup \overline{B} >,$$
$$< \underline{A}, \overline{A} >^c = < \underline{A}', \overline{A}' >$$

where $\cap$, $\cup$, $'$ are ordinary union, intersection and complementation of sets. To validate the first two identities, it is required to prove that there are sets $P \subseteq X$, $Q \subseteq X$ such that

$$\underline{P} = \underline{A} \cap \underline{B}, \overline{P} = \overline{A} \cap \overline{B} \text{ and}$$
$$\underline{Q} = \underline{A} \cup \underline{B}, \overline{Q} = \overline{A} \cup \overline{B}.$$

Existence of such P and Q has been shown in [4] viz.

$$P = (A \cap B) \cup (A \cap \overline{B} \cap (\overline{A \cap B})^c),$$
$$Q = (A \cup B) \cup (A \cup \underline{B} \cup (\underline{A \cup B})^c).$$

The algebra thus generated is called a pre-rough algebra which has many equivalents. These are all qBas with additional properties.

An abstract pre-rough algebra $< \mathcal{A}, \sqcap, \sqcup, \neg, L, 0, 1 >$ is defined as follows:

1. $< \mathcal{A}, \sqcap, \sqcup, 0, 1 >$ is a bounded distributive lattice, $\neg$ and L are two unary operators satisfying
2. $\neg\neg a = a$
3. $\neg(a \sqcup b) = \neg a \sqcap \neg b$
4. $La \leq a$
5. $L(a \sqcap b) = La \sqcap Lb$
6. $LLa = La$
7. $L1 = 1$
8. $MLa = La$
9. $\neg La \sqcup La = 1$
10. $L(a \sqcup b) = La \sqcup Lb$
11. $La \leq Lb$ and $Ma \leq Mb$ imply $a \leq b$ where $Ma = \neg L \neg a$ and $a, b \in \mathcal{A}$.

This definition has been simplified later in [47]. If an algebraic structure satisfies axioms 1–8, it is called a topological quasi-Boolean algebra (tqBa). Other algebras can also be built on rough set structures.

For a survey of algebras generated out of rough set structures, we refer to [3, 6]. With respect to the above definitions of fuzzy sets and rough sets, both turn out to be quasi-Boolean algebras. But the generalized varieties of the two concepts may give rise to different algebras.

1.3 Generalizations

There have been many generalizations of both the notions, fuzzy set as well as rough set.

1.3.1 Fuzzy Sets

It may be observed that since the algebraic operations $\cap$, $\cup$, ^{c} of fuzzy sets are defined by the corresponding operations in the membership-value set pointwise, the algebra of the value set induces a similar algebra in the set of all fuzzy sets in the universe. As per above definition, the algebra generated in the set [0, 1] with respect to the operations ($\wedge$) min, ($\vee$) max and $1-(\cdot)$ turns out to be the same algebra as of $(\mathcal{F}(X), \cap, \cup, ^c, \phi, X)$, $\mathcal{F}(X)$ being the set of all fuzzy sets in X. The algebra of fuzzy sets is a quasi-Boolean algebra [46] since $([0, 1], \wedge, \vee, 1-(\cdot), 0, 1)$ is a qBa.

Instead of min ($\wedge$), max ($\vee$), a host of other operators, called t-norms and s-norms, are taken in [0, 1] and corresponding fuzzy set operations are defined. With respect to these operations, [0, 1] may not turn into a qBa, and hence, the corresponding fuzzy sets algebra will change. Below we present one such algebra which is called Łukasiewicz (- Moisil) algebra. Here, the t-norm is $t = \max(0, a+b-1)$ and s-norm is $s = \min(1, a+b)$, $a, b \in [0, 1]$. These operators take care of intersection and union, respectively. The complementation of a fuzzy set is defined, as before, by the operator $1-(\cdot)$. Though the laws of contradiction and excluded middle also hold here, it is a different algebra since $t(a, a) < a$ and $s(a, a) > a$ for all $0 < a < 1$, and hence, unlike the min-max case, idempotence law does not hold. Another such pair of t-norm, s-norm is $(a \cdot b, a+b-a \cdot b)$.

For an extensive list of t-norms and s-norms, readers are referred to [29].

The next step of generalization consists in taking an arbitrary algebraic structure having operations corresponding to $\cap, \cup, ^c$ of fuzzy sets. The minimal structure that is expected of the value set is that of a lattice with an involution operator. Long back in 1967 Goguen publish the first paper in this direction, fuzzy sets thus obtained are called lattice-valued fuzzy sets or simply L-fuzzy sets [23]. Of course, some special kind of lattices has gained importance such as MV-algebra and residuated lattices [16, 17, 19, 25]. Residuated lattice is used in fuzzy logic to compute implication operator, in particular, and thus to define fuzzy consequence relations in the theory of graded consequence. We give the definition below [25].

A residuated lattice is an algebraic structure $(\mathcal{A}, \wedge, \vee, *, \rightarrow, 0, 1)$ such that

- $(\mathcal{A}, \wedge, \vee, 0, 1)$ is a bounded lattice,
- $(\mathcal{A}, *, 1)$ is a commutative monoid,
- $a \leq b$ implies $a * c \leq b * c$ and
- $(*, \rightarrow)$ forms an adjoint pair, i.e.,
  $a * b \leq c$ iff $a \leq b \rightarrow c$
  for all $a, b, c \in \mathcal{A}$.

It is, in fact, necessary to take the lattice complete, and then, the algebraic structure is called a complete residuated lattice. We shall see the use of the implication operator $\rightarrow$ in section 1.5 for the computation of the degree (grade) of fuzzy MP rule.

The above cases of t and s norms and complementation are instances of MV-algebra which is defined below. An MV-algebra $(\mathcal{A}, \oplus, \neg, 0, 1)$ where $\oplus$ and $\neg$ are binary and unary operations and 0 is a constant is an algebra satisfying the following conditions :

MV 1. $a \oplus b = b \oplus a$
MV 2. $a \oplus (b \oplus c) = (a \oplus b) \oplus c$
MV 3. $a \oplus 0 = a$
MV 4. $\neg\neg a = a$
MV 5. $a \oplus 1 = 1$ (where $1 = \neg 0$)
MV 6. $\neg(\neg a \oplus b) \oplus b = \neg(\neg b \oplus a) \oplus a$.

Defining another dual binary operator $\odot$ by $a \odot b = \neg(\neg a \oplus \neg b)$, one gets the dual properties of MV 1–MV 3, MV 5–MV 6. $\odot$ and $\oplus$ serve as the operators for intersection and union respectively of $\mathcal{A}$-valued fuzzy sets on any universe X. Complementation is computed by $\neg$. An MV-algebra turns into a lattice by defining the order relation $\leq$ by $a \leq b$ if and only if $\neg a \oplus b = 1$ (equivalently $a \odot \neg b = 0$. The above set of axioms MV 1–MV 6 are proposed much later in [17]. The original notion is due to Chang with an elaborate set of axioms for $\oplus$ and $\odot$ (see [16]).

Yet another direction of generalization of fuzzy sets consists in the so called 'intuitionistic' fuzzy sets (IFS) where a pair of functions (A^+, A^-) from the universe X to [0, 1] are taken giving the belongingness degree and non-belongingness degree [1]. Thus, the intuitionistic fuzzy set $\tilde{A}^I = (A^+, A^-)$ such that $A^+, A^- : X \to [0, 1]$ and $A^+(x) + A^-(x) \leq 1$. $A^+(x)$ and $A^-(x)$ are respectively the measures of belongingness and non-belongingness of the object $x \in X$ in the IFS $\tilde{A}^I$. Though the word 'intuitionistic' is somewhat misleading (about which point we will say a few words later), the idea of measuring the degrees of belongingness and non-belongingness by two separate functions (of course, bounded by a condition) seems to be a very natural extension and acceptable from the standpoint of the philosophy of vagueness. Intersection, union, complementation and inclusion are then defined as follows.

Let $\tilde{A}^I = (A^+, A^-)$, $\tilde{B}^I = (B^+, B^-)$. Then
$\tilde{A}^I \cap \tilde{B}^I = (A^+ \wedge B^+, A^- \vee B^-)$,
$\tilde{A}^I \cup \tilde{B}^I = (A^+ \vee B^+, A^- \wedge B^-)$,
$(\tilde{A}^I)^c = (A^-, A^+)$

and $\tilde{A}^I \subseteq \tilde{B}^I$ iff $A^+ \leq B^+$, $B^- \leq A^-$ (both pointwise).

There may be another way of defining an IFS that shows the algebraic structure more directly. Instead of taking two functions A^+ and A^-, one can first take the product $[0, 1] \times [0, 1]$ and consider $\tilde{A}^I$ as a mapping from X to $[0, 1]^2$. Thus, $\tilde{A}^I(x)$ is a pair (a, b) such that $a + b \leq 1$, the first and second components are respectively the degrees of belongingness and non-belongingness of x in the intuitionistic fuzzy set $\tilde{A}^I$. Intersection, union and complementation are defined w.r.t. suitable t-norms, s-norms and negation operators in the base set [0, 1]. For the first two operators, min and max are generally used. How can other t-norms and s-norms maybe chosen has been aptly discussed in [18]. We, however, present below the algebraic structure to develop IFS-algebra w.r.t. min and max operators. The value set now is a subset K of $[0, 1]^2$ viz. $K = \{(a, b) : a + b \leq 1, a, b \in [0, 1]\}$. $\wedge, \vee, \neg$ are defined by,

$$(a, b) \wedge (c, d) = (\min(a, c), \max(b, d))),$$
$$(a, b) \vee (c, d) = (\max(a, c), \min(b, d))) \text{ and}$$
$$\neg(a, b) = (b, a).$$

$(K, \wedge, \vee)$ turns out to be a distributive lattice in which the order relation becomes $(a, b) \leq (c, d)$ iff $a \leq c, d \leq b$. In fact, this lattice is complete with $(0, 1)$ and $(1, 0)$ as the least and top elements, respectively. The following properties for $\neg$ hold.

- $\neg\neg(a, b) = (a, b)$ (double negation),
- $\neg(0, 1) = (1, 0)$, $\neg(1, 0) = (0, 1)$,
- De Morgan laws and
- the law of contradiction (equivalently, the law of excluded middle) does not hold.

Since the law 'double negation' holds in K, calling the structure of functions from X to K 'intuitionistic' is not justified.

Besides, topological operators L an M may be defined by $L(a, b) = (a, 1 - a)$ and $M(a, b) = (1 - b, b)$. It is easily verified that L, M satisfy

$$L(a, b) = \neg M\neg(a, b), M(a, b) = \neg L\neg(a, b),$$
$$L(a, b) \leq (a, b) \leq M(a, b), LL(a, b) = L(a, b),$$
$$MM(a, b) = M(a, b), LM(a, b) = M(a, b), ML(a, b) = L(a, b).$$

Hence, K with the above operators is a topological quasi-Boolean algebra. In fact, the algebraic structure satisfies all the axioms 1–11 of pre-rough algebra except axiom 9. In [47], there are some discussions on this type of algebra.

From the definition of L and M, it follows that if $\tilde{A}^{I}$ is an IFS, then $L(\tilde{A}^{I})$ and $M(\tilde{A}^{I})$ are ordinary fuzzy sets. Ordinary fuzzy sets form a proper subclass of the class of intuitionistic fuzzy sets both over X as the universe.

1.3.2 Rough Sets

In the case of rough sets also, we observe generalizations from various angles of which two types are quite natural. The first one is by taking an arbitrary relation R in the universe instead of an equivalence relation. With respect to the generalized approximation space (X, R) thus obtained, the lower and upper approximations of a set $A \subseteq X$ are defined by $\underline{A}_R = \{x \in X : R_x \subseteq A\}$ and $\overline{A}^R = \{x \in X : R_x \cap A \neq \phi\}$ where $R_x = \{y \in X : xRy\}$. Depending upon various properties (such as reflexivity, symmetry, transitivity, seriality, etc.), various properties of the approximations are generated (see Table 1.1). For an extensive list of such properties, we refer to [48, 52]. It should be clear from the definitions above and the table that modal logic systems can be given rough set semantics by interpreting the modality 'necessity' as lower approximation and the modality 'possibility' as the upper approximation.

Table 1.1 Properties of relation-based approximations

	R	R_r	R_s	R_t	R_{rs}	R_{rt}	R_{st}	R_{rst}	R_{ser}
Duality of $\underline{A}, \overline{A}$	Y	Y	Y	Y	Y	Y	Y	Y	Y
$\underline{\phi} = \phi$	N	Y	N	N	Y	Y	N	Y	Y
$\phi = \overline{\phi}$	Y	Y	Y	Y	Y	Y	Y	Y	Y
$\underline{X} = X$	Y	Y	Y	Y	Y	Y	Y	Y	Y
$X = \overline{X}$	N	Y	N	N	Y	Y	N	Y	Y
$\underline{A \cap B} \subseteq \underline{A} \cap \underline{B}$	Y	Y	Y	Y	Y	Y	Y	Y	Y
$\underline{A} \cap \underline{B} \subseteq \underline{A \cap B}$	Y	Y	Y	Y	Y	Y	Y	Y	Y
$\overline{A \cup B} \subseteq \overline{A} \cup \overline{B}$	Y	Y	Y	Y	Y	Y	Y	Y	Y
$\overline{A} \cup \overline{B} \subseteq \overline{A \cup B}$	Y	Y	Y	Y	Y	Y	Y	Y	Y
$A \subseteq B$ implies $\underline{A} \subseteq \underline{B}$	Y	Y	Y	Y	Y	Y	Y	Y	Y
$A \subseteq B$ implies $\overline{A} \subseteq \overline{B}$	Y	Y	Y	Y	Y	Y	Y	Y	Y
$\underline{A} \subseteq A$	N	Y	N	N	Y	Y	N	Y	N
$A \subseteq \overline{A}$	N	Y	N	N	Y	Y	N	Y	N
$\underline{A} \subseteq \overline{A}$	N	Y	N	N	Y	Y	N	Y	Y
$A \subseteq \underline{(\overline{A})}$	N	N	Y	N	Y	N	Y	Y	N
$\overline{(\underline{A})} \subseteq A$	N	N	Y	N	Y	N	Y	Y	N
$\underline{A} \subseteq \underline{(\underline{A})}$	N	N	N	Y	N	Y	Y	Y	N
$\overline{(\overline{A})} \subseteq \overline{A}$	N	N	N	Y	N	Y	Y	Y	N
$\overline{A} \subseteq \underline{(\overline{A})}$	N	N	N	N	N	N	N	Y	N
$\overline{(\underline{A})} \subseteq \underline{A}$	N	N	N	N	N	N	N	Y	N

It may be mentioned that the most important of all these cases is the one when R is reflexive and transitive. The second natural generalization is obtained by taking a covering $\mathcal{C}$ on the universe X in place of the partition. $(X, \mathcal{C})$ is called the covering space. A covering $\mathcal{C}$ is a collection $\{C_i\}_{i \in I}$ of subsets of X such that $\bigcup_{i \in I} C_i = X$. Given a covering $\mathcal{C} = \{C_i\}$, for each $x \in X$, the following subsets of X are taken:

$$N_x = \bigcup \{C_i \in \mathcal{C} : x \in C_i\},$$
$$N(x) = \bigcap \{C_i \in \mathcal{C} : x \in C_i\} \text{ and}$$
$$P_x = \{y \in X : x \in C_i \text{ iff } y \in C_i \text{ for all } i \in I\}.$$

Various types of lower and upper approximations of a set A have been defined using the above sets (often called 'granules'). Among these lower–upper approximation pairs, some are non-duals yet they possess many, often almost all, important properties of these approximation operators. For an elaborate survey of covering based rough sets, we refer again to [48, 52]. Various kinds of abstract algebraic structures have come into existence from studies of generalized rough set [6, 7, 9, 27, 28, 30, 36, 45].

For topological approaches to covering rough sets, we refer [36, 52].

We shall not, however, delve into the topological aspects of both the theories which, in itself is a fascinating area of research.

1.4 Fuzzy Sets Vis a Vis Rough Sets

Although, in some early papers, there had been efforts to see the connections between fuzzy sets and rough sets [50], at least from the mathematical angle, it did not advance much. In [39] Pawlak states that neither fuzzy set nor rough set may be subsumed in the other. There were, rather, attempts to make a fusion between the two. The seminal work in this direction was by Dubios and Parade [20]. However, there had been other significant publications also [34]. In [20], notions of fuzzy rough set and rough fuzzy set are introduced. In the first, a fuzzy equivalence relation is taken on the universe X. Then a fuzzy subset $\tilde{A}$ (in particular a crisp subset) of X is approximated by two fuzzy subset $\tilde{A}_*$ and $\tilde{A}^*$, while in the second, in the universe X, a crisp equivalence relation R is taken and a fuzzy subset $\tilde{A}$ of X is approximated by two fuzzy sets, $\underline{R}(\tilde{A})$ and $\overline{R}(\tilde{A})$. This was in 1990. Definitions of the above mentioned pairs in a little modified form are given below.

1.4.1 *Fuzzy Rough Set*

It will be easier to comprehend the definitions by re-writing Pawlakian lower/upper approximations in the following way:

(1) $x \in \underline{A}$ iff for all $y \in X$, $y \in [x]$ implies $y \in A$

and

(2) $x \in \overline{A}$ iff there exists $y \in X$ such that $y \in [x] \cap A$.

Let $\tilde{R}$ be a fuzzy equivalence relation on X. Then the fuzzy equivalence class of $x \in X$ denoted by $[x]_{\tilde{R}}$ is a fuzzy set in X defined by

$$[x]_{\tilde{R}}(y) = \tilde{R}(x, y) \text{ for all } y \in X.$$

Now, lower and upper approximations of a fuzzy set $\tilde{A}$ w.r.t. $\tilde{R}$ denoted by $\tilde{R}_*(\tilde{A})$ and $\tilde{R}^*(\tilde{A})$, respectively, are fuzzy sets obtained by generalizing (1) and (2) which are given by

$$\begin{aligned}\tilde{R}_*(\tilde{A})(x) &= \inf_{y\in X}(y \in [x]_{\tilde{R}} \rightarrow y \in \tilde{A}) \\ &= \inf_{y\in X}((1 - [x]_{\tilde{R}}(y)) \vee \tilde{A}(y))\end{aligned}$$

and

$$\tilde{R}^*(\tilde{A}) = \sup_{y\in X}([x]_R(y) \wedge \tilde{A}(y)).$$

The pair $(\tilde{R}^*(\tilde{A}), \tilde{R}^*(\tilde{A}))$ is called a fuzzy rough set.

1.4.2 Rough Fuzzy Set

Let X be the universe with the crisp equivalence relation R and $\tilde{A} : X \longrightarrow [0, 1]$ a fuzzy set in X. Then the lower and upper approximations $\underline{R}(\tilde{A})$ and $\overline{R}(\tilde{A})$ of $\tilde{A}$ are given by fuzzy sets in X as defined below.

$$\underline{R}(\tilde{A})(x) = \inf\{\tilde{A}(x) : x \in [x]_R\}$$

and

$$\overline{R}(\tilde{A})(x) = \sup\{\tilde{A}(x) : x \in [x]_R\}.$$

1.4.3 Some Other Approaches

We would like to present briefly some other efforts to put fuzzy and rough sets together.

(a) In 1992, a short paper was published in Fuzzy Sets and Systems [34] bearing the title 'fuzzy rough set' but it meant something different. By this term the authors meant a pair of $< \tilde{A}_L, \tilde{A}_U >$ of L-fuzzy sets $\tilde{A}_L : X_L \longrightarrow L$, $\tilde{A}_U : X_U \longrightarrow L$ where $< X_L, X_U >$ forms a generalized version of standard rough set and such that $\tilde{A}_L(x) \leq \tilde{A}_U(x)$ for all $x \in X_U$ where L is a lattice. It appears that these authors and the reviewers as well were unaware of the paper by Dubois and Prade [20] published two years ago.

(b) A notion of rough fuzzy sets based on soft sets was introduced in [22]. In order to understand this notion, the idea of soft sets is to be discussed. Soft set is a very close associate of rough set. It was introduced by Molodtsov in 1999 [33]. A soft set in the universe X is a pair $(F, \mathcal{A})$ where $\mathcal{A} \subseteq E$ (called the set of parameters) and $F : \mathcal{A} \to \mathcal{P}(X)$. Parameters may be interpreted as properties or attribute of the objects of X. In rough set theory, the equivalence relation emerges from the attribute-value information of the objects of the universe, formally called 'information system.' So, for every attribute $e \in \mathcal{A}$, $F(e)$ may be considered as the subset of all objects in X having the same value with respect to the attribute e. In fact, soft set and information system are mutually definable and in this sense equivalent. But the two theories diverge in formalism just after this initial overlap. They converge again on many domains of application however.

Binary operations of soft sets are defined with respect to the binary operations in $\mathcal{P}(X)$. If $\oplus$ is an operation in $\mathcal{P}(X)$ and $(F, \mathcal{A})$ and $(G, \mathcal{B})$ are two soft sets in X then a binary operation $\oplus'$ between them is defined by

$$(F, \mathcal{A})\oplus'(G, \mathcal{B}) = (H, \mathcal{A} \times \mathcal{B})$$

where $\mathcal{A} \times \mathcal{B}$ is the cartesian product of $\mathcal{A}$ and $\mathcal{B}$ and for $(e_1, e_2) \in \mathcal{A} \times \mathcal{B}$, $H(e_1, e_2) = F(e) \oplus F(e')$. A soft set $(F, \mathcal{A})$ in X is said to be full if $\bigcup_{a\in\mathcal{A}} F(a) = X$, that is $\{F(a) : a \in \mathcal{A}\}$ forms a covering of X. In such a case, $(X, F, \mathcal{A})$ is called a soft approximation space. Now with respect to a soft approximation space $S = (X, F, \mathcal{A})$, a fuzzy subset $\tilde{A}$ of X is approximated following Dubois and Prade by

$$\underline{S}(\tilde{A})(x) = \inf_{y\in X}\{\tilde{A}(y) : \{x, y\} \subseteq F(a) \text{ for some } a \in \mathcal{A}\}$$

and

$$\overline{S}(\tilde{A})(x) = \sup_{y\in X}\{\tilde{A}(y) : \{x, y\} \subseteq F(a) \text{ for some } a \in \mathcal{A}\}.$$

$\underline{S}(\tilde{A})$ and $\overline{S}(\tilde{A})$ are called the lower and upper approximation of $\tilde{A}$ w.r.t. the soft approximation space S. In case the subset $\tilde{A}$ is crisp, say A, then the above two equations reduce to

$$\underline{S}(A) = \bigcup_{a\in\mathcal{A}}\{F(a) : F(a) \subseteq A\}$$

and

$$\overline{S}(A) = \bigcup_{a\in\mathcal{A}}\{F(a) : F(a) \cap A \neq \phi\}.$$

For some other interesting results, see [22].

(c) Based on soft sets, one also gets fuzzy rough sets [32]. The notion of fuzzy soft sets was already introduced by Maji et al. [31]. Let X be the universe and E the set of parameters. Then a pair $(F, \mathcal{A})$ is called a fuzzy soft set if $\mathcal{A} \subseteq E$ and $F : \mathcal{A} \to \mathcal{F}$, the set of all fuzzy subsets of X, i.e., F maps an attribute to a fuzzy subset of X. In this sense, the soft set is fuzzy. For any fuzzy subset $\tilde{A}$ of X, fuzzy lower and upper approximations of it w.r.t. the pair $(F, \mathcal{A})(\equiv S)$ are defined respectively by the equations (1) and (2) below.

(1) $\underline{S}(\tilde{A})(x) = \inf_{a \in \mathcal{A}}(1 - F(a)(x)) \vee (\inf_{y \in X}((1 - F(a)(y)) \vee \tilde{A}(y)))$.
(2) $\overline{S}(\tilde{A})(x) = \sup_{a \in \mathcal{A}}(F(a)(x) \wedge (\sup_{y \in X}(F(a)(y) \wedge \tilde{A}(y))))$.

The interpretations of the above two equations in the case when $\tilde{A}$ is a crisp set (say A) and F(a) is also a crisp subset are the following:

$$x \in \underline{S}(A) \text{ iff } x \in F(a) \text{ implies } F(a) \subseteq A \text{ for all } a \in \mathcal{A}$$

and

$$x \in \overline{S}(A) \text{ iff there is some } a \in \mathcal{A} \text{ such that } x \in F(a) \text{ and } F(a) \cap A \neq \phi.$$

We have observed above several attempts to put fuzzy, rough and soft sets together. In the following section, the question of reducibility of one to another, in particular rough sets into fuzzy sets via rough membership functions, will be discussed.

1.4.4 Rough Membership Function-Based Approach

We shall present here a direct correspondence between the two kinds of mathematical entities. The notion of rough membership function is quite old; it was introduced by Pawlak and Skowron [43] and developed in [44].

Definition 2 Rough membership function: given the approximation space (X, R) and a set $A \subseteq X$, the rough membership function $f_A : X \longrightarrow [0, 1]$ is defined by:

$$f_A(x) = \frac{\mid [x]_R \cap A \mid}{\mid [x]_R \mid}$$

where $\mid P \mid$ denotes the cardinality of the set P.

Originally, the notion was defined on the finite universe but later in [11], it was extended over any universe with the assumption that the equivalence classes generated by R are of finite cardinalities. It is immediately observed that

$$f_A(x) = \begin{cases} 1 & \text{if } x \in \underline{A}_R \\ 0 & \text{if } x \in (\overline{A}^R)^c \\ 0 < f_A(x) < 1 & \text{if } x \in Bd(A). \end{cases}$$

The following properties of rough membership function are established [51]:

- $\max[0, f_A(x) + f_B(x) - 1] \leq f_{A\cap B}(x) \leq \min[f_A(x), f_B(x)]$,
- $\max[f_A(x), f_B(x)] \leq f_{A\cup B}(x) \leq \min[1, f_A(x) + f_B(x)]$,
- $f_{A\cup B}(x) = f_A(x) + f_B(x) - f_{A\cup B}(x)$.

There may be two distinct sets A, B such that $f_A = f_B$.

Though these functions resembled fuzzy subsets of X, they could not be considered as fuzzy sets proper until [11] was published, where the following properties had been established: for any two rough membership functions, f_A and f_B the functions defined by $f_A \wedge f_B$ and $f_A \vee f_B$ pointwise are also rough membership functions. This means, for any sets A, B, there exist sets P and Q such that $f_A \wedge f_B = f_P$ and $f_A \vee f_B = f_Q$. This observation in a sense, may be considered as a breakthrough in establishing the relation between fuzzy and rough sets.

Let X be a universe with a partition. Let $\mathcal{F}(X)$ denote the collection of all fuzzy subsets of X. On the other side, by $MF(X)$, we denote the set of all rough membership functions viz.

$$\{f_A \ : \ A \in \mathcal{P}(X), \text{ the power set of } X\}.$$

A correspondence between these two sets may be established. For the cluster of subsets $\{A' : f_A = f'_A\}$, the corresponding fuzzy set is f_A. This correspondence from $MF(X)$ to $\mathcal{F}(X)$ may not be surjective (onto). The set $MF(X)$is now partitioned in terms of the above mentioned clusters. More formally, in the power set $\mathcal{P}(X)$, an equivalence relation $\equiv$ is defined by $A \equiv B$ if and only if $f_A = f_B$. Each equivalence class $[A]_\equiv$ has a unique fuzzy set in $\mathcal{F}(X)$ viz. f_A. It may be recalled that a Pawlakian rough set $< \underline{A}, \overline{A} >$ is equivalently defined as the equivalence class $[A]_\approx$ where $\approx$ is the equivalence relation defined in $\mathcal{P}(X)$ by $A \approx B$ if and only if $\underline{A} = \underline{B}$ and $\overline{A} = \overline{B}$. Obviously, $\equiv$ generates a finer partition than $\approx$ in $\mathcal{P}(X)$. Since all members of $[A]_\equiv$ have the same membership function, $[A]_\equiv$ is called a membership function based rough set or m-f rough set which corresponds to a unique member of $\mathcal{F}(X)$. In fact, the fuzzy set f_A is itself named as an m-f rough set and intersection, union and complementation of m-f rough sets are defined exactly as those of fuzzy sets viz. $f_A \wedge f_B$, $f_A \vee f_B$ and $1 - f_A = f_{A^c}$, respectively. Thus, $MF(X)$ constitutes a substructure of $(\mathcal{F}(X), \wedge, \vee, 1 - \cdot)$.

The converse question is the following.

Given a fuzzy set $f : X \longrightarrow [0, 1]$, does there exist a partition of X and a subset $A \subseteq X$ such that the rough membership function $f_A = f$? The answer to this question is 'no,' though we are not providing here a formal proof of the above claim. However, in a special case, the answer is positive. The statement of that assertion may be formulated as follows: given a universe with a partition (X, R), the equivalence classes being of finite cardinality, any fuzzy set $f : X \longrightarrow [0, 1]$ such that for any $x \in X$, $f(x) = 0/\frac{1}{n}/\frac{2}{n}/\ldots/\frac{n-1}{n}/1$ where $\mid [x]_R \mid= n$ is the rough membership function f_A for some $A \subseteq X$. It is immediately observed that the A is obtained by picking up $0/1/2/\ldots/n-1/n$ elements respectively from the equivalence class $[x]_R$. Obviously, A can be constructed in more than one ways. Thus, in this special case, there can be a matching between the fuzzy sets and m-f rough sets.

1.5 Logics

In this last section, we shall discuss some core issues of fuzzy logic and rough logic. In fact, it would be proper to say 'logics' instead of 'logic' in both the cases. Here is a big divergence: while fuzzy logics are based on many-valued logics, rough logics are based on modal logics. In almost all logics, classical or otherwise, the rule of inference called Modus Ponens (MP) plays a key role. Informally speaking, the rule is:

(Σ): from two statements (or formulae) 'α' and 'if α then β' one is allowed to infer 'β' since when 'α' and 'if α then β' are 'true,' 'β' is true as well, whatever notion of truth one is inclined to adopt.

In the following subsections, we present fuzzy and rough MP rules, both generalize the classical MP rule Σ.

1.5.1 *Fuzzy MP Rule*

Fuzzy logics are broadly of two kinds viz. fuzzy logic in narrow sense (or mathematical fuzzy logic) [25] and the theory of graded consequence [15]. Pavelka in [37] introduced the idea of fuzzy rule of inference that follows Goguen's [24] notion where the fuzzy MP rule is intended to capture the following idea:

'If you know α is true at least to the degree a and $\alpha \rightarrow \beta$ at least to the degree b, you conclude that β is true at least to the degree $a \cdot b$' ($\cdot$ being the multiplication in the interval [0.1]). As mentioned before, the logic is based on many-valued semantics. Formally written, the fuzzy MP rule may be expressed as

$$\frac{\begin{array}{c}(\alpha, a)\\(\alpha \rightarrow \beta, b)\end{array}}{(\beta, a \cdot b)}$$

The product $(\cdot)$ may be generalized to any t-norm [24].

Pavelka generalized the above schema over to a complete residuated lattice as the truth set and in his notation the rule stands as

$$\begin{array}{c} \alpha \quad (a) \\ \alpha \to \beta \quad (b) \\ \hline \beta \quad (a * b) \end{array}$$

$*$ being the product operator of the residuated lattice. In fact, in Pavelka's presentation, a fuzzy rule of inference r consists of two components (r', r'') where the first (grammatical) component r' operates on formulas and the second (evaluation) r'' operates on the values. The above MP rule is one such rule of inference.

On the other hand, in the context of graded consequence, the fuzzy rule Modus Ponens is a fuzzy relation $|\!\sim$ assigning a fixed value to $\{\alpha, \alpha \to \beta\} \mid\!\sim \beta$ for all α, β. This value denoted by $gr(\{\alpha, \alpha \to \beta\} \mid\!\sim \beta)$ is determined with respect to a set $\{T_i\}_{i \in I}$ of fuzzy sets on the set of formulas For, i.e., T_i : For $\longrightarrow [0, 1]$, $i \in I$. T_i's constitute a set of fuzzy valuations of the set of formulas. The determining expression is

$$gr(\{\alpha, \alpha \to \beta\} \mid\!\sim \beta) = \inf_{\alpha,\beta}(\inf_i[(T_i(\alpha) \wedge (T_i(\alpha) \to_o T_i(\beta))) \to_m T_i(\beta)]).$$

The above expression, though looks complicated, is in fact, value of the following sentence:

'For all α, β and for all valuations T_i, if α and $\alpha \to \beta$ are true with respect to T_i, then β is also true w.r.t. T_i.' This is a generalization of Σ in the many-valued context.

It is to be noted that there are involved two implication operators $\to_o$ and $\to_m$ defined in [0, 1], the first for the object language and the second for the meta language. The meta language implication $\to_m$ has to be a residuation in a residuated lattice in particular the set [0, 1] with a residuation operator. (For details see [15]). The gross difference between the fuzzy MP rule and graded MP rule lies in that while the first is a crisp relation that holds between the set of pairs $\{(\alpha, a), (\alpha \to \beta, b)\}$ and the pair $(\beta, a * b)$ for all α, β, a and b, i.e., a crisp relation between fuzzy sets, the second is a fuzzy relation between the set $\{\alpha, \alpha \to \beta\}$ and β. The formalism and various notions such as the notion of consistent set in the two logical systems are quite different. Outside the logic systems proper, application of the two MP rules in addressing the Sorites paradox is intriguing [15].

1.5.2 *Rough Modus Ponens Rules*

As mentioned before, Rough Modus Ponens (RMP) pre-supposes an underlying modal logical system. We shall not delve into modal systems here for which the reader may consult [26]. It may only be mentioned that the basic modal logic systems are K, T, S_4, B and S_5 all having the same language but different sets of axioms. Since we shall deal with the system S_5 only, its axioms and rules are given below. It is to note that the propositional modal language is the classical propositional language enhanced by two more unary operators L (necessity) and M (possibility) where L and M are interdefinable.

Axioms:

All propositional logic axioms and
Ax K: $L(\alpha \rightarrow \beta) \rightarrow (L\alpha \rightarrow L\beta)$
Ax T: $L\alpha \rightarrow \alpha$
Ax S_5: $M\alpha \rightarrow LM\alpha$.

Rules:

$$\frac{\alpha,\ \alpha \rightarrow \beta}{\beta}\ \text{(MP)} \quad \frac{\alpha}{L\alpha}\ \text{(N)}$$

Just to recall that in the system S_5, $L\alpha \rightarrow LL(\alpha)$ (Ax S_4) and $\alpha \rightarrow LM\alpha$ (Ax B) can be derived. The system S_5 has a direct relationship with the original rough set theory as proposed by Pawlak. The standard interpretation of the formulas of S_5 is made in the Kripke frame (X, R) [26] where X is a set of 'possible worlds' and R is an 'accessibility relation' which is reflexive, symmetric and transitive. So, the Kripke frame, in this case, is the approximation space of rough set theory. Also, the wffs $L\alpha$ and $M\alpha$ admit a rough set theoretic interpretation by $v(L\alpha) = \underline{v(\alpha)}_R$ and $v(M\alpha) = \overline{v(\alpha)}^R$ where v is a Kripke frame-based interpretation [35]. We have mentioned this point in Sect. 1.3.2. With this very brief introductory remarks, we shall now present rough MP rules and logics developed thereby.

In the more general setup, classical MP rule may be written as

$$\frac{\Gamma \vdash \alpha, \Gamma \vdash \alpha \rightarrow \beta}{\Gamma \vdash \beta}$$

where Γ is a set of wffs denoting the premise set and $\vdash$ denotes classical consequence relation. In rough logics, the above MP rule is generalized by

$$\frac{\Gamma \mid\!\sim \alpha,\ \Gamma \mid\!\sim \beta \rightarrow \gamma, \vdash_{S_5} \aleph(\alpha, \beta)}{\Gamma \mid\!\sim \gamma}$$

where $\mid\!\sim$ is the rough consequence relation and $\aleph(\alpha, \beta)$ is anyone of the following list of wffs:

(i) $L\alpha \rightarrow L\beta$ (iv) $\alpha \rightarrow L\beta$ (vii) $M\alpha \rightarrow L\beta$ (x) $M(\alpha \rightarrow \beta)$

(ii) $L\alpha \rightarrow \beta$ (v) $\alpha \rightarrow \beta$ (viii) $M\alpha \rightarrow \beta$ (xi) $L(\alpha \rightarrow \beta)$

(iii) $L\alpha \rightarrow M\beta$ (vi) $\alpha \rightarrow M\beta$ (ix) $M\alpha \rightarrow M\beta$ (xii) $\alpha \Rightarrow \beta$ (xiii) $\alpha \Leftrightarrow \beta$.
[$\alpha \Rightarrow \beta$ standing for $(L\alpha \rightarrow L\beta) \wedge (M\alpha \rightarrow M\beta)$ and $\alpha \Leftrightarrow \beta$ for $(\alpha \Rightarrow \beta) \wedge (\beta \Rightarrow \alpha)$]. These thirteen formulas are partitioned in the following equivalence classes under the base modal logic S_5.

{(i), (ii)}, {(iii), (x)}, {(iv), (vii), (viii)}, {(v), (xi)}, {(vi), (ix)}, {(xii)}, {(xiii)}. This means that any two formulas belonging to the same class are mutually derivable in the system S_5.

Thus, seven RMP rules would be available picking up one from each class viz.

$$\text{RMP}_1 \ \frac{\Gamma|\!\sim\alpha,\ \Gamma|\!\sim\beta\rightarrow\gamma,\ \vdash_{S_5} L\alpha\rightarrow L\beta}{\Gamma|\!\sim\gamma}$$

$$\text{RMP}_2 \ \frac{\Gamma|\!\sim\alpha,\ \Gamma|\!\sim\beta\rightarrow\gamma,\ \vdash_{S_5} L\alpha\rightarrow M\beta}{\Gamma|\!\sim\gamma}$$

$$\text{RMP}_3 \ \frac{\Gamma|\!\sim\alpha,\ \Gamma|\!\sim\beta\rightarrow\gamma,\ \vdash_{S_5} M\alpha\rightarrow\beta}{\Gamma|\!\sim\gamma}$$

$$\text{RMP}_4 \ \frac{\Gamma|\!\sim\alpha,\ \Gamma|\!\sim\beta\rightarrow\gamma,\ \vdash_{S_5} \alpha\rightarrow\beta}{\Gamma|\!\sim\gamma}$$

$$\text{RMP}_5 \ \frac{\Gamma|\!\sim\alpha,\ \Gamma|\!\sim\beta\rightarrow\gamma,\ \vdash_{S_5} M\alpha\rightarrow M\beta}{\Gamma|\!\sim\gamma}$$

$$\text{RMP}_6 \ \frac{\Gamma|\!\sim\alpha,\ \Gamma|\!\sim\beta\rightarrow\gamma,\ \vdash_{S_5} \alpha\Rightarrow\beta}{\Gamma|\!\sim\gamma}$$

$$\text{RMP}_7 \ \frac{\Gamma|\!\sim\alpha,\ \Gamma|\!\sim\beta\rightarrow\gamma,\ \vdash_{S_5} \alpha\Leftrightarrow\beta}{\Gamma|\!\sim\gamma}.$$

The interpretation of RMP_1 is the following:

if α and $\beta \rightarrow \gamma$ roughly follow from Γ and in the semantics of the underlying system S_5, $\underline{v(\alpha)} \subseteq \underline{v(\beta)}$, then γ roughly follows from Γ. Similarly, the other rules may be interpreted.

Axiomatic presentation of the notion 'roughly follows' is given in Sect. 1.5.3.

The above RMP rules are related by the following hierarchical relations:

$\text{RMP}_1 \Rightarrow \text{RMP}_4 \Rightarrow \text{RMP}_3$,
$\text{RMP}_2 \Rightarrow \text{RMP}_5$ and
$\text{RMP}_1 \Rightarrow \text{RMP}_6 \Rightarrow \text{RMP}_7$.

where $\text{RMP}_i \Rightarrow \text{RMP}_j$ means that the ith rule implies the jth one, $i \leq j$. Besides these, two more rules, also called rough MP, had been proposed in [2, 8] which are as follows:

$$\text{R}_1 \ \frac{\Gamma|\!\sim\alpha,\ \vdash_{S_5} M\alpha\rightarrow M\gamma}{\Gamma|\!\sim\gamma} \quad \text{and}$$

$$\text{R}_2 \ \frac{\Gamma|\!\sim M\alpha,\ \Gamma|\!\sim M\gamma}{\Gamma|\!\sim M\alpha\wedge M\gamma}$$

Depending on various RMP rules, various rough logics are obtained. For more detail, see [49].

1.5.3 Rough Logics

Let S be a modal system with consequence relation $\vdash_S$. Based on S, system L_r is defined axiomatically by using Rough consequence relation $|\!\sim$ as follows:

$\boldsymbol{L_r}$:

(i) $\vdash_S \alpha$ implies $\Gamma \mid\!\sim \alpha$.
(ii) $\{\alpha\} \mid\!\sim \alpha$.
(iii) $\Gamma \mid\!\sim \alpha$ implies $\Gamma \cup \Delta \mid\!\sim \alpha$.
(iv) RMP may be applied.

So we have rough logic systems L_{r_1}–L_{r_7} corresponding to the rules RMP_1–RMP_7 and two more systems LR_1 and LR_2 corresponding to the rules R_1 and R_2. However, first coinage of the term 'rough consequence' was most probably in [10], since then lot of modifications have taken place.

The relation between the corresponding logics L_{r_i} and L_{r_j} will be reverse inclusion, $L_{r_j} \preceq L_{r_i}$ for $i \leq j$.

A detailed study of the systems L_{r_i} when S is S_5 with rules RMP$_i$ is done in [8, 49].

We present below a diagram (Fig. 1.1) depicting the relevant portion of that study after making a few modifications.

In the following diagram, (Fig. 1.1) $\sim$ means equivalence, connection by a line means the lower logical system is proper subsystem of the upper one and connection by dotted line means the corresponding systems are mutually independent.

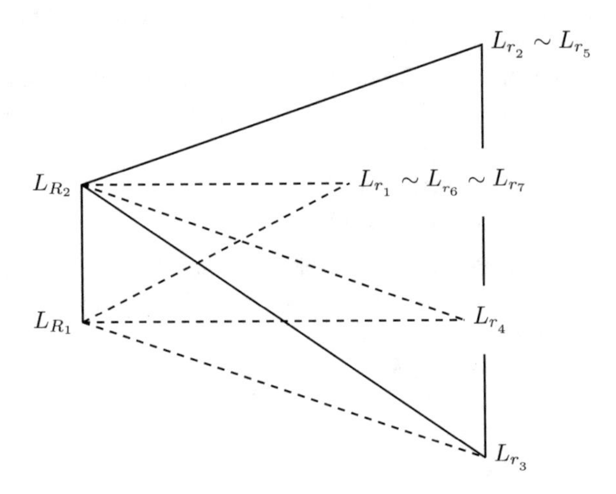

Fig. 1.1 Hierarchy of rough logics

It is to be noted that we have presented only tiny fragments of fuzzy logics and rough logics. Both Zadeh and Pawlak were interested in logic and the notion of truth. In Zadeh's perception, a statement may be 'true,' 'very true,' 'more or less true' and so on. He proposed linguistic calculi with these values to model approximate reasoning [54]. On the other hand, Pawlak proposed for rough truth [40] which was formalized in [2]. For an earlier survey of rough logics, we refer to [5].

1.6 Concluding Remarks

In this chapter, we have presented only glimpses of a few aspects of fuzzy set theory and rough set theory and made a comparative analysis. Besides the presented aspects, there is a number of others. However, both the theories have had wide range of applications that overlap. This aspect has not been touched upon at all. In our opinion, it is high time to do a comprehensive research on the similarities and differences of the two theories from all aspects, viz., mathematical import, range of applications and philosophy of vagueness and indiscernibility.

Acknowledgements We are thankful to the referee for some helpful suggestions.

References

1. Atanassov, K.T.: Intuitionistic fuzzy sets. Fuzzy Sets Syst. **20**(1), 87–96 (1986)
2. Banerjee, M.: Logic for rough truth. Fundam. Inform. **71**(2,3), 139–151 (2006)
3. Banerjee, M., Chakraborty, M.K.: Rough algebra. Bull. Pol. Acad. Sci. (Math.) **41**(4), 293–297 (1993)
4. Banerjee, M., Chakraborty, M.K.: Rough sets through algebraic logic. Fundam. Inform. **28**(3–4), 211–221 (1996)
5. Banerjee, M., Chakraborty, M.K.: Rough logics: a survey with further directions. In: Orloska, E. (ed.) Incomplete Information: Rough Set Analysis. Studies in Fuzziness and Soft Computing, vol. 13, pp. 579–600. Physica-Verlag (1998)
6. Banerjee, M., Chakraborty, M.K.: Algebras from rough sets. In: Pal, S.K., Polkowski, L., Skowron, A. (eds.) Rough-Neural Computing-Techniques for Computing with Words. Springer, Berlin, Heidelberg, New York (2004)
7. Bonikowski, Z.: A certain conception of the calculus of rough sets. Notre Dame J. Formal Logic **33**, 412–421 (1992)
8. Bunder, M.W., Banerjee, M., Chakraborty, M.K.: Some rough consequence logics and their interrelations. In: Peters, J.F., Skowron, A. (eds.) Transactions on Rough Sets VIII. LNCS, vol. 5084, pp. 1–20. Springer, Berlin, Heidelberg (2008)
9. Cattaneo, G., Ciucci, D.: Lattices with interior and closure operators and abstract approximation spaces. In: Peters, J., Skowron, A. (eds.) Foundations of Rough Sets X, pp. 67–116. Springer (2009)
10. Chakraborty, M.K., Banerjee, M.: Rough consequence. Bull. Pol. Acad. Sci. (Math.) **41**(4), 299–304 (1993)
11. Chakraborty, M.K.: Membership function based rough set. Int. J. Approx. Reas. **55**, 402–411 (2014)

12. Chakraborty, M.K.: On some issues in the foundation of rough sets: the problem of definitions. Fundam. Inform. (2016)
13. Chakraborty, M.K., Das, M.: Studies on fuzzy relations over fuzzy subsets. Fuzzy Sets Syst. **9**, 79–89
14. Chakraborty, M.K., Das, M.: On fuzzy equivalence. I, II. FSS **11**, 185–193, 299–307 (1983)
15. Chakraborty, M.K., Soma, D.: Theory of Graded Consequence. A General Framework for Logic of Uncertainty. Springer (2019)
16. Chang, C.C.: Algebraic analysis of many-valued logics. Trans. Am. Math. Soc. **88**, 467–490 (1958)
17. Cignoli, R.L., d'Ottaviano, I.M., Mundici, D.: Algebraic Foundations of Many-Valued Reasoning. Kluwer Academic Publishers, Dordrecht, Boston, London (2000)
18. Cornelis, C., De Cock, M., Serve, E.E.: Intuitionistic fuzzy rough sets: at the crossroads of imperfect knowledge. Expert Syst. **20**(5), 260–269 (2003)
19. Di Nola, A.: MV algebras in the treatment of uncertainty. In: Lowen, R. (ed.) Fuzzy Logic, Proceedings of the International Congress IFSA, Brucellas, pp. 123–131. Kluwer, Dordrecht (1991)
20. Dubois, D., Prade, H.: Rough fuzzy sets and fuzzy rough sets. Int. J. Gen. Syst. **17**, 191–209 (1990)
21. Soma, D., Chakraborty, M.K.: Fuzzy relation and fuzzy function over fuzzy sets: a retrospective. Soft Comput. **19**, 99–112 (2015)
22. Feng, F., Li, C., Davvaz, B., Irfan Ali, M.: Soft sets combined with fuzzy sets and rough sets: a tentative approach. Soft Comput. 899–911 (2010)
23. Goguen, J.A.: L-fuzzy sets. J. Math. Anal. Appl. **18**(1), 145–174 (1967)
24. Goguen, J.A.: The logic of inexact concept. Synthese **19**, 325–373 (1968)
25. Hajek, P.: Mathematics of Fuzzy Logic. Kluwer Academic Publisher, Dordrecht (1998)
26. Hughes, G.E., Cresswell, M.J.: A New Introduction to Modal Logic. Routledge (1996)
27. Jarvinen, J.: Approximations and rough sets based tolerances. In: Ziarko, W., Yao, Y. (eds.) RSCTC 2000. LNAI, vol. 2005, pp. 182–189. Springer, Berlin, Heidelberg (2001)
28. Järvinen, J.: The ordered set of rough sets. In: Tsumoto, S. et al. (eds.) RSCTC 2004. LNAI, vol. 3066. Springer, Berlin, Heidelberg, pp. 49–58 (2004)
29. Klir, G.J., Yuan, B.: Fuzzy Sets and Fuzzy Logic: Theory and Applications. Prentice-Hall, India (1997)
30. Kumar, A., Banerjee, M.: Definable and rough sets in covering-based approximation spaces. In: Li, T., et al. (eds.) RSKT 2012. LNAI, vol. 7414, pp. 488–495. Springer, Berlin, Heidelberg (2012)
31. Maji, P.K., Biswas, R., Roy, A.R.: Fuzzy Math. **9**, 589–602 (2001)
32. Dan, M., Xiaohong, Z., Keyun, Q.: Soft rough fuzzy sets and soft fuzzy rough set. Comput. Math. Appl. **62**, 4635–4645 (2011)
33. Molodtsov, D.: Soft set theory-first results. Comput. Math. Appl. **37**, 19–31 (1999)
34. Nanda, S., Majumdar, S.: Fuzzy rough sets. FSS **45**, 157–160 (1992)
35. Orlowska, E.: Kripke semantics for knowledge representation logics. Stud. Logica **XLIX**, 255–272 (1990)
36. Pagliani, P., Chakraborty, M.K.: A Geometry of Approximation—Rough Set Theory: Logic, Algebra and Topology of Conceptual Patterns. Springer (2008)
37. Pavelka, J.: On fuzzy logic I, II, III. Z. Math. Logic Grundlagen Math. **25**, 45–52, 119–134, 447–464 (1979)
38. Pawlak, Z.: Rough sets. Int. J. Comput. Inf. Sci. **11**(5) (1982)
39. Pawlak, Z.: Rough sets and fuzzy Sets. Institute of Computer Science, Polish Academy of Sciences Report 540 (ICSPAS Reports) (1984)
40. Pawlak, Z.: Rough logic. Bull. Pol. Acad. Sci. (Tech. Sci.) **35**(5–6), 253–258 (1987)
41. Pawlak, Z.: Rough Sets—Theoretical Aspects of Reasoning About Data. Kluwer Academic Publisher (1991)
42. Pawlak, Z.: Some issues on rough sets. Transactions on Rough Sets I. LNCS, vol. 3100, pp. 1–58 (2004)

43. Pawlak, Z., Skowron, A.: Rough membership function. In: Yager, R., Fedrizzi, M., Kacprzyk, J. (eds.) Advances in Dempster-Shafer Theory of Evidence, pp. 251–271. Wiley, New York (1994)
44. Polkowski, L., Skowron, A.: Rough mereology: A new paradigm for approximate reasoning. Int. J. Approx. Reas. **15**, 81–91 (1996)
45. Pomykala, J., Pomykala, J.A.: The stone algebra of rough sets. Bull. Pol. Acad. Sci. Math. **36**(7–8), 495–508 (1988)
46. Rasiowa, H.: An Algebraic Approach to Non-classical Logics. North Holland Publishing Co., Amsterdam (1974)
47. Saha, A., Sen, J., Chakraborty, M.K.: Algebraic structure in the vicinity of pre-rough algebras and their logics. Inf. Sci. **282**, 296–320 (2014)
48. Samanta, P., Chakraborty, M.K.: Covering based approaches to rough sets and implication lattices. In: Sakai, H., et al. (eds.) RSFDGrC 2009. LNAI, vol. 5908, pp. 127–134. Springer, Berlin, Heidelberg (2009)
49. Samanta, P., Chakraborty, M.K.: Interface of rough set systems and modal logics: a survey. In: Peters, J.F., Skowron, A., Slezak, D., Nguyen, H.S., Bazan, J.G. (eds.) Transactions on Rough Sets XIX, pp. 114–137 (2015)
50. Wygralak, M.: Some remarks on rough and fuzzy sets. BUSEFAL **21**, 43–49 (1985)
51. Yao, Y.Y.: Semantics of fuzzy sets in rough set theory. In: Peters, J.F. et al. (eds.) Transactions on Rough Sets II. LNCS, vol. 3135, pp. 297–318 (2004)
52. Yao, Y.Y., Yao, B.: Covering based rough set approximations. Inf. Sci. **200**, 91–107 (2012)
53. Zadeh, L.A.: Fuzzy sets. Inf. Control **8**, 338–353 (1965)
54. Zadeh, L.A.: Fuzzy Logic and approximate reasoning. Synthese **30**, 407–428 (1975)
55. Zhu, W.: Topological approaches to covering rough sets. Inf. Sci. **177**, 1499–1508 (2007)

Chapter 2
Enhancing the Prediction of Anti-cancer Peptides by Suitable Feature Extraction and FRFS with ACO Search Followed by Resampling

Rakesh Kumar Pandey, Anoop Kumar Tiwari, Shivam Shreevastava, and Tanmoy Som

2.1 Introduction

Fermentation is one of the well-known techniques of food-processing that uses biochemical transformations to improve the qualities of food products. Fermentation produces enzymatic hydrolysates that imparts to taste characteristics [1]. Enzymatic hydrolysates available in food proteins are usually an anti-cancer taste. In the literature [2, 3], the association between anti-cancerness and chemical structure is widely discussed and found that anti-cancer-tasting peptides are usually isolated from numerous sources. Anti-cancer taste [4, 5] is a Japanese notion, which means "savoury" or "broth-like" and is produced by several peptides. This concept is specified by the taste of monosodium L-glutamate and prototypical stimulus, and sodium lactate or peptides. Anti-cancer peptides [6], especially tripeptides and dipeptides, have been segregated and specified from soy sauce, cheese, and miso, which can play vital role in various health advantages, including weight gain, reducing fat

R. K. Pandey
Dr. K.N. Modi University, Niwai, Rajasthan 304021, India

A. K. Tiwari (✉)
Department of Computer Science and Information Technology, Central University of Haryana, Mahendergarh 123031, India
e-mail: anoop.phd2014@gmail.com

S. Shreevastava
Department of Mathematics, School of Basic & Applied Sciences, Harcourt Butler Technical University, Kanpur, Uttar Pradesh 208002, India

T. Som
Department of Mathematical Sciences, Indian Institute of Technology (BHU), Varanasi 221005, India

T. Som et al. (eds.), *Fuzzy, Rough and Intuitionistic Fuzzy Set Approaches for Data Handling*, Forum for Interdisciplinary Mathematics,
https://doi.org/10.1007/978-981-19-8566-9_2

removal and plasma leptin levels. Anti-cancer ingredients are used to control gastrointestinal operations and to reduce the threat of stroke as well as coronary heart disease. Moreover, anti-cancer ingredients play key role for food seasoning and are extensively utilized in food production. However, identification and characterization of anti-cancer peptides are always time consuming and expensive. So, computer-aided methods are always effective to complete this task by expediting the experimental techniques. In the literature, very few investigations have been reported for building computational models to predict anti-cancer peptides. Wei et al. [4] have given one of the most interesting machine learning-based approach, which was developed specifically for discriminating anti-cancer and non-anti-cancer peptides.

There are variety of factors that can directly degrade the average performances of the learning algorithms to predict the anti-cancer peptides. Extraction of informative features [7], selection of relevant and/or non-redundant features from large number of features [8, 9], resampling of imbalanced datasets [10], and selection of suitable learning technique [11, 12] are key factors among them. In case of peptide sequences with variety of lengths, feature extraction can play a key role in designing well-performed predictors. Feature extraction process produces a constant length of feature vectors from the different lengths of peptide sequences that reflects the necessary correlation with the target to establish a potential classifier. Feature extraction techniques can provide various characteristics of the data points to the machine learning algorithms as it produces different interesting representative features, which leads to improve the average performances of the learned models.

Dimensionality of the datasets is increasing expeditiously by procuring and amassing an increasing number of features from peptide sequences. Some of the features are responsible for creation of noise to the target. Noisy features can be irrelevant and/or redundant and degrade the performance of the learning algorithms. Irrelevant features do not have any direct association with the decision features but adversely influence the overall performances of the learning techniques. Redundant features do not produce any supplementary information to the decision feature. Therefore, prior to use a dataset containing noisy features, it is the foremost necessity to pre-process the dataset for eliminating redundant and irrelevant features. Feature selection is an efficacious and extensively used tool to remove redundant and/or irrelevant features. The main objective is to ascertain a subset of optimal features with powerful classification capability based on specific evaluation criteria. Moreover, feature selection provides better interpretation ability as high-dimensional characteristics can be obtained by analysing low-dimensional datasets. Hence, feature selection can be applied as a tool for simplification of data and reduction of computational complexity for learning algorithms.

Class imbalance is the frequent and common challenge for the researchers in the field of data mining [13, 14]. Area. Class imbalance is concerned with the fact that the samples related to the one class is much higher in number than that of samples in other class. Conventional learning approaches are often biased towards the class with higher number of samples, while the class of importance is usually the class with the less number of samples. Consequently, lower sensitivity and higher specificity are reported for these datasets. The available solutions for class imbalance problems can be divided into four categories namely: cost-sensitive techniques, algorithm-driven methods, data-level methods, and integration approaches. Data-level methods are extensively implemented among other techniques as these methods include easy operations and found to be independent of algorithms. The main notion of data-level methods are associated with resampling concept. Resampling can be further divided into two major concepts namely: oversampling and undersampling. In the last few years, oversampling techniques have produced excellent results for imbalanced biological datasets.

After applying the above mentioned pre-processing approaches, selection of effective learning algorithms always plays a vital role in improving the prediction performances. Predominantly, ensemble learning algorithms produce a better sensitivity and accuracy when compared to conventional learning algorithms in case of biological datasets.

In the current study, we employ iFeature web server to extract diverse features of fixed-length feature vectors from variety length of peptide sequences containing anti-cancer and non-anti-cancer. Then, fuzzy rough feature selection with harmony search and particle swarm optimization (PSO) search is implemented to produce non-redundant and relevant features. Next, SMOTE is applied on reduced datasets to generate balanced datasets to avoid the biasedness of learning algorithms. Moreover, eight classifiers are used to evaluate their performances on highly balanced reduced datasets based on the evaluation parameters. Finally, we observe that the prediction of anti-cancer and non-anti-cancer peptides by using pre-processing techniques, such as feature extraction, feature selection, and data balancing techniques followed by selection of an appropriate machine learning algorithm. Entire methodology is displayed through schematic framework (see Fig. 2.1).

2.2 Material and Methods

2.2.1 Dataset

An authentic and informative benchmark dataset is always a key aspect to develop suitable and robust models. In the present work, we have completed the experimental study by using a benchmark dataset ACP500, which was obtained by performing instance selection in Wei et al. [4] dataset. ACP500 was comprised of 200 positive class samples, i.e. anti-cancer peptides and 300 negative class samples, i.e. non-anti-cancer peptides after applying instance selection. This dataset (ACP500) is consisted of the ubiquitous class imbalance issue as the ratio of positive to negative samples is different from 1:1. Wei et al. [4] exercised three steps to construct ACP500 dataset. Firstly, various literatures and BIOPEPUWM databases were utilized to collect different anti-cancer peptides, which were already validated through a series of experiments. Moreover, anti-cancer peptides presented by Wei et al. were accumulated as non-anti-cancer peptides as anti-cancer taste contains the intuitively aversive

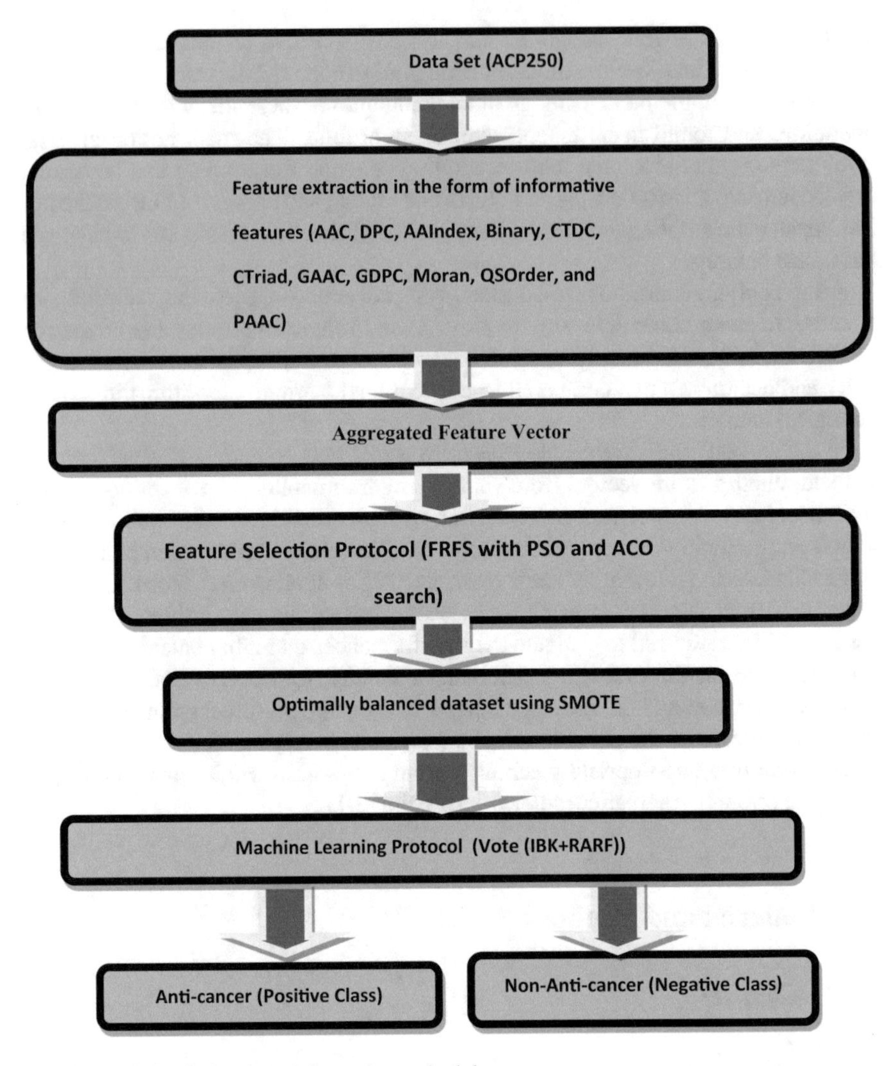

Fig. 2.1 Schematic framework for entire methodology

taste when compared to anti-cancer taste. Secondly, peptides comprising of non-standard letters such as "B", "X", "U", or "Z" were eliminated. Thirdly, peptide sequences having redundancy were also deleted for further processing.

2.2.2 *Feature Extraction*

The feature extraction from peptide samples is the foremost and essential step for developing a strong predictor. Peptide sequences consisting of anti-cancer and non-anti-cancer classes can be formulated with an extensive and suitable feature vector that demonstrates an effective correlation with decision class for creating a powerful predictor. Amino acid sequences carry diverse aspects, including composition, permutation, physicochemical properties, profiles, and combination modes of amino acids, which can be used for differentiating positive and negative samples. In this paper, feature vector of fixed length from various protein sequences has been generated by using iFeature web server [15]. Here, we have extracted 11 types of features [16, 17] namely: Amino acid composition (AAC), dipeptide composition (DPC), binary composition (binary), Moran correlation (Moran), composition/transition/distribution (CTD), pseudo-amino acid composition (PAAC), conjoint triad (CTF), quasi-sequence order (QSO), amino acid index (AAI), grouped dipeptide composition (GDPC), grouped amino acid composition (GAAC). Hence, we have applied feature extraction process on this dataset to extract appropriate features for conducting the entire experiments. After applying feature extraction, we obtain 489 suitable features from different categories.

2.2.3 *Feature Selection*

Feature selection is an effective and widely used method to remove negative effects caused by irrelevant and/or redundant features [18, 19]. Feature selection approach is preferred over other reduction techniques as it maintains the original sense of the available features. Data generated from various sources usually include the problems of imprecision, vagueness, and uncertainty. Conventional feature selection methods cannot address these issues in effective and efficient way. Moreover, traditional feature selectors can lead to information loss as they require additional information to eliminate redundant and/irrelevant features. In the recent years, rough set theory [20] has been efficaciously utilized as a mathematical notion to handle aforementioned issues in data analysis. Three aspects are responsible for the successful implementation of rough set theory. First, it produces depiction of knowledge in a minimal form for the available data. Second, this facilitates to analyse only the hidden facts present in the datasets. Third, no extra information such as threshold value or expert knowledge about particular domain is needed for data analysis. Various feature selection algorithms based on rough set theory have been introduced and successfully applied for high-dimensional datasets. However, classical rough set theory is based on equivalence relation concept and can handle nominal attributes only. So, it is not applicable on real-life datasets as these contain continuous features.

One of the attainable solution to deal with numerical attributes is to discretize these attributes, which is an important cause of information loss. In response to this issue, fuzzy rough set theory [21] was utilized to avoid the problem of discretization to deal with real-valued datasets. In the last few years, different fuzzy rough set-based feature selection [22] have been developed and effectively applied to the datasets containing numerical or continuous features. Fuzzy rough sets summarize connected but distinct notions of indiscernibility and fuzziness. In fuzzy rough sets assisted feature selection, fuzzy binary relations are employed to characterize the similarity between data points. So, continuous feature values are no longer required to be discretized. They are transformed to membership grades of samples corresponding to decisions. Hence, information about numerical or continuous feature values can be preserved.

Dataset with numerical or continuous feature values is normalized to fuzzy information system (U, C, D). Here, U indicates a non-empty set of data points, C denotes collection of conditional features, and D represents collection of decision features. Fuzzy lower and upper approximations are computed based on similarity between features. By calculating union of lower approximations, we can obtain positive region. This positive region indicates degree of certainty of a sample belong to a category. Now, the degree of dependency of target feature over conditional feature is assessed based on positive region value. In this approach, features are added iteratively to potential reduct set and evaluation is done based on degree of dependency. Finally, the algorithm provides reduct set when some convergence criteria is attained. Various extensions of fuzzy rough feature selection concept have been presented in the recent years [23–29]. In this study, we have applied FRFS with ACO and PSO to obtain the reduct sets [30, 31].

2.2.4 Balancing Protocol

Due to development of IoT and advanced data computing technologies, various researchers have access these advancement to achieve more data with various types and large amount. However, imbalanced data problem leads to information loss for artificial intelligence models built on these data and results in extremely poor performances [32]. Essentially, an imbalanced dataset refers to those samples in the dataset which fails to provide approximately equal representations for all patterns. Oversampling is well-known efficient technique to cope with class imbalance ubiquitous issue by reduplicating or generating the minority class samples, resulting in balance between the samples of the majority and minority class. Chawla et al. [33] introduced an interesting and massively applied oversampling technique, which is known as synthetic minority oversampling technique (SMOTE). In this method, synthetic samples of minority class are generated by employing the idea of nearest neighbour. These samples are created artificially by randomly choosing minority class samples and its nearest sample from the same class. For each instance m related to minority class, n nearest neighbours of the identical class are selected. If the available sampling rate is j, then j instances from n nearest neighbours are selected randomly. Here, a new object can be generated by using the information from each object m_i ($i = 1, 2, \ldots, j$) and the original instance m as below:

$$m_{\text{new}} = m + \text{rand}(0, 1) \times (m_i - m)$$

where rand(0, 1) indicates a random number in the interval (0, 1) and plays vital role in the generation of synthetic samples. After combining the generated artificial samples with the original available samples, we obtain a new balanced dataset for further processing.

2.2.5 *Machine Learning Protocol*

A comprehensive experimental study has been performed in the current study to justify the domination of our methodology. Nine classifiers namely (1) random forest [34], (2) vote-based classifier [35, 36], (3) JRip [37], (4), boosted random forest (RARF) [38], (5) sequential minimization optimization (SMO) [39], (6) PART [40], (7) rotation forest [41], (8) IBK [42], and (9) Naïve Bayes [43], have been employed on original, reduced, and optimally balanced reduced versions of benchmark dataset (ACP500). These classifiers are extensively utilized learning algorithms on biological datasets to perform the prediction tasks effectively. Vote-based classifier has reported the best results in the current study. A brief depiction of vote-based classifier can be given as follows.

2.2.5.1 Vote-Based Classifier

For the biological datasets, the conventional classifiers have their own limitations and advantages while carrying out the prediction tasks. The vote-based classifier comes in the category of ensemble learning techniques, which can be applied to overcome the inadequacy of individual classifiers. In case of vote-based classification technique, different base classifiers are trained with training datasets. Thereafter, their predictions are combined to achieve better precision and recall, which results in the enhanced value of F-measure and overall accuracy when compared to individual models. Here, individual experts assign class labels to each pattern and vote-based classifiers operate on these class labels. The maximum value selectors are utilized to obtain the class labels, which can be achieved by adjusting the soft decision outputs. Specific expert produces the votes for each class and the vote rule output can be expressed as a function of votes. We always select the choice containing maximum value wherein the decision consisting of multiple choices. In this paper, IBK and RARF are used as base learner for vote-based classifier.

2.2.6 *Performance Measures*

There are various validation techniques such as jackknife test, independent dataset test, and *k*-fold cross validation for learning models evaluation. In this paper, tenfold cross validation (CV) and percentage split of 80:20 validation techniques have been used to evaluate our entire methodology for the prediction of anti-cancer and non-anti-cancer. In tenfold CV, entire dataset is randomly divided into 10 identical sets or folds, where all the 10 sets contain similar number of positive as well as negative samples. Then, 9 sets or folds are employed for training purpose and remaining 1 for testing. Entire procedure iterates 10 times, and we obtain average values of the evaluation metrics for different learning algorithms. In percentage split of 80:20 validation method, 80% of the samples are randomly divided for constructing a prediction model and rest 20% of the samples are applied for evaluation of the model.

In addition to efficient and feasible evaluation method for conducting the experiments, effective evaluation indicators are also required to assess the predictive potential of the model. In the experimental study, five basic metrics are examined for evaluating the performance of the nine learning algorithms. These metrics are the combination of both threshold-dependent and threshold-independent. Confusion matrix provides a way to compute evaluation parameters. There are four elements of the confusion matrix namely: true positive (tp), false positive (fp), true negative (tn), and false negative (fn). Here, tp denotes count of correctly predicted anti-cancer peptides, fp represents count of incorrectly predicted anti-cancer peptides, tn is the representation of correctly predicted non-anti-cancer peptides, whereas incorrectly predicted non-anti-cancer peptides are indicated by fn. Threshold-dependent and threshold-independent evaluation parameters can be briefly describes as follows:

Sensitivity: This parameter indicates the percentage of rightly predicted anti-cancer peptides and is calculated as follow:

$$\text{Sensitivity} = \frac{\text{tp}}{(\text{tp} + \text{fn})} \times 100$$

Specificity: This parameter denotes the percentage of rightly predicted non-anti-cancer peptides and is computed by the following equation:

$$\text{Specificity} = \frac{\text{tn}}{(\text{tn} + \text{fp})} \times 100$$

Accuracy: The percentage of rightly predicted anti-cancer and non-anti-cancer peptides is denoted by this metric and is expressed by the following equation:

$$\text{Accuracy} = \frac{\text{tp} + \text{tn}}{\text{tp} + \text{fp} + \text{tn} + \text{fn}} \times 100$$

AUC: This evaluation metric is the abbreviation of area under the receiver operating characteristic curve (AUROC). For a better performing predictor, AUC gives its value closer to 1.

MCC: Mathew's correlation coefficient is extensively used parameter for binary classifications and can be expressed by the following equation:

$$\text{MCC} = \frac{\text{tp} \times \text{tn} - \text{fp} \times \text{fn}}{\sqrt{(\text{tp} + \text{fp})(\text{tp} + \text{fn})(\text{tn} + \text{fp})(\text{tn} + \text{fn})}}$$

MCC value nearer to 1 can be contemplated as better predictor when compared to others.

2.3 Experimentation

In this paper, we have done a complete observational study to determine the strength, abstraction, and applicable authentication of our suggested methodology. In Beginning, iFeature web server has applied to excerpt various feature vectors of nailed length from 500 peptide sequences of ACP500 for a given each peptide sequence. Attributes were extracted from 11 various compositions, which encompasses of various properties for making competent predictors to identify anti-cancer and non-anti-cancer peptides. Thereafter, we imposed distinct traditional and fuzzy [44] concept-based attribute selection approach to rule out superfluous and extraneous features. Filter-based attribute selection [45] as classic crisp approach and FRFS along with PSO and ACO were used to give rise to useful and significant attributes. In our concern, FSS, FRFS along with PSO, and ACO has been used to get useful and significant attributes. In this FSS, FRFS with PSO and FRFS along with ACO has extensively produced 145, 73, and 117 attributes, respectively. Complete data with respect to ACP500 dataset after doing attribute extraction and attribute selection operations have recorded in Table 2.1. The reduced datasets which were given by various FRFS techniques had been changed into rationally balanced dataset with the help of SMOTE. In continuation of this positive to negative sample ratios are also changed into 1:1.

Further, we used majorly nine machine learning models to analyse their realization over reduced and unreduced datasets. Over fitting as well as unbiasedness of our given methodology have also been avoided by using tenfold CV and percentage split in ratios of 80:20 validation methods.

Attainment of all AI models for integral datasets which are being provided by FFS has also been represented in Tables 2.2 and 2.3. Estimation measures for diminished dataset as given by FFS have also been represented in Tables 2.4 and 2.5. Datasets

Table 2.1 Benchmark datasets characteristics and reduct size

Dataset	Instances	Attributes	Reduct size		
			FSS	FRFS with PSO	FRFS with ACO
ACP500	500	489	145	73	117

having more reduced attributes as given by FRFS using PSO and ACO and optimally uniform datasets by SMOTE had been used to give insight performance of various learning models and tabulated in Tables 2.6, 2.7, 2.8, and 2.9. With basis of experimental datasets, one can easily make conclusions that vote-based classifier is better performing in comparison of other machine learning algorithms. The finest results had been achieved with specificity, sensitivity, accuracy, AUC, and MCC of 99.1, 97.3%, 98.2%, 0.983, and 0.888, respectively, along with vote-based classifiers which are based upon percentage split in ratios of 80:20 for rationally balanced reduced dataset given by FRFS with ACO search followed by SMOTE.

The overall statistics have shown that our representational methodology (approach) is the best approach for favouring the anti-cancer and non-anti-cancer peptides as the lastly reported best analysis was produced with specificity, sensitivity, accuracy, AUC, and MCC 97.5%, 75.4%, 91.5%, 0.902, and 0.683, respectively. As per tenfold cross validation, our supposed methodology has been reported

Table 2.2 Performance results of different learning algorithms based on evaluation metrics for original ACP500 dataset using percentage split of 80:20

IBK	Specificity	Sensitivity	Accuracy	AUC	MCC
Naïve Bayes	86.4	81.6	83.8	0.842	0.609
IBK	91.4	72.7	86.5	0.805	0.602
JRip	91.7	69.6	84.7	0.826	0.604
Random forest	92.7	74.4	86.6	0.902	0.618
Rotation forest	92.4	75.9	88.8	0.904	0.624
PART	89.1	66.7	81.6	0.804	0.504
SMO	88.1	78.4	86.1	0.808	0.625
Vote	93.1	77.3	82.7	0.914	0.656
RARF	92.4	75.1	87.8	0.902	0.619

Table 2.3 Performance of various learning algorithms based on evaluation metrics for original ACP500 dataset using tenfold CV

Learning algorithm	Specificity	Sensitivity	Accuracy	AUC	MCC
Naïve Bayes	84.8	82.0	88.8	0.826	0.645
IBK	91.1	66.7	88.1	0.851	0.656
JRip	86.5	64.7	82.4	0.762	0.551
Random forest	93.4	61.3	84.8	0.885	0.592
Rotation forest	88.2	68.3	79.4	0.895	0.529
PART	75.3	65.0	74.7	0.769	0.363
SMO	78.0	66.3	77.0	0.742	0.474
Vote	91.1	68.7	86.1	0.865	0.686
RARF	90.1	68.3	87.0	0.896	0.689

Table 2.4 Performance results of different learning algorithms based on evaluation metrics using percentage split of 80:20 for reduced ACP500 dataset generated by conventional filter subset selection method

Learning algorithm	Specificity	Sensitivity	Accuracy	AUC	MCC
Naïve Bayes	87.4	78.4	81.2	0.893	0.659
IBK	82.7	69.1	80.6	0.800	0.598
JRip	86.7	74.0	80.8	0.823	0.599
Random forest	89.7	71.3	81.6	0.889	0.655
Rotation forest	88.4	68.1	79.6	0.802	0.659
PART	87.7	69.4	80.9	0.834	0.653
SMO	88.4	71.6	86.4	0.798	0.666
Vote	91.1	72.0	84.1	0.827	0.686
RARF	88.1	71.0	81.7	0.892	0.686

Table 2.5 Performance results of different learning algorithms based on evaluation metrics using tenfold CV for reduced ACP500 dataset produced by conventional filter subset selection method

Learning algorithm	Specificity	Sensitivity	Accuracy	AUC	MCC
Naïve Bayes	83.0	81.3	82.8	0.816	0.651
IBK	88.7	52.3	71.3	0.727	0.451
JRip	90.4	64.7	81.0	0.718	0.650
Random forest	86.7	61.3	82.7	0.827	0.517
Rotation forest	87.7	72.0	81.0	0.814	0.624
PART	83.0	71.7	74.7	0.716	0.314
SMO	84.2	72.7	82.0	0.824	0.604
Vote	88.9	63.5	76.5	0.814	0.525
RARF	84.3	64.3	88.7	0.827	0.517

Table 2.6 Performance results of different learning algorithms based on evaluation metrics using percentage split of 80:20 for reduced ACP500 dataset generated by FRFS with PSO search

Learning algorithm	Specificity	Sensitivity	Accuracy	AUC	MCC
Naïve Bayes	71.0	76.5	75.9	0.816	0.517
IBK	92.4	86.2	94.9	0.918	0.829
JRip	81.9	83.7	85.8	0.824	0.715
Random forest	92.2	81.7	84.4	0.910	0.729
Rotation forest	90.8	82.3	95.1	0.920	0.812
PART	82.2	81.0	84.6	0.813	0.624
SMO	91.6	82.3	85.6	0.826	0.714
Vote	93.8	91.4	94.7	0.919	0.827
RARF	92.8	90.7	95.7	0.904	0.809

Table 2.7 Performance results of different learning algorithms based on evaluation metrics using tenfold CV for reduced ACP500 dataset produced by FRFS with PSO search

IBK	Specificity	Sensitivity	Accuracy	AUC	MCC
Naïve Bayes	77.5	84.8	78.6	0.898	0.574
IBK	88.7	96.4	93.1	0.957	0.872
JRip	84.4	84.5	84.9	0.862	0.689
Random forest	89.7	92.4	88.6	0.978	0.771
Rotation forest	89.4	95.1	87.7	0.972	0.786
PART	87.1	88.8	87.9	0.892	0.739
SMO	87.8	89.4	88.1	0.891	0.772
Vote	89.7	96.0	93.4	0.979	0.849
RARF	88.7	95.7	92.7	0.980	0.835

Table 2.8 Performance results of different learning algorithms based on evaluation metrics using percentage split of 80:20 for reduced ACP500 dataset generated by FRFS with ACO search

Learning algorithm	Specificity	Sensitivity	Accuracy	AUC	MCC
Naïve Bayes	89.5	82.0	86.3	0.877	0.698
IBK	95.4	95.3	95.4	0.957	0.888
JRip	87.2	83.7	86.5	0.892	0.690
Random forest	97.1	93.7	95.4	0.982	0.888
Rotation forest	97.1	88.7	92.9	0.982	0.871
PART	92.2	88.7	89.4	0.897	0.789
SMO	89.5	92.0	88.3	0.883	0.795
Vote	99.1	97.3	98.2	0.983	0.888
RARF	95.4	92.0	93.7	0.959	0.845

Table 2.9 Performance results of different learning algorithms based on evaluation metrics using tenfold CV for reduced ACP500 dataset produced by FRFS with ACO search

IBK	Specificity	Sensitivity	Accuracy	AUC	MCC
Naïve Bayes	83.5	88.4	86.4	0.950	0.680
IBK	88.1	97.7	91.4	0.952	0.870
JRip	83.8	89.1	83.9	0.872	0.740
Random forest	88.4	92.4	89.9	0.975	0.788
Rotation forest	89.7	95.1	86.9	0.971	0.778
PART	88.4	89.1	89.2	0.872	0.775
SMO	87.8	93.1	89.9	0.875	0.790
Vote	97.6	99.8	99.7	0.999	0.902
RARF	89.7	95.1	90.9	0.932	0.789

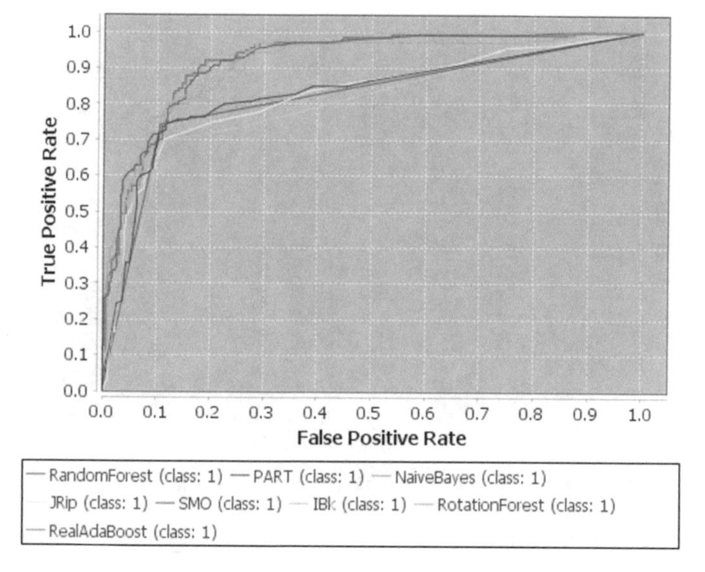

Fig. 2.2 AUC of eight learning algorithms for original ACP500 dataset

the best outcomes with specificity of 97.6%, sensitivity of 99.8%, accuracy of 99.7%, AUC of 0.999, and MCC of 0.902 as the lastly reported best results were 95.0% for specificity, sensitivity for 95.5%, accuracy for 95.5%, AUC for 0.965, and MCC for 0.888. In Tables 2.1, 2.2, 2.3, 2.4, 2.5, 2.6, 2.7, 2.8, and 2.9, it is reported that fuzzy-based attribute selection is more active and adequate than traditional methods used for making of powerful predictor for foretelling anti-cancer and non-anti-cancer peptides. The capacious approach called ROC is more applicable to visualize the realization of all nine learning algorithms. In Figs. 2.2, 2.3, 2.4, and 2.5, there is a plot of ROC for decreased and non-decreased ACP500 datasets. Extraction of attributes was done at the beginning with iFeature web server, while experiments like as attribute selection, classification, and visualization using various validation approaches were accomplished in Weka 3.8 [46] on Hardware platform using Intel® Core™ i5-8265U CPU @ 1.60 GHz, 1.80 GHz with 8.00 GB RAM.

2.4 Conclusion

Anti-cancer taste plays vital role in the field of food industry as well as medical science. Anti-cancer ingredients have been successfully utilized for treatment of various critical diseases such as stroke and food seasoning. So, it is prerequisite to present an efficient computational methodology to improve the discriminating ability of machine learning algorithms for anti-cancer and non-anti-cancer peptides. Informative features related to different composition, relevant and non-redundant features,

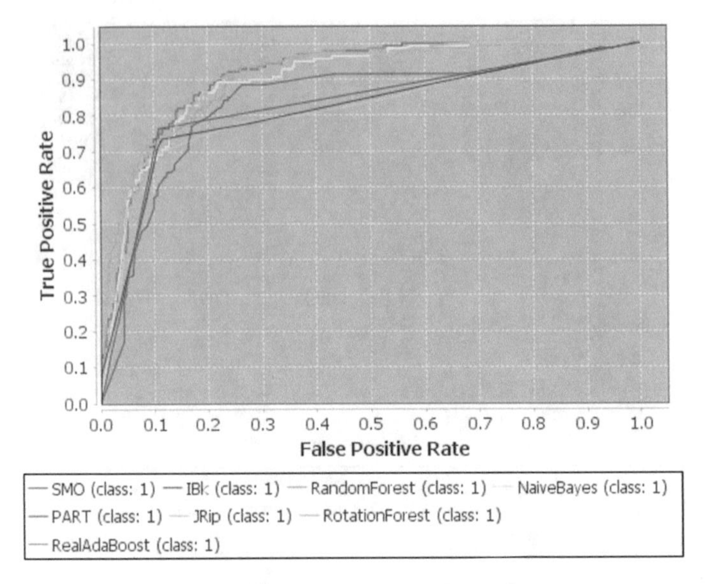

Fig. 2.3 AUC of eight learning algorithms for reduced ACP500 dataset generated by conventional filter subset selection method

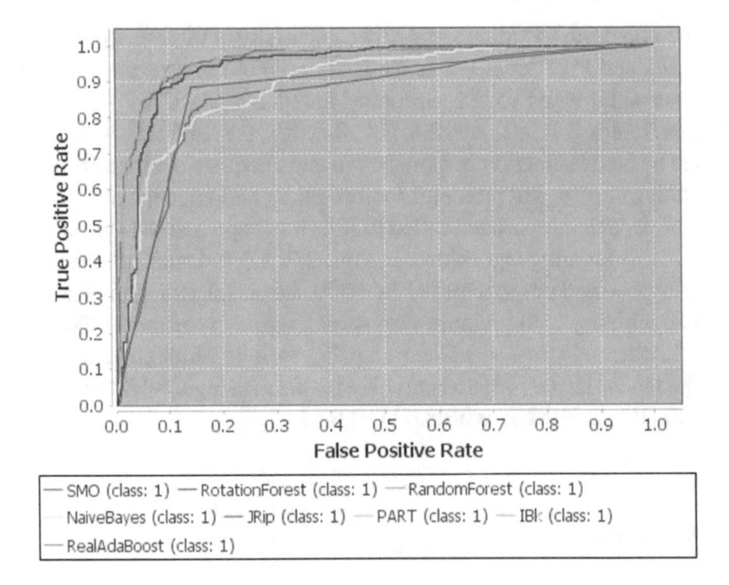

Fig. 2.4 AUC of eight learning algorithms for reduced ACP500 dataset produced by FRFS with PSO search

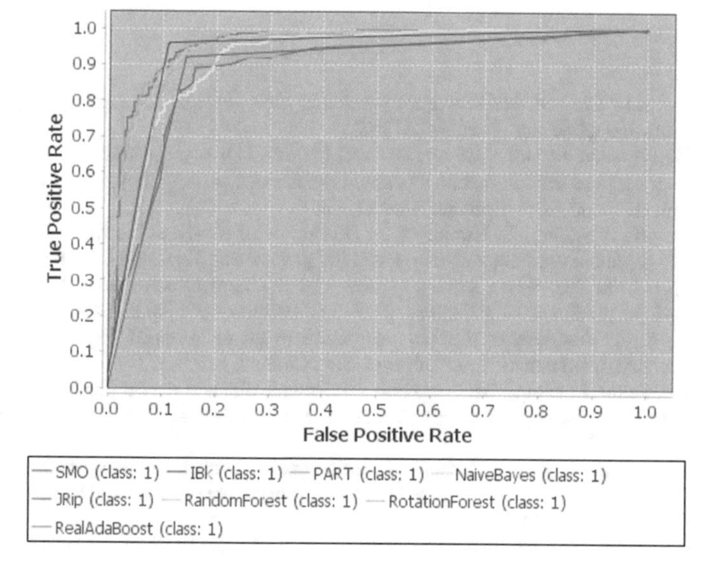

Fig. 2.5 AUC of eight learning algorithms for reduced ACP500 dataset produced by FRFS with ACO search

optimal balancing, and effective learning algorithms are the key factors that can lead to upgrade the overall performances of the classification techniques. In this paper, we focussed on all the aforementioned factors and developed an efficient procedure to enhance the prediction of anti-cancer and non-anti-cancer peptides. Firstly, peptide sequences consisted of anti-cancer and non-anti-cancer peptides were loaded on iFeature web server and 489 features were extracted based on AAC, CTD, DPC, PAAC, binary, Moran, AAI, QSO, GDPC, CTF, and GAAC compositions. Secondly, we obtained a reduced feature set of 117 features after eliminating irrelevant and redundant features by using FRFS with ACO search. Thirdly, SMOTE was applied on the imbalanced dataset with reduced feature set and optimally balanced reduced dataset was produced. Now, the performances of various classifiers were explored with the optimally balanced reduced dataset by using percentage split of 80:20 and tenfold CV. Then, a comprehensive comparative experimental work was performed. Finally, we achieved the best results by using vote-based classifier with sensitivity of 97.1%, specificity of 99.1%, accuracy of 98.2%, MCC of 0.888, and AUC of 0.983 based on percentage split of 80:20 validation method. Entire experimental study indicated that results provided by our proposed approach are better than existing results till date.

References

1. Xu, D., et al.: ACHP: A web server for predicting anti-cancer peptide and anti-hypertensive peptide. Int. J. Peptide Res. Ther. 1–12 (2021)
2. Lane, N., Kahanda, I.: DeepACPpred: a novel hybrid CNN-RNN architecture for predicting anti-cancer peptides. In: International Conference on Practical Applications of Computational Biology & Bioinformatics. Springer, Cham (2020)
3. Sakamoto, K., Masutani, T., Hirokawa, T.: Generation of KS-58 as the first K-Ras (G12D)-inhibitory peptide presenting anti-cancer activity in vivo. Sci. Rep. **10**(1), 1–16 (2020)
4. Wei, L., et al.: ACPred-FL: a sequence-based predictor using effective feature representation to improve the prediction of anti-cancer peptides. Bioinformatics **34**(23), 4007–4016 (2018)
5. You, H., et al.: Anti-cancer peptide recognition based on grouped sequence and spatial dimension integrated networks. Interdiscipl. Sci.: Comput. Life Sci. 1–13 (2021)
6. Chantawannakul, J., et al.: Virtual screening for biomimetic anti-cancer peptides from *Cordyceps militaris* putative pepsinized peptidome and validation on colon cancer cell line. Molecules **26**(19), 5767 (2021)
7. Jain, P., Tiwari, A.K., Som, T.: Enhanced prediction of anti-tubercular peptides from sequence information using divergence measure-based intuitionistic fuzzy-rough feature selection. Soft. Comput. **25**(4), 3065–3086 (2021)
8. Jensen, R., Shen, Q.: Computational intelligence and feature selection: rough and fuzzy approaches (2008)
9. Shang, C., Barnes, D.: Fuzzy-rough feature selection aided support vector machines for mars image classification. Comput. Vis. Image Underst. **117**(3), 202–213 (2013)
10. Nath, A., Subbiah, K.: Maximizing lipocalin prediction through balanced and diversified training set and decision fusion. Comput. Biol. Chem. **59**, 101–110 (2015)
11. Eyers, C.E., Lawless, C., Wedge, D.C., Lau, K.W., Gaskell, S.J., Hubbard, S.J.: CONSeQuence: prediction of reference peptides for absolute quantitative proteomics using consensus machine learning approaches. Mol. Cell. Proteomics **10**(11), M110.003384 (2011)
12. Nath, A.A., Subbiah, K.: Probing an optimal class distribution for enhancing prediction and feature characterization of plant virus-encoded RNA-silencing suppressors. 3 Biotech **6**(1), 93 (2016)
13. Nath, A., Karthikeyan, S.: Enhanced prediction and characterization of CDK inhibitors using optimal class distribution. Interdiscipl. Sci.: Comput. Life Sci. **9**(2) (2017)
14. Pirizadeh, M., Alemohammad, N., Manthouri, M., Pirizadeh, M.: A new machine learning ensemble model for class imbalance problem of screening enhanced oil recovery methods. J. Petrol. Sci. Eng. **198**, 108214 (2021)
15. Chen, Z., Zhao, P., Li, F., Leier, A., Marquez-Lago, T.T., Wang, Y., Webb, G.I., Smith, A.I., Daly, R.J., Chou, K.-C.: iFeature: a python package and web server for features extraction and selection from protein and peptide sequences. Bioinformatics **34**(14), 2499–2502 (2018)
16. Manavalan, B., Basith, S., Shin, T.H., Wei, L., Lee, G.: AtbPpred: a robust sequence-based prediction of anti-tubercular peptides using extremely randomized trees. Comput. Struct. Biotechnol. J. **17**, 972–981 (2019)
17. Usmani, S.S., Bhalla, S., Raghava, G.P.: Prediction of antitubercular peptides from sequence information using ensemble classifier and hybrid features. Front. Pharmacol. **9**, 954 (2018)
18. Khalid, S., Khalil, T., Nasreen, S.: A survey of feature selection and feature extraction techniques in machine learning. In: 2014 Science and Information Conference. IEEE (2014)
19. Liu, H., Motoda, H. (eds.): Computational Methods of Feature Selection. CRC Press (2007)
20. Pawlak, Z.: Rough sets: Theoretical Aspects of Reasoning About Data. Springer (2012)
21. Dubois, D., Prade, H.: Putting rough sets and fuzzy sets together. Intelligent Decision Support, pp. 203–232. Springer (1992)
22. Jensen, R., Shen, Q.: Fuzzy–rough attribute reduction with application to web categorization. Fuzzy Sets Syst. **141**(3), 469–485 (2004)
23. Chen, J., Mi, J., Lin, Y.: A graph approach for fuzzy-rough feature selection. Fuzzy Sets Syst. **391**, 96–116 (2020)

24. Jensen, R., Shen, Q.: Semantics-preserving dimensionality reduction: rough and fuzzy-rough-based approaches. IEEE Trans. Knowl. Data Eng. **16**(12), 1457–1471 (2004)
25. Jensen, R., Shen, Q.: Fuzzy-rough sets assisted attribute selection. IEEE Trans. Fuzzy Syst. **15**(1), 73–89 (2007)
26. Jensen, R., Shen, Q.: New approaches to fuzzy-rough feature selection. IEEE Trans. Fuzzy Syst. **17**(4), 824–838 (2008)
27. Sheeja, T., Kuriakose, A.S.: A novel feature selection method using fuzzy rough sets. Comput. Ind. **97**, 111–116 (2018)
28. Wang, C., Huang, Y., Ding, W., Cao, Z.: Attribute reduction with fuzzy rough self-information measures. Inf. Sci. **549**, 68–86 (2021)
29. Wang, C., Huang, Y., Shao, M., Fan, X.: Fuzzy rough set-based attribute reduction using distance measures. Knowl.-Based Syst. **164**, 205–212 (2019)
30. Jensen, R., Shen, Q.: Fuzzy-rough data reduction with ant colony optimization. Fuzzy Sets Syst. **149**(1), 5–20 (2005)
31. Son, H., Kim, C.: Forecasting short-term electricity demand in residential sector based on support vector regression and fuzzy-rough feature selection with particle swarm optimization. Procedia Eng. **118**, 1162–1168 (2015)
32. Mukherjee, M., Khushi, M.: SMOTE-ENC: a novel SMOTE-based method to generate synthetic data for nominal and continuous features. Appl. Syst. Innov. **4**(1), 18 (2021)
33. Chawla, N.V., Bowyer, K.W., Hall, L.O., Kegelmeyer, W.P.: SMOTE: synthetic minority over-sampling technique. J. Artif. Intell. Res. **16**, 321–357 (2002)
34. Breiman, L.: Random forests. Mach. Learn. **45**(1), 5–32 (2001)
35. Kuncheva, L.I.: Combining Pattern Classifiers: Methods and Algorithms. Wiley (2014)
36. Ashraf, M., Zaman, M., Ahmed, M.: To ameliorate classification accuracy using ensemble vote approach and base classifiers. Emerging Technologies in Data Mining and Information Security, pp. 321–334. Springer (2019)
37. Leon, F., Zaharia, M.H., Gâlea, D.: Performance analysis of categorization algorithms
38. Breiman, L.: Bagging predictors. Mach. Learn. **24**(2), 123–140 (1996)
39. Platt, J.: Sequential Minimal Optimization: A Fast Algorithm for Training Support Vector Machines (1998)
40. Frank, E., Witten, I.H.: Generating Accurate Rule Sets Without Global Optimization (1998)
41. Rodriguez, J.J., Kuncheva, L.I., Alonso, C.J.: Rotation forest: a new classifier ensemble method. IEEE Trans. Pattern Anal. Mach. Intell. **28**(10), 1619–1630 (2006)
42. Kumar, V., Zinovyev, R., Verma, A., Tiwari, P.: Performance Evaluation of Lazy and Decision Tree Classifier: A Data Mining Approach for Global Celebrity's Death Analysis, pp. 1–6
43. Mukherjee, S., Sharma, N.: Intrusion detection using naive Bayes classifier with feature reduction. Procedia Technol. **4**, 119–128 (2012)
44. Zadeh, L.A.: Fuzzy Sets, Fuzzy Sets, Fuzzy Logic, and Fuzzy Systems: Selected Papers by Lotfi A Zadeh, pp. 394–432. World Scientific (1996)
45. Ambusaidi, M.A., He, X., Nanda, P., Tan, Z.: Building an intrusion detection system using a filter-based feature selection algorithm. IEEE Trans. Comput. **65**(10), 2986–2998 (2016)
46. Hall, M., Frank, E., Holmes, G., Pfahringer, B., Reutemann, P., Witten, I.H.: The WEKA data mining software: an update. ACM SIGKDD Expl. Newslett. **11**(1), 10–18 (2009)

Chapter 3
New Methods of Vagueness and Uncertainty Quantification in Lattice Boltzmann Method-Based Solute Transport Model

Tushar Kanti Pal, Debabrata Datta, and R. K. Bajpai

3.1 Introduction

Mathematical models of various science and engineering problems are associated with parameters which cannot be treated as crisp parameters because of scarcity in data collected and heterogeneity of the material/media involved with the process. For example, the parameters permeability, porosity, etc., of a geological material associated with groundwater flow models are imprecise in nature. Vagueness is another word generally used to describe data uncertainty refers to lack of definite or sharp distinction in data collected in the form of linguistic variable from various sources such as literatures and expert opinions [1]. The model outcomes become highly uncertain because of imprecision and vagueness present in the data of model parameters. These types of problems with imprecise and vague data require estimation of plausible range of the model outputs. Application of uncertainty analysis is useful for enhancing confidence in model estimates and also useful for weighting the various model estimates. Quantification of uncertainty and vagueness is therefore an unavoidable and inevitable component of such studies. There are generally two types of parametric uncertainty analysis techniques: the first one is called aleatory uncertainty analysis where natural variability or randomness are the source of uncertainty in the parameters, whereas in the second one which is called epistemic uncertainty analysis, lack of knowledge and insufficient information are the source of uncertainty [2]. Traditional Monte Carlo simulation (MCS) is used to quantify aleatory uncertainty whereas various mathematical tools, such as fuzzy sets, interval-valued

T. K. Pal (✉) · R. K. Bajpai
Technology Development Division, Bhabha Atomic Research Centre, Mumbai, India
e-mail: tkpal@barc.gov.in

D. Datta
Radiological Physics and Advisory Division, Bhabha Atomic Research Centre, Mumbai, India

T. Som et al. (eds.), *Fuzzy, Rough and Intuitionistic Fuzzy Set Approaches for Data Handling*, Forum for Interdisciplinary Mathematics,
https://doi.org/10.1007/978-981-19-8566-9_3

fuzzy sets, fuzzy-valued (type-2) fuzzy sets, intuitionistic fuzzy sets, rough sets, hybridization of rough and fuzzy sets, are used to quantify epistemic uncertainty.

Fuzzy set, proposed by Zadeh [3], is used to model non-statistical uncertainty due to imprecise data which are spread around a most likely measured value or data with a most likely measured value that will lie in between its minimum and maximum. These types of impression or fuzziness of model parameters are easily represented by fuzzy numbers [4]. Intuitionistic fuzzy set (IFS), which was introduced by Atanassov [5, 6] as a generalized version for the existing fuzzy sets, is more efficient than fuzzy sets in tackling vagueness and representing imperfect knowledge of model parameters. Parametric uncertainty analysis of a mathematical model having closed form solution is easily carried out using various techniques such as MCS, fuzzy set, IFS theory. However, in many practical situations, it become difficult to obtain analytical solution of a physical model which represents a nonlinear process and also for models with complicated boundary interfaces. In such cases, application of numerical techniques is inevitable. Simulation of migration of dissolved contaminants through groundwater which is widely used for environmental assessment of a contaminated site often requires numerical techniques for solving the governing solute transport equation. Since the medium (geological rock, soil, etc.) though which migration of solute takes place is highly heterogeneous, deterministic solute transport equation is not applicable here. In fact, the parameters of the solute transport equation are imprecise and vague due to scarcity of experimental data and/or use of experts' opinion in linguistic forms. Therefore, uncertainty and vagueness analysis of the solute transport model outcomes in the form of solute concentration at a given spatio-temporal location are invaluable for appropriate decision making with confidence. Researchers have successfully applied traditional numerical techniques such as finite difference method (FDM), finite element method (FEM) as well as advanced numerical techniques such as lattice Boltzmann (LB) method, differential quadrature method (DQM) for uncertainty analysis of solute transport model with imprecise model parameters which are represented by fuzzy numbers [7–11]. IFS theory has not been used to elaborate uncertainty and vagueness associated with solute transport model with vague parameters. This may be due to lack of attempt to construct intuitionistic fuzzy numbers for the model parameters.

In this chapter, an attempt has been made to explore LB technique [12, 13] for numerically solving solute transport equation with fuzzy as well as vague. Here, imprecise and vague parameters are represented by fuzzy and intuitionistic fuzzy numbers, and LB technique is used to solve the deterministic equation obtained using fuzzy and intuitionistic fuzzy vortex methods [14]. These newly proposed numerical techniques are therefore named as fuzzy lattice Boltzmann (FLB) and intuitionistic fuzzy lattice Boltzmann (IFLB) methods. The chapter presents the detailed numerical formulation of the LB scheme to solve a governing equation with fuzzy and vague parameters as its coefficients. The developed schemes are applied for vagueness and uncertainty analysis of one-dimensional model of solute transport in saturated porous media. The remaining part of the chapter is organized in the following way. Section 3.2 presents the mathematical model of solute transport through saturated porous media in presence of imprecise and vague model parameters. Section 3.3

presents preliminaries of fuzzy sets and IFS including fuzzy vertex method and intuitionistic fuzzy vertex method used to solve the governing solute transport equation. Details of the numerical schemes, FLB and IFLB are provided in Sect. 3.4. Construction of fuzzy and intuitionistic fuzzy numbers using experts' opinion and development of LB solver and use of the solver for analysis of uncertainty and vagueness of the solute transport model output for the presence of imprecise and vague parameters, hydrodynamic dispersion coefficient and groundwater velocity are reported in Sect. 3.5.

3.2 Solute Transport Model with Imprecise and Vague Parameters

Transport of solutes through saturated porous media is governed by advection dispersion equation (ADE). These types of models are used to compute concentration of solute (contaminant) moving through an aquifer at a specified location away from the point of discharge and at any given time. Here, one-dimensional form of ADE as given in Eq. (3.1) is considered for uncertainty analysis of the estimated solute concentration. The parameters involved with the ADE are groundwater flow velocity and hydrodynamic dispersion coefficient. For a saturated porous media, the governing 1D ADE can be written as

$$\frac{\partial C(x,t)}{\partial t} = D_L \frac{\partial^2 C(x,t)}{\partial x^2} - u \frac{\partial C(x,t)}{\partial x} \tag{3.1}$$

where $C(x, t)$ is solute concentration (mg/L), u represents groundwater flow velocity (m/day), D_L stands for longitudinal hydrodynamic dispersion coefficient (m^2/day), x and t are downstream distance (m) and time of observation (days), respectively. The above equation is subjected to following initial and boundary conditions.

$$C(x,t) = \begin{cases} 0 \quad \forall x \geq 0 \text{ at } t = 0 \\ C_0 \; \forall t \geq 0 \;\text{ at } x = 0 \\ 0 \quad \forall t \geq 0 \text{ at } x = \infty \end{cases}$$

For the above initial and boundary conditions, closed form solution of Eq. (3.1) can be written as [8]

$$C(x,t) = \frac{C_0}{2}\left[\operatorname{erfc}\left(\frac{x - ut}{\sqrt{4D_L t}}\right) + \exp\left(\frac{ux}{D_L}\right)\operatorname{erfc}\left(\frac{x + ut}{\sqrt{4D_L t}}\right)\right] \tag{3.2}$$

where

$$\operatorname{erf}(\beta) = \frac{2}{\sqrt{\pi}} \int_{0}^{\beta} e^{-t^2} \mathrm{d}t \quad \text{and} \quad \operatorname{erfc}(\beta) = 1 - \operatorname{erf}(\beta)$$

Here, $\operatorname{erf}(x)$ and $\operatorname{erfc}(x)$ are mathematical error function and complementary error functions, respectively. In order to model uncertainty and vagueness associated with the estimated concentration of solute by solving an ADE with imprecise or vague parameters, it is required to represent the model parameters in a proper way. When there is most likely value of the measured data and others data are around the most probable value, triangular fuzzy number is generally used to maintain the measurement uncertainty in their representation, and when the numerical data are collected from experts, triangular intuitionistic fuzzy number is used to represent vagueness of these measures and to incorporate uncertainty due to hesitation.

When the parameters of the 1D ADE are fuzzy variables, Eq. (3.1) can be rewrite as

$$\frac{\partial \tilde{C}}{\partial t} = \widetilde{D}_L \frac{\partial^2 \tilde{C}}{\partial x^2} - \tilde{u} \frac{\partial \tilde{C}}{\partial x} \tag{3.3}$$

where $\widetilde{D}_L$ and $\tilde{u}$ are two fuzzy variables representing longitudinal hydrodynamic dispersion coefficient and groundwater flow velocity, respectively. The tilde, "~", sign is used to distinguish these fuzzy parameters from their classical (crisp) representation. Similarly considering vagueness of the parameters of the governing Eq. (3.1), we can rewrite it as

$$\frac{\partial \overset{\sim i}{C}}{\partial t} = \overset{\sim i}{D}_L \frac{\partial^2 \overset{\sim i}{C}}{\partial x^2} - \overset{\sim i}{C} \frac{\partial \overset{\sim i}{C}}{\partial x} \tag{3.4}$$

where $\overset{\sim i}{D}_L$ and $\overset{\sim i}{u}$ are two intuitionistic fuzzy variables representing longitudinal hydrodynamic dispersion coefficient and groundwater flow velocity, respectively. Here, the tilde sign and i, "~i" sign is used to distinguish these intuitionistic fuzzy variables from their classical (crisp) and fuzzy representation. In order to obtain the numerical solutions of Eqs. (3.3) and (3.4), the fuzziness and vagueness of the model parameters need to be removed by implementing the measure in the form a closed interval which can be represented by the α-cut values of the fuzzy numbers, $\widetilde{D}_L$, $\tilde{u}$ and intuitionistic fuzzy numbers, $\overset{\sim i}{D}_L$, $\overset{\sim i}{u}$. In order to complete this chapter, a short description pertaining to the definition of a fuzzy set, intuitionistic fuzzy set and their α-cut representations and some algebraic properties which are required for LBM scheme presented in the subsequent paragraphs. Detailed description of fuzzy set and intuitionistic fuzzy set can be found elsewhere in [4, 15–17].

3.3 Preliminaries of Fuzzy Set and Intuitionistic Fuzzy Set

In this section, some basic definitions of fuzzy set and intuitionistic fuzzy set are presented before utilizing them for solving solute transport equation with imprecise and vague parameters.

3.3.1 *Fuzzy Set*

A generalized version of classical set theory, where elements of a set have binary membership, was introduced by Zadeh in the form of fuzzy set theory [3]. A fuzzy set is a collection of objects with graded membership value which can be any number from 0 to 1.

A fuzzy set, $\tilde{A}$, drawn from the universe of discourse X, is represented as

$$\tilde{A} = \{x, \mu_{\tilde{A}}(x) | x \in X\}$$

where $\mu_{\tilde{A}}(x)$ is degree of membership of elements $x \in X$ to a fixed set $A \subset X$, defined as $\mu_{\tilde{A}}(x) : X \to [0, 1]$. Fuzzy set theory provides a platform to address uncertainty caused by imprecise information. The details of fuzzy set theory can be found from available literatures [4]. Here, only the fuzzy vertex method [14], generally used for computing function of fuzzy variables, is highlighted in the followings.

3.3.2 *Basic Concept of Intuitionistic Fuzzy Set*

Intuitionistic fuzzy set (IFS) can be viewed as a generalized version of conventional fuzzy set. The inefficiency of the later one in dealing with vague information leads to the development of IFS theory by Atanassov [5, 6]. In fuzzy set theory, the grade of membership of an element x_i in the universe of discourse $X = \{x_1, x_2, \ldots, x_n\}$ is represented by a real value between 0 and 1, and it provides evidence for $x_i \in X$ but does not say anything against $x_i \in X$. The problem faced due to the phrase "not indicating evidence against $x_i \in X$" is removed by bringing intuitionistic fuzzy set (IFS). An IFS $\overset{\sim i}{A}$ defined in X is bounded by two functions, one is called membership function, $\mu_{\overset{\sim i}{A}}(x)$ and another is non-membership function $\nu_{\overset{\sim i}{A}}(x)$, each possess a real number lying between 0 and 1. Higher grade of membership and lower grade of non-membership of x means the value of membership function is closer to unity and value of non-membership function is closer to zero. The element x belongs to $\overset{\sim i}{A}$ is defines by $\mu_{\overset{\sim i}{A}}(x) = 1$ and $\nu_{\overset{\sim i}{A}}(x) = 0$ and does not belong to $\overset{\sim i}{A}$ is defined by $\mu_{\overset{\sim i}{A}}(x) = 0$ and $\nu_{\overset{\sim i}{A}}(x) = 1$. The intuitionistic fuzzy set $\overset{\sim i}{A}$ in X, is defined as

$$\overset{\sim i}{A} = \{x, \mu_{\overset{\sim i}{A}}(x), \nu_{\overset{\sim i}{A}}(x) | x \in X\}$$

where $\mu_{\overset{\sim i}{A}}(x)$ and $\nu_{\overset{\sim i}{A}}(x)$ are degrees of membership and non-membership of elements $x \in X$ to a fixed set $A \subset X$, defined as $\mu_{\overset{\sim i}{A}}(x) : X \rightarrow [0, 1]$ and $\nu_{\overset{\sim i}{A}}(x) : X \rightarrow [0, 1]$. For every element $x \in X$, $0 \leq \mu_{\overset{\sim i}{A}}(x) + \nu_{\overset{\sim i}{A}}(x) \leq 1$. Furthermore, we have the term $\pi(x) = 1 - \mu_{\overset{\sim i}{A}}(x) - \nu_{\overset{\sim i}{A}}(x)$ called the intuitionistic fuzzy set index or hesitation margin of x in X. $\pi(x)$ is the degree of indeterminacy of $x \in X$ to the IFS $\boldsymbol{A}$ and defined as $\pi_A(x) : X \rightarrow [0, 1]$. For every element $x \in X$, $0 \leq \pi_A(x) \leq 1$. $\pi_A(x)$ expresses the lack of knowledge or uncertainty level of the element x belongs to IFS. Therefore, the degree of membership of x in the IFS is characterized by the interval $\left[\mu_{\overset{\sim i}{A}}(x), 1 - \nu_{\overset{\sim i}{A}}(x)\right]$. For every fuzzy set, $\nu_{\overset{\sim i}{A}}(x) = 1 - \mu_{\overset{\sim i}{A}}(x)$ and hence $\pi_A(x) = 0$ which implies fuzzy sets are not able to model hesitancy in data.

3.3.3 *Fuzzy Vertex Method*

Fuzzy vertex method (FVM) introduced by Dong and Shah (1987) has been used by various researchers to solve fuzzy differential equation [1–3]. In this method, concept of α-cut of a fuzzy set and interval analysis for computation of fuzzy number are used to solve the fuzzy and intuitionistic fuzzy solute transport model [1]. The α-cut of a fuzzy set is another fuzzy set where minimum membership value of any element is α. The α-cut of a fuzzy set $\tilde{A} = \left\{x, \mu_{\tilde{A}}(x)\right\}$ is defined as $\widetilde{A}_\alpha = \{x | \mu_{\tilde{A}}(x) \geq \alpha\}$. In FVM, the deterministic model corresponding to the fuzzy model is executed for each lower and upper bounds obtained from the α-cut representation of a fuzzy number, and the same step is repeated for a set of values of α. The FVM used for the present study can be written in following algorithmic steps:

1. Construct fuzzy numbers for all the uncertain input parameters (here in this study triangular fuzzy numbers are constructed).
2. Obtain finite number of discrete α-cut levels from the continuous range of membership grade [0, 1].
3. Formulate the representative intervals for each fuzzy number at the specified α-cut levels.
4. Take one end point from each of the intervals (under an α-cut level). Note that, there are 2^n combinations for n fuzzy sets.
5. Solve 2^n deterministic sub-models for each lower and upper bound of the model output function.
6. Integrate the solutions of the sub-models by max–min rule and report the final solution for the output function at any α-cut level as [lower bound, upper bound].

3.3.4 Fuzzy Vertex Method for IFS

The α-cut and interval analysis used in the fuzzy vertex method described in the previous section is extended to intuitionistic fuzzy set. Here, the same techniques are applied on the membership function, $\mu_{\underset{A}{\sim i}}(x)$ as well as on the membership function with hesitancy, $1 - \nu_{\tilde{A}^i}(x)$. The α-cut of an intuitionistic fuzzy set $\overset{\sim i}{A} = \left\{x, \mu_{\underset{A}{\sim i}}(x), \nu_{\underset{A}{\sim i}}(x)\right\}$ in the universe of discourse X is similar to that of fuzzy set and mathematically represented as $\underset{\alpha}{\overset{\sim i}{A}} = \{x \in X | \mu_{\underset{A}{\sim i}}(x) \geq \alpha, 1 - \nu_{\underset{A}{\sim i}}(x) \geq \alpha\}$. The FVM for IFS used for the present study can be written in following algorithmic steps:

1. Construct intuitionistic fuzzy numbers for all the vague input parameters (here in this study triangular IFNs are constructed).
2. Obtain finite number of discrete α-cut levels from the continuous range of membership grade $[0, \mu]$ and $[0, 1 - \nu]$
3. Formulate the representative intervals for $\mu_{\underset{A}{\sim i}}(x)$ and $1 - \nu_{\underset{A}{\sim i}}(x)$ for each IFS at the specified α-cut levels.
4. Take one end point from each of the intervals (under an α-cut level). There are 2^n combinations for $\mu_{\underset{A}{\sim i}}(x)$ as well as for $1 - \nu_{\underset{A}{\sim i}}(x)$ for n IFSs and therefore total of 2×2^n combinations for n IFSs.
5. Solve 2×2^n deterministic sub-models for each lower and upper bound of membership function and membership function with hesitation margin of the model output function.
6. Integrate the solutions of the sub-models by max–min rule and report the final solution for the output function with and without hesitation margin at any α-cut level as [lower bound, upper bound].

3.4 Formulation of Fuzzy and Intuitionistic Fuzzy Lattice Boltzmann Scheme

In this section, the fuzzy lattice Boltzmann (FLB) and intuitionistic fuzzy lattice Boltzmann (IFLB) schemes used to solve Eqs. (3.3) and (3.4) numerically are derived. Single relaxation time (SRT)-based collision term is used in the formulation of the numerical scheme. The formulation starts with α-cut representation of fuzzy parameters ($\widetilde{D}_L$ and $\tilde{u}$) and intuitionistic fuzzy parameters ($\underset{L}{\overset{\sim i}{D}}$ and $\overset{\sim i}{u}$) and corresponding α-cut representation of Eqs. (3.3) and (3.4) as given in Eq. (3.5).

$$\frac{\partial C_\alpha}{\partial t} = D_{L\alpha} \frac{\partial^2 C_\alpha}{\partial x^2} - u_\alpha \frac{\partial C_\alpha}{\partial x} \tag{3.5}$$

Equation (3.5) is similar to the ADE as given in Eq. (3.1) with defined parameters. Equation (3.5) can be solved using conventional numerical techniques such as FEM, FDM, and FVM where the algebraic form of Eq. (3.5) is solved, whereas in LB method, discrete velocity Boltzmann equation is solved in a discretized special and temporal domain. In this study, the standard SRT LB scheme for mass/energy transport equations is utilized to solve it numerically [13]. The governing equation of standard SRT LB scheme is a discrete velocity LB equation which can be written as [12]

$$f_i\left(\vec{r} + \vec{e_i}\,\Delta t, t + \Delta t\right) = f_i(\vec{r}, t) + \Omega_i^{\mathrm{BGK}}(\vec{r}, t)$$

$$\Omega_i^{\mathrm{BGK}}(\vec{r}, t) = \frac{1}{\tau}\left[f_i^{\mathrm{eq}}(\vec{r}, t) - f_i(\vec{r}, t)\right] \tag{3.6}$$

where $f_i(\vec{r}, t)$ and $f_i^{\mathrm{eq}}(\vec{r}, t)$ represent single particle distribution function and its equilibrium form at a spatio-temporal location $(\vec{r}, t)$ along the discrete velocity direction $\vec{e_i}$, $\Omega_i^{\mathrm{BGK}}(\vec{r}, t)$ is SRT BGK collision operator along ith direction at same spatio-temporal coordinate, Δt and τ are simulation time step and relaxation coefficient, respectively. α-cut representation of the above LB Eqs. (3.6) can be written as

$$f_{i\alpha}\left(\vec{r} + \vec{e_i}\,\Delta t, t + \Delta t\right) = f_{i\alpha}(\vec{r}, t) + \Omega_{i\alpha}^{\mathrm{BGK}}(\vec{r}, t)$$

$$\Omega_{i\alpha}^{\mathrm{BGK}}(\vec{r}, t) = \frac{1}{\tau}\left[f_{i\alpha}^{\mathrm{eq}}(\vec{r}, t) - f_{i\alpha}(\vec{r}, t)\right] \tag{3.7}$$

The same LB Eq. (3.6) can be used to solve various kinds of physical problems, and the form of the single particle equilibrium distribution function (EDF) decides the type of problem being solved. Therefore, EDF has important role in solving a physical problem using LBM. Here, the EDF is basically the truncated form of the famous Boltzmann equilibrium distribution function which for the ADE [Eq. (3.5)] takes the following form [18]

$$f_{i\alpha}^{\mathrm{eq}}(\vec{r}, t) = w_i C_\alpha(\vec{r}, t)\left(1 - \frac{\vec{e_i}\,.\overrightarrow{u_{x\alpha}}}{e_s^2}\right) \tag{3.8}$$

where w_i is weight factor for the single particle distribution function along the discrete velocity direction $\vec{e_i}$ and e_s is called "pseudo-sound speed" [18]. Based on the type of problem being solved, lattices for LB technique are selected, for example, D1Q2 and D1Q3 lattices are generally used for 1-D equations for diffusive and advective–diffusive type of problems, for the same types of problems in 2-D, D2Q4, D2Q5 are used and similarly for 3-D, D3Q7, D3Q9, D3Q15 lattices are generally used. The factor n in the symbol of a general lattice DnQm represents the dimension of the problem and m is the number of discrete velocity directions. Schematic views of D1Q3, D2Q5 and D3Q15 lattices are shown in Fig. 3.1a, b and c, respectively.

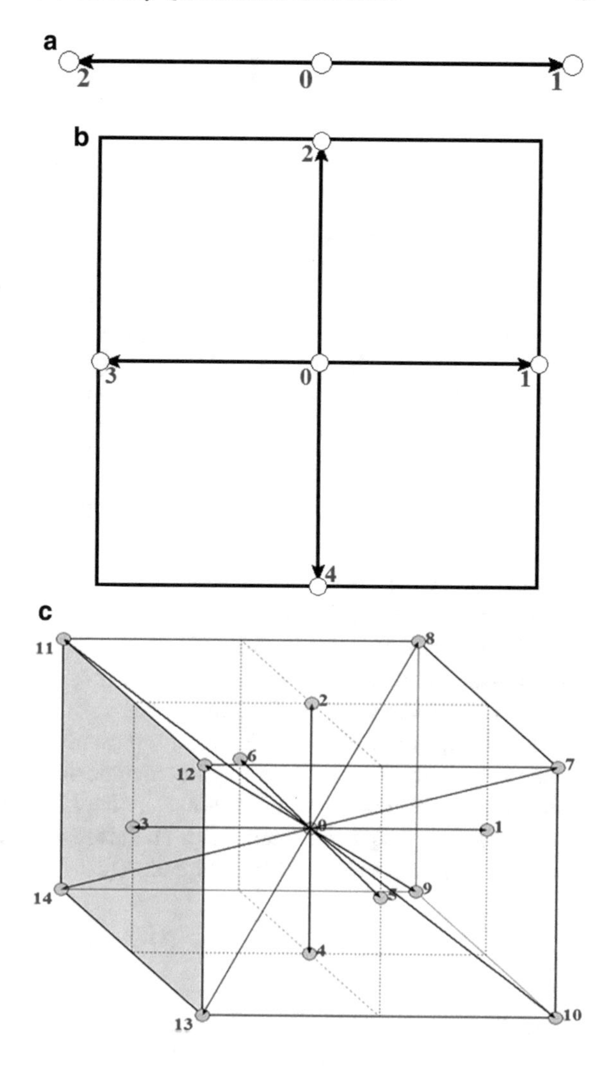

Fig. 3.1 **a** D1Q3 lattice. **b** D2Q5 lattice. **c** D3Q15 lattice

The macroscopic dependent variable such as particle density is calculated from the zero-order velocity moment of the particle distribution function, and therefore, the solution of Eq. (3.5) in the form of solute concentration is written as

$$C_\alpha(\vec{r}, t) = \sum_i f_{i\alpha}(\vec{r}, t) \tag{3.9}$$

The following moments of the EDF are used to calculate EDF and weight factors for the LB scheme.

$$\sum_i w_i = 1 \tag{3.10}$$

$$\sum_{i} f_{i\alpha}^{\mathrm{eq}}(\vec{r}, t) C_{\alpha}(\vec{r}, t) \tag{3.11}$$

$$\sum_{i} e_{ix} f_{i\alpha}^{\mathrm{eq}}(\vec{r}, t) = u_{x\alpha} C_{\alpha}(\vec{r}, t) \tag{3.12}$$

$$\sum_{i} e_{ix} e_{iy} f_{i\alpha}^{\mathrm{eq}}(\vec{r}, t) = e_s^2 C_{\alpha}(\vec{r}, t) \delta_{xy} \tag{3.13}$$

where the Kronecker delta function, δ_{xy} is equal to 1 when $x = y$ and equal to 0 when $x \neq y$. The recovery of 1D ADE governing solute transport process [Eq. (3.5)] from the LBE [Eq. (3.7)] is carried out using Chapman–Enskog multi-scale analysis technique [18], and in this process of derivation, we get the expression

$$D_{L\alpha} = e_s^2 \left(\tau - \frac{1}{2} \right) \tag{3.14}$$

Equation (3.14) correlates macroscopic parameter, hydrodynamic dispersion coefficient, with the mesoscopic parameter, relaxation coefficient. It can be observed from Eq. (3.14) that hydrodynamic dispersion coefficient has a linear relationship with relaxation coefficient; therefore, it is appropriate to write τ_α for τ when LB Eq. (3.7) is used to simulate the same process for different α-cut values of hydrodynamic dispersion coefficient. Here, τ_α represents the relaxation coefficient corresponding to a given α-cut value of hydrodynamic dispersion coefficient ($D_{L\alpha}$). From stability point of view, it is well established that LB solutions with lattice units are more stable than the physical units [19]. Therefore, in this study, we have used lattice units for the developed LB scheme. Equation (3.14) can be written in lattice units as

$$D_L^* = e_s^2 \left(\tau - \frac{1}{2} \right) \tag{3.15}$$

where D_L^* is lattice hydrodynamic dispersion coefficient. The time step (Δt^*) and lattice length (Δx^*) in lattice units are taken as $\Delta t^* = 1$ and $\Delta x^* = 1$. Therefore, magnitude of lattice velocity ($\vec{e}$) is

$$\overrightarrow{|e|} = \frac{\Delta x^*}{\Delta t^*} = 1$$

The value of the "pseudo-sound speed" is $e_s = \overrightarrow{|e|}/\sqrt{3}$ (valid for D1Q3, D2Q5, D2Q9 and similar lattices [18]). Therefore, the lattice hydrodynamic dispersion coefficient as given in Eq. (3.15) can be rewritten for a D1Q3 lattice as

$$D_L^* = \frac{1}{3} \left(\tau - \frac{1}{2} \right) \tag{3.16}$$

The lattice dispersion coefficient (D_L^*) is linked with physical hydrodynamic dispersion coefficient (D_L) by the following equation.

$$D_L^* = \frac{D_L}{\frac{\Delta x^2}{\Delta t}} \tag{3.17}$$

From the above two Eqs. (3.16) and (3.17), physical time step can be written in the following form

$$\Delta t = \frac{\Delta x^2}{3D_L}\left(\tau - \frac{1}{2}\right) \tag{3.18}$$

For $\tau = 1$, Eq. (3.18) can be written as

$$\Delta t = \frac{\Delta x^2}{6D_L} \tag{3.19}$$

Equation (3.19) shows that the physical time step (Δt) is inversely proportional to the hydrodynamic dispersion coefficient, and therefore, for different values of the parameter obtained from the α-cuts, time step value (Δt) will vary. In the form of α-cuts, Eq. (3.19) can therefore be written as

$$\Delta t_\alpha = \frac{\Delta x^2}{6D_{x\alpha}} \tag{3.20}$$

There are four combinations of input parameters (hydrodynamic dispersion coefficient and groundwater flow velocity), such as, $\left[D_L^{\text{Low}}, u^{\text{Low}}\right]_\alpha$, $\left[D_L^{\text{Low}}, u^{\text{High}}\right]_\alpha$, $\left[D_L^{\text{High}}, u^{\text{Low}}\right]_\alpha$, $\left[D_L^{\text{High}}, u^{\text{High}}\right]_\alpha$ at each α-cut value (except α-cut = 1). Our task here is to simulate the process of solute transport using the LBE [Eq. (3.7)] for each combination of these input parameters. It can be observed from Eq. (3.8) that groundwater flow velocity is associated with the EDF, and therefore, its change can be directly incorporated into the LBE by changing the EDF accordingly. The physical hydrodynamic dispersion coefficient is not directly linked with the LBE, but its effect can be incorporated into the LB equation by the following way; change in hydrodynamic dispersion coefficient modifies the physical time step value via Eq. (3.20) and physical time step value determines the numbers of iterations need to be executed for a given simulation time. Following the standard LB algorithm, the algorithm for the fuzzy LB equation as provided in Eq. (3.7) can be decomposed in the following two processes [18].

Collision process

In this process, the single particles distribution function relaxes towards the local EDF, which can be expressed mathematically as

$$f_{i\alpha}^{*}(\vec{r}, t + \Delta t) = f_{i\alpha}(\vec{r}, t) + \Omega_{i\alpha}^{\mathrm{BGK}}(\vec{r}, t) \tag{3.21}$$

where $f_i^*(\vec{r}, t + \Delta t)$ is called post-collision single particle distribution function, and the collision operators is the same BGK collision operator as given in Eq. (3.7).

Streaming process

Streaming is basically the process of hopping of particles from one lattice location to its nearest neighbour lattice location along the lattice velocity direction. In this process, no algebraic operation is carried out only swapping in storage memory takes place. The process can be written in following algorithmic form

$$f_{i\alpha}\left(\vec{r} + \vec{e_i}\,\Delta t, t + \Delta t\right) = f_{i\alpha}^{*}(\vec{r}, t + \Delta t) \tag{3.22}$$

Dirichlet, Neumann and Cauchy type boundary conditions are easily incorporated in the LB framework and a special type of boundary condition called bounce-back boundary conditions [20] where particles reverse their velocity directions after colliding with obstacles or boundary walls are widely used for fluid flow simulation required for estimation of ground water flow velocity.

3.5 Vagueness and Uncertainty Analysis of Solute Transport Model

In this section, vagueness and uncertainty analysis of the solute transport model (STM) discussed in Sect. 3.3 is carried in three steps. In the first step, fuzzy and intuitionistic fuzzy numbers for the input variables such as groundwater velocity and hydrodynamic dispersion coefficient are constructed using inputs from expert opinion. In the second step, stability analysis of the developed FLB and IFLB schemes is carried out for a given set of fuzzy and intuitionistic fuzzy numbers. Finally, fuzzy membership and intuitionistic fuzzy membership and non-membership functions of solute concentration are computed using the developed FLB and IFLB schemes.

3.5.1 Generation of Fuzzy and Intuitionistic Fuzzy Number

The input data in the form of lower bound, most likely, upper bound for construction of triangular fuzzy and intuitionistic fuzzy numbers for the solute transport model parameters such as groundwater velocity and hydrodynamic dispersion coefficient are collected from expert for a specific site. The initial imprecise data were in the form of a real number a and its upper and lower uncertainties in the range of $h_1\%$ and $h_2\%$. Using these data, triangular fuzzy numbers are constructed in the form of $\left[a - a \cdot \frac{h_2}{100}, a, a + a \cdot \frac{h_1}{100}\right]; \mu$ with an α-cut $a^\alpha = \left[a_L^\alpha, a_U^\alpha\right]$. The lower and upper

Table 3.1 Input fuzzy parameters

Parameters	Lower bound	Most likely	Upper bound
Velocity of flow (m/day)	0.1	0.3	0.6
Hydrodynamic dispersion coefficient (m^2/day)	0.8	3.0	9.0

bounds and most probable values of the imprecise parameters, ground water velocity and hydrodynamic dispersion coefficient are as shown in Table 3.1, and their fuzzy membership functions are as shown in Fig. 3.2a and b, respectively. Similarly, triangular intuitionistic fuzzy numbers are constructed by considering degree of acceptance of the data for groundwater velocity and hydrodynamic dispersion coefficient as 0.6 and 0.7 and equal degree of rejection as 0.1. So, the degree of hesitation or degree of uncertainty is $1 - (0.6 + 0.1) = 0.3$ for groundwater velocity and $1 - (0.7 + 0.1) = 0.2$ for hydrodynamic dispersion coefficient. The constructed triangular intuitionistic fuzzy membership functions are as shown in Fig. 3.2c and d.

3.5.2 *Development of LB Solver*

The numerical simulation is carried out using an in-house developed LB solver in Python programming language. For verification and validation of the LB solver, the results are compared with the closed form solution provided in Eq. (3.2). For a given set of crisp input parameters, hydrodynamic dispersion coefficient as 1.0 m^2/day and groundwater velocity as 0.3 m/day, the LB simulation of the solute transport process is carried out for four different lattice sizes (2.5, 5, 10 and 20 m). The special profiles of solute concentrations after 800 days are shown in Fig. 3.3a–d for these four different lattice lengths, respectively. It can be observed from the results that the stability of the LB solution dependent on lattice size. The solution becomes unstable for a lattice length $\geq$ 10 m. From this study, a lattice length of 5 m was initially selected. In the next step, the ground water flow velocity is uniformly varied from the lower bound (0.1 m/day) to the upper bound (0.6 m/day) and hydrodynamic dispersion coefficient is fixed at the lower limit (0.8 m^2/day). Two different lattice lengths (5 and 2.5 m) are used in this study, and the results as shown in Fig. 3.4a and b indicate that the solution for the upper bound of velocity is unstable for a lattice length of 5 m. It can be observed from the results shown in Fig. 3.4c that LB solution is highly stable for all the velocity values even at a lattice length of 20 m when the value of hydrodynamic dispersion coefficient is the upper bound value, i.e. 9.0 m^2/day. All these studies can be concluded by the result shown in Fig. 3.4d where it can be observed that stability of the LB solution is dependent on the Peclet (Pe) number. The figure shows that for a stable LB solution, the Peclet number should be kept below 2.0. The Peclet number 1.88 corresponds to the upper bound of velocity and lower bound of dispersion coefficient with a lattice length of 2.5 m. Therefore, this value of lattice length ensures stability for all the LB solutions need to be obtained for vagueness and uncertainty analysis.

3.5.3 LB-Based Numerical Solution

At the beginning of the LB simulation, preprocessing of the fuzzy number and intuitionistic fuzzy number of the hydrodynamic dispersion coefficient and groundwater velocity is carried out to obtain α-cuts in the form of closed intervals. Then for every

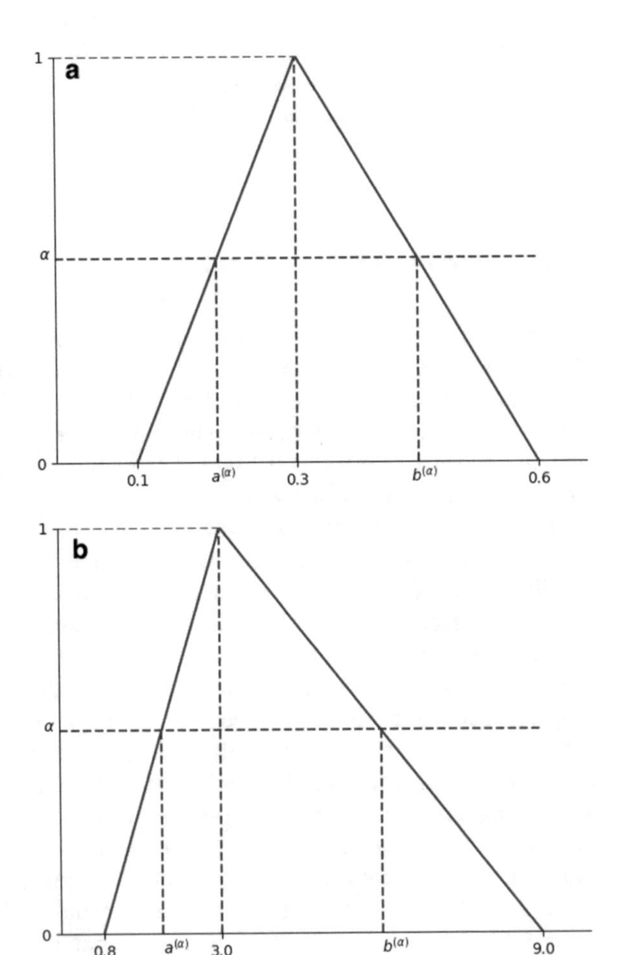

Fig. 3.2 **a** Triangular fuzzy velocity: α-cut representation. **b** Triangular fuzzy dispersion coefficient: α-cut representation. **c** Triangular intuitionistic fuzzy velocity: α-cut representation. **d** Triangular intuitionistic fuzzy velocity: α-cut representation

Fig. 3.2 (continued)

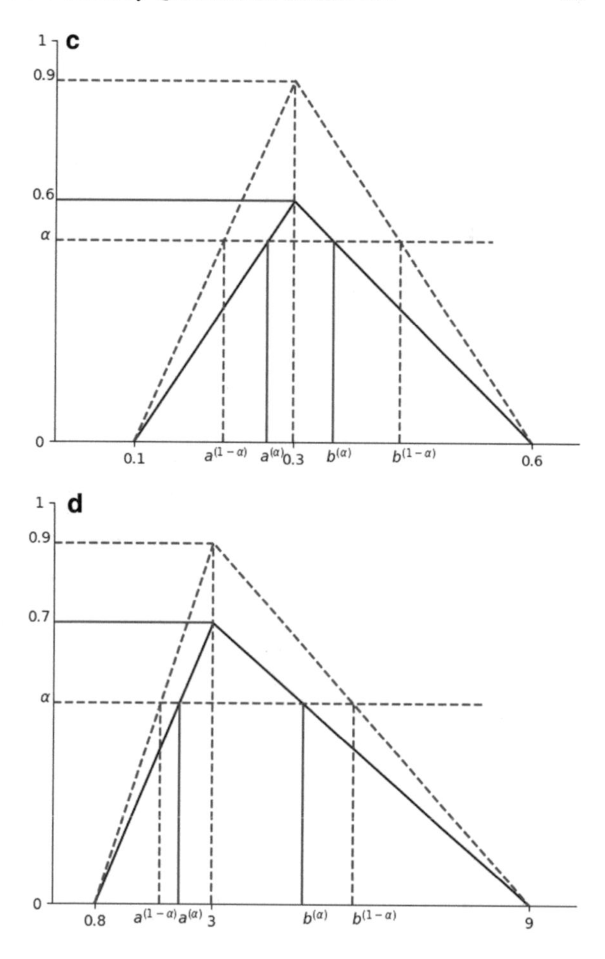

α-cut, vertices are constructed using fuzzy vertex method. For the solute transport model with two parameters, $2^2 - 4$ vertices are constructed at each α-cut for the fuzzy number and $2 \times 2^2 = 8$ vertices for IFN, where four vertices for membership and four for non-membership function. These vertices are used as inputs for the LB simulations. Therefore, a set of four numerical solutions of solute concentration at any spatio-temporal location are obtained for a given α-cut. Minimum and maximum values of solute concentrations are then extracted from these four solutions. This step is repeated for other values of alpha with an increment of 0.1 for estimation of lower and upper bounds of the solute concentration. For the case of IFN, lower and upper bounds of the solute concentration with and without hesitation margin are computed.

The domain length of 1000 m and 401 uniform grid points are used in the LB simulation. Therefore, physical lattice length is $\Delta x = 2.5$ m. In the actual LB simulation, unit lattice units, i.e. $\Delta x^* = 1$ lbu, and $\Delta t^* = 1$ lbu as discussed in Sect. 3.4 are considered. The time step (Δt) for different values of hydrodynamic dispersion coefficients, $D_{L\alpha}$, taken from the α-cuts, is calculated using Eq. (3.20). The

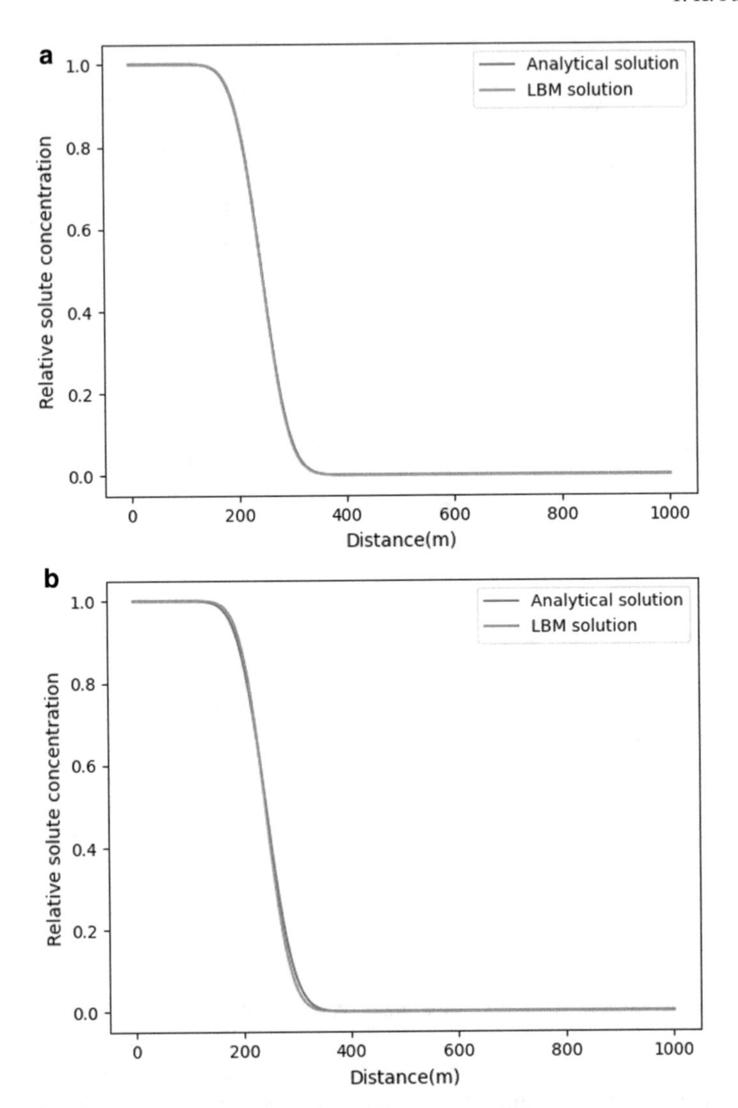

Fig. 3.3 **a** Spatial profile of solute concentration after 800 days with 2.5 m lattice length. **b** Spatial profile of solute concentration after 800 days with 5 m lattice length. **c** Spatial profile of solute concentration after 800 days with 10 m lattice length. **d** Spatial profile of solute concentration after 800 days with 20 m lattice length

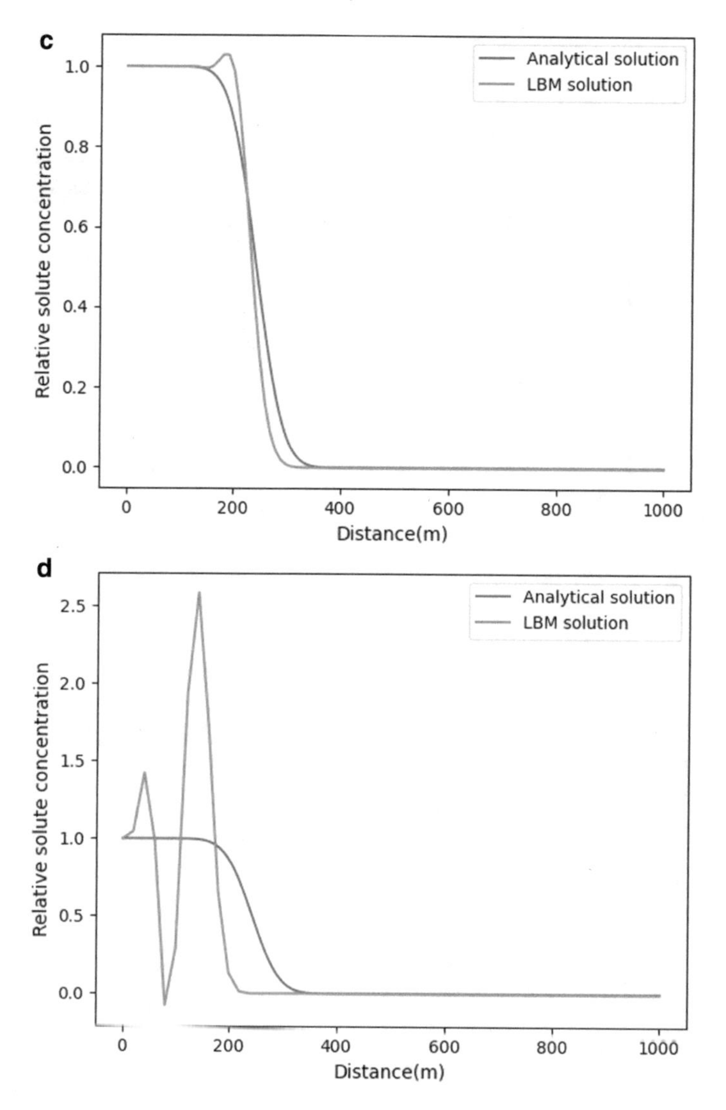

Fig. 3.3 (continued)

solute concentration at source position, i.e. at $x = 0$ m, is taken as $C_0 = 100$ mg/l for any time $t > 0$. The LB-based numerical results are calculated in the form of solute concentration at a spatial distance of 250 m from the source position after 600, 700, 800 and 900 days. The fuzziness of the solute concentration in the form of fuzzy membership function and vagueness of the solute concentration in form of membership function and non-membership function are shown in Fig. 3.5a–d. The computation was carried out on a platform having Intel 1.2 GHz i3 processor, and the required computation time to obtain any one of the following figures was about

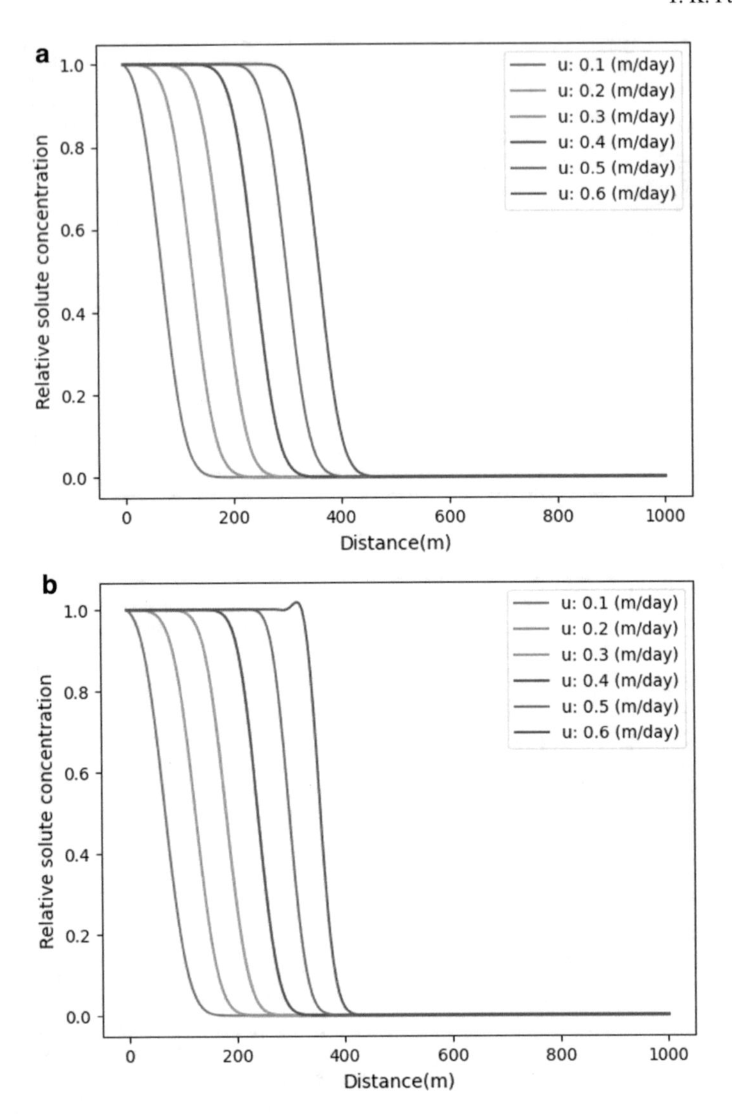

Fig. 3.4 **a** Velocity dependent spatial profiles of solute concentration after 600 days with 2.5 m lattice length and lower bound of dispersion coefficient. **b** Velocity dependent spatial profiles of solute concentration after 600 days with 5 m lattice length and lower bound of dispersion coefficient. **c** Velocity dependent spatial profiles of solute concentration after 600 days with 20 m lattice length and upper bound of dispersion coefficient. **d** Spatial profile of solute concentration after 600 days with different Peclet numbers

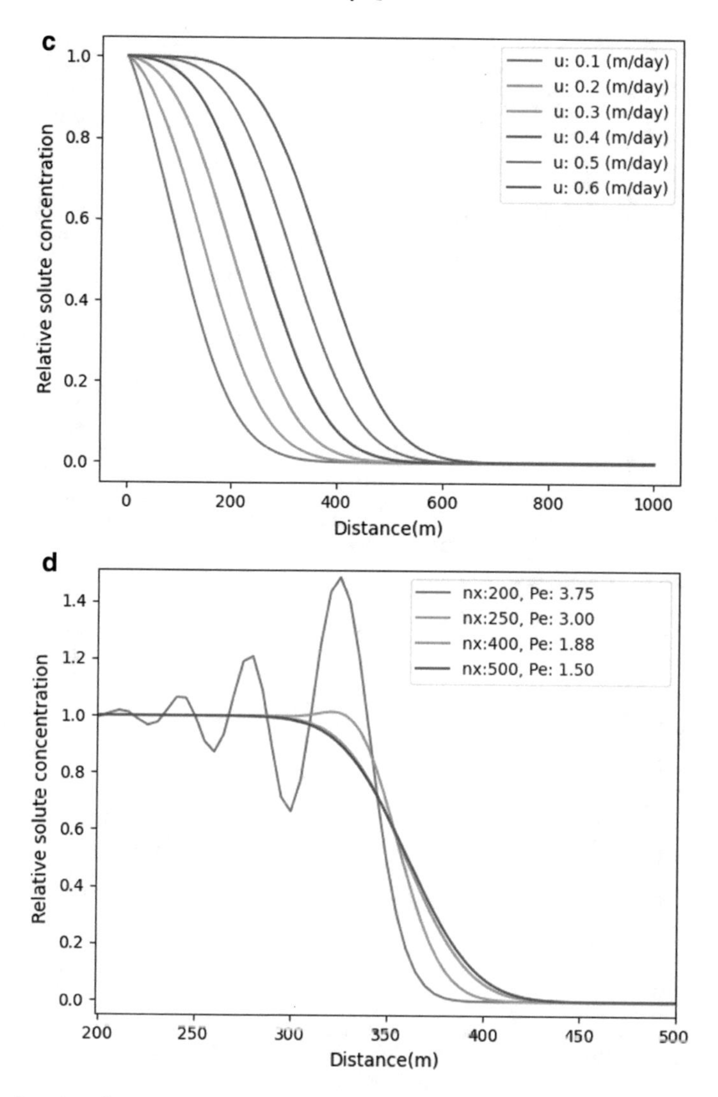

Fig. 3.4 (continued)

50 s. Uncertainty and vagueness of solute concentration can be calculated using any defuzzification technique generally used for fuzzy number and intuitionistic fuzzy number [21, 22].

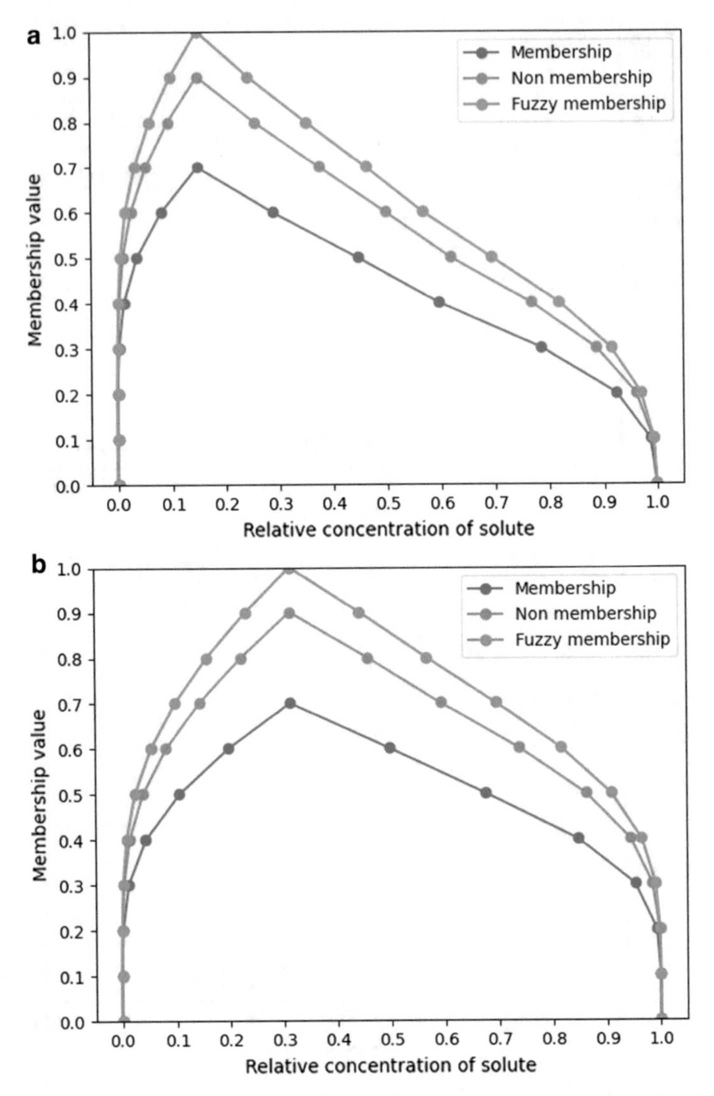

Fig. 3.5 **a** Solute concentration at 250 m after 600 days. **b** Solute concentration at 250 m after 700 days. **c** Solute concentration at 250 m after 800 days. **d** Solute concentration at 250 m after 900 days

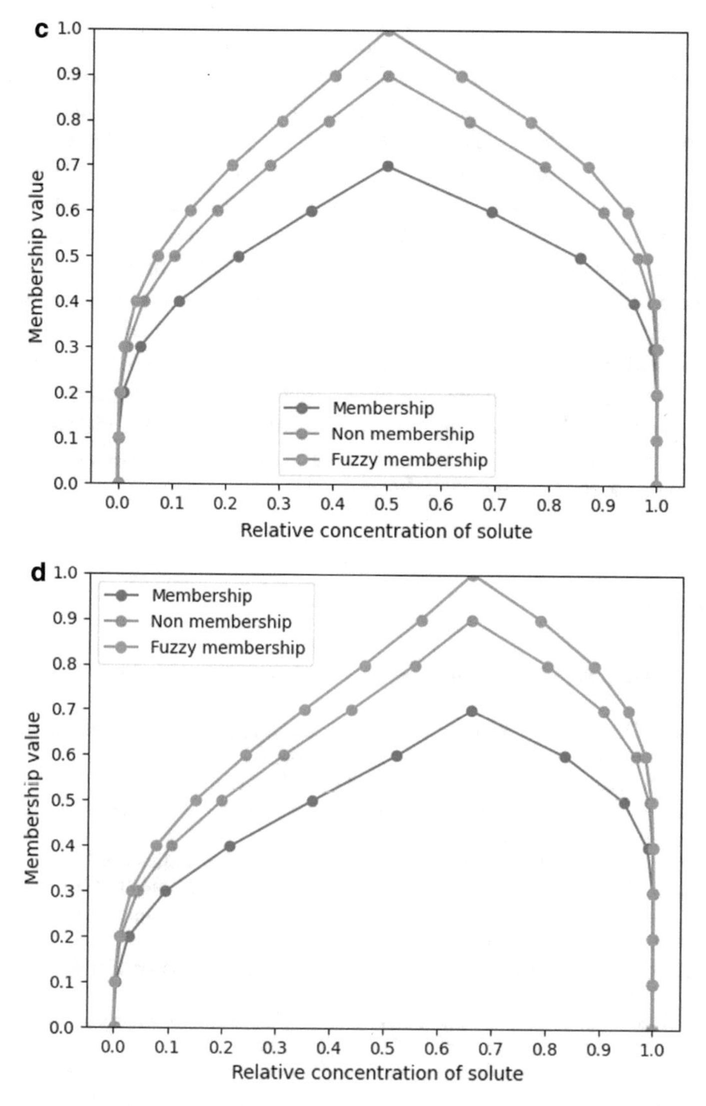

Fig. 3.5 (continued)

3.6 Conclusions

In this chapter, advanced numerical technique such as LB method is used to develop numerical schemes for solving solute transport equation in presence of imprecise and vague parameters. Detailed formulation of FLB and IFLB schemes is provided which can be used to solve other related problems. The developed schemes are basically amalgamation of LB technique with fuzzy set and IFS. The simple collision and streaming operations of LB algorithm which are not directly linked with the macroscopic parameters of the solute transport equation make the schemes very flexible for implementation. The chapter highlighted how the required fuzzy and intuitionistic fuzzy numbers can be generated using experts' opinion. The stability analysis of the LB scheme shows that the Peclet number must be kept below 2, and therefore, corresponding lattice length can be calculated using the upper bound of ground water velocity and lower limit of hydrodynamic dispersion coefficient. The comparison of LB solution with corresponding closed form solution shows that LB technique provides very accurate results when stability criterion is fulfilled. The membership functions of the fuzzy and intuitionistic fuzzy solute concentration at a given spatial location are constructed for four different simulation times using the developed FLB and IFLB schemes. The computation time is very minimal about 50 s for construction of fuzzy and intuitionistic fuzzy solute concentration at given spatio-temporal location. The solutions of the solute transport equation are presented in the form of membership functions. The results can also be presented in crisp form using standard defuzzification techniques.

References

1. Klir, G.J.: Principles of uncertainty: what are they? Why do we need them? Fuzzy Sets Syst. **74**, 15–31 (1995)
2. Ross, J.L., Ozbek, M.M., Pinder, G.F.: Aleatoric and epistemic uncertainty in groundwater flow and transport simulation. Water Resour. Res. **45**, W00B15 (2009)
3. Zadeh, L.A.: Fuzzy sets. Inform. Control **8**, 338–353 (1965)
4. Klir, G.J., Yuan, B.: Fuzzy sets and fuzzy logic: theory and applications. Possib. Theory Versus Probab. Theory **32**, 207–208 (1996)
5. Atanassov, K.T.: Intuitionistic fuzzy sets. VII ITKR's Session, Sofia (1983)
6. Atanassov, K.T.: Intuitionistic fuzzy sets. Fuzzy Sets Syst. **20**, 87–96 (1986)
7. Datta, D.: Non-probabilistic uncertainty analysis of analytical and numerical solution of heat conduction. Int. J. Energy Inf. Commun. **2**, 143--156 (2011)
8. Dou, C., Woldt, W., Bogardi, I., Dahab, M.: Numerical solute transport simulation using fuzzy sets approach. J. Contam. Hydrol. **27**, 107–126 (1997)
9. Prasad, R.K., Mathur, S.: Groundwater flow and contaminant transport simulation with imprecise parameters. J. Irrig. Drain. Eng. **133**, 61–70 (2007)
10. Pal, T.K., Datta, D.: Parametric uncertainty analysis of solute transport process using fuzzy lattice Boltzmann scheme. Life Cycle Reliab. Saf. Eng. **6**, 239–248 (2017)
11. Datta, D., Pal, T.K.: Development of fuzzy differential quadrature numerical method and its application for uncertainty quantification of solute transport model. Life Cycle Reliab. Saf. Eng. **6**, 249–256 (2017)
12. Chen, S., Doolen, G.D.: Lattice Boltzmann method for fluid flows. Annu. Rev. Fluid Mech. **30**, 329–364 (1998)
13. Succi, S.: The Lattice Boltzmann Equation for Fluid Dynamics and Beyond. Oxford University Press, New York (2001)
14. Dong, W., Shah, H.: Vertex method for computing functions of fuzzy variables. Fuzzy Sets Syst. **24**, 65–78 (1987)
15. Atanassov, K.: Intuitionistic Fuzzy Sets. Springer, Heidelberg (1999)
16. Atanassov, K.: On Intuitionistic Fuzzy Sets Theory. Springer, Berlin (2012)
17. Atanassov, K.: More on intuitionistic fuzzy sets. Fuzzy Sets Syst. **33**, 37–46 (1989)

18. Mohamad, A.: Lattice Boltzmann Method Fundamentals and Engineering Applications with Computer Codes. Springer, London (2011)
19. Perko, J., Patel, R.A.: Single-relaxation-time lattice Boltzmann scheme for advection-diffusion problems with large diffusion-coefficient heterogeneities and high-advection transport. Phys. Rev. E. **89**, 053309 (2014)
20. Zeigler, D.P.: Boundary conditions for the lattice Boltzmann simulations. J. Stat. Phys. **71**, 1171–1177 (1993)
21. Zhang, Q., Xiao, Y., Wang, G.: A new method for measuring fuzziness of vague set (or intuitionistic fuzzy set). J. Intell. Fuzzy Syst. **25**, 505–515 (2013)
22. Liang, J., Chin, K.S., Dang, C., Yam, R.C.: A new method for measuring uncertainty and fuzziness in rough set theory. Int. J. Gen. Syst. **31**, 331–342 (2002)

Chapter 4
Fuzzy Rough Set-Based Feature Selection for Text Categorization

Ananya Gupta and Shahin Ara Begum

4.1 Introduction

With recent trends in digital revolution, accumulation of information in digital repositories have grown exponentially. Tools for knowledge acquisition, recovery, storage and maintenance must progress at an equal pace so as to combat this exponential growth. Knowledge Discovery in Databases (KDD) is increasingly gaining popularity [16, 38] as proper KDD aids in using knowledge effectively and efficiently. Traditional process of KDD is based on manual investigation and interpretation of data. Manual form of searching data has many limitations and thus evolved automated process of knowledge discovery. The KDD process consists of several stages represented in Fig. 4.1 [20].

Data Mining: This stage involves mining relevant information from databases based on the domain under consideration. The choice of mining algorithm is dependent on many factors, including but not limited to dataset source and values in it.

Interpretation/Evaluation: After knowledge discovery, it is assessed in terms of its validity, simplicity and usefulness.

Feature selection aims at dimensionality reduction, which may result in significant data loss. There are two methods of dimensionality reduction. These are the feature selection and feature extraction methods [8, 14, 17]. Feature selection method focuses on reducing dimensionality of original feature space by selecting feature subset without any transformation. It preserves the physical interpretability of the selected

A. Gupta · S. A. Begum (✉)
Department of Computer Science, Assam University, Assam, India
e-mail: shahin.ara.begum@aus.ac.in

A. Gupta
e-mail: gupta.ananya77@gmail.com

T. Som et al. (eds.), *Fuzzy, Rough and Intuitionistic Fuzzy Set Approaches for Data Handling*, Forum for Interdisciplinary Mathematics,
https://doi.org/10.1007/978-981-19-8566-9_4

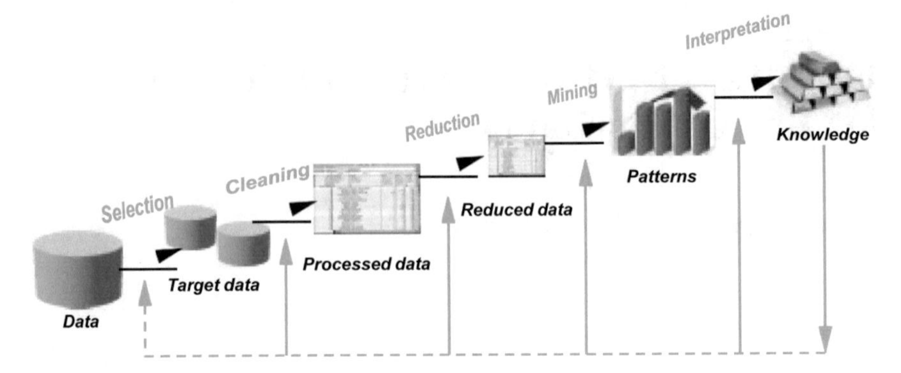

Fig. 4.1 Knowledge discovery process

features as in the original space. Feature extraction methods reduce the dimensionality by linear transformation of the input features into a completely different space. The linear transformation involved in feature extraction causes the features to be altered, making their interpretation difficult. Features in the transformed space lose their physical interpretability and their original contribution becomes difficult to ascertain [8].

The choice of the dimensionality reduction method is completely application specific and depends on the type of data. Feature selection is advantageous especially as features keep their original physical meaning because no transformation of data is made. Figure 4.2 depicts a schematic representation of data dimension reduction methods.

Selection-based approach encompasses dimensionality reduction on the basis of preserving semantics. The paper highlights the potential of fuzzy rough feature selection in the domain of text categorization. This paper proposes a hybrid feature selection technique called landmark-based fuzzy rough feature selection (LBFRFS) for

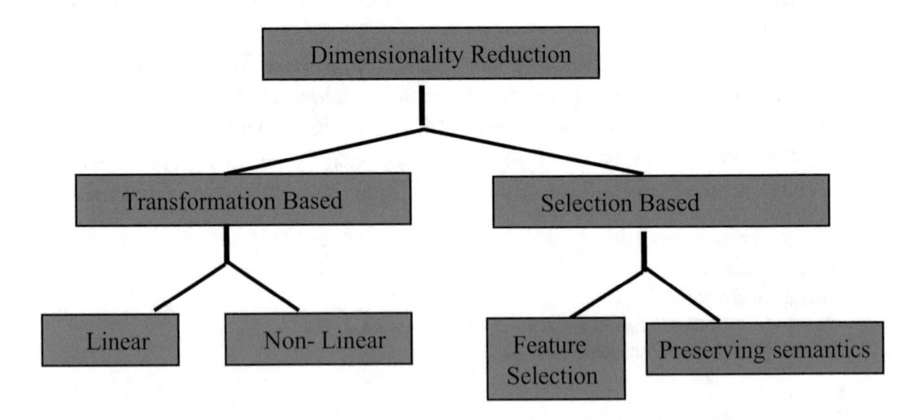

Fig. 4.2 Taxonomy of dimension reduction approaches

data dimensionality reduction of large-scale datasets such as text datasets for memory short environments. The proposed technique (LBFRFS) hybridizes large-scale spectral clustering [5] with landmark-based representation [10] with fuzzy rough set-based feature selection and its efficiency is evaluated on three text datasets using k-means clustering.

The remainder of the paper is organized as follows: Sects. 4.2 and 4.3 describe the theoretical background of feature selection and fuzzy rough feature selection, Sect. 4.4 is on related work pertaining to fuzzy rough feature selection, Sect. 4.5 is on potential of fuzzy rough feature selection on text categorization, and Sect. 4.6 concludes the paper.

4.2 Feature Selection

Feature selection selects discriminative features from high dimensional feature space toward dimensionality reduction, which in turn reduces the computational cost, eliminates redundant and irrelevant features thereby improving prediction performance. Principle of parsimony (or Occam's Razor) is the basis of feature selection. It aims to enhance classification performance and produces fast and cost-effective predictors. Moreover, feature selection enables simplification of classification problems as it provides better understanding of the process that generates the data. Different authors define relevancy of features in different ways. Almuallim and Dietterich [3] define features to be relevant if they appear in any logical formula that defines the target concept. Gennari et al. [18] consider features to be relevant if they affect the conditional distribution of the labels. John et al. [25] demonstrate the notion of strong and weak relevance. A feature (or a set of features) is said to be strongly relevant if they cannot be replaced by a feature (or set of features), whereas a feature is considered weakly relevant when they are useful yet can be replaced by other feature (or set of features). The authors emphasize that relevance does not imply optimality and conversely optimality does not imply relevance. Yu and Liu [42] in their paper demonstrate that relevance of feature is not the absolute parameter for feature selection.

It is hard to apply the principle of parsimony to problems involving feature selection. The selection of best features is a NP-complete problem [2]. The task of feature selection is challenging because (i) a single feature may not appear relevant yet in combination with other features may become highly relevant. Real-world datasets are characterized by multiple interdependencies among its features. Consequently, a weak association of a feature with a prediction (or classification) can enhance prediction accuracy if complemented with other features and (ii) some of the relevant features may be redundant as there can be multi-way redundancy in the feature space. Removal of such features will reduce complexity of the prediction problem.

The subset of best features can be obtained by an exhaustive search of subsets of features. However, an exhaustive search is practically infeasible even for a medium sized database with n features as the number of possible outcomes is 2^n. The current

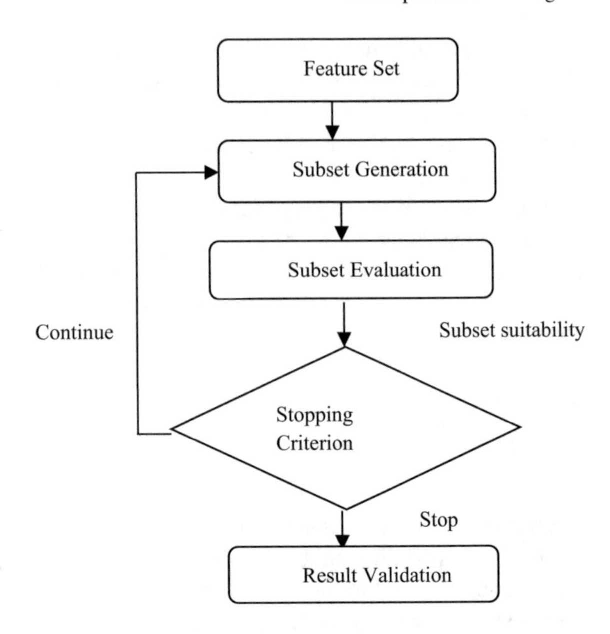

Fig. 4.3 Feature selection process

research trends rely on an optimal subset of features. A trade-off has to be worked out between the quality of the feature subset generated and its computational time.

A feature selection process typically comprises four basic stages, viz., generating subset, evaluating subset, criterion for stopping, and validation of results [12] as shown in Fig. 4.3.

Subset Generation

Subset generation process generates the next subset for evaluation. On the basis of a given search strategy, candidate features are produced that are subsequently evaluated. Candidate subsets produced are compared with the previous best subset on the basis of certain evaluation criterion. If the new subset produced is found better than the previous one, then the new subset replaces the previous best subset.

Two issues that are of utmost importance in this stage are the search starting points and the search strategy. There exist different generation procedures such as the sequential, exhaustive, and random search. Sequential algorithms can be easily implemented and are computationally fast due to their linear complexity. Features are added or removed one at a time. In case of exhaustive search strategy, all probable combinations of features are considered for evaluation.

However, it is practically infeasible even for small sized feature sets due to its high complexity. Even small sized feature sets have complexity as large as O (2^n), where n denotes the cardinality of the entire set of features. In the case of random search algorithms, search begins with a randomly selected subset of features, to which feature sets are randomly inserted or deleted. Their performance degrades with large feature sets. The generation procedure can begin with all features, no features or with random feature subset. In the first two cases, features are removed iteratively or added, whereas features either are iteratively added/removed or randomly produced thereafter in the last case [13].

Subset Evaluation

Subset evaluation evaluates the goodness of a generated subset using a given criterion function. The value obtained by the function is compared to the previous best subset. If the current subset has an evaluation value better than the previous value, then the newly generated subset replaces the previous one. A subset is considered optimal subject to a specific evaluation function (i.e., an optimal subset produced by a given evaluation function may not be the same as the optimal subset produced by another). Evaluation functions can be independent or dependent on the basis of the dependency criterion of the mining algorithms.

Stopping Criteria

An appropriate stopping criterion is necessary for the feature selection process, as otherwise it may exhaustively run through the space of subsets. The choice of the stopping criterion relies on the generation process and evaluation functions. Stopping criteria for a generation process can be set as a predefined number of features selected. Alternatively, a predefined number of iterations can also be considered as stopping criteria. For an evaluation function, the stopping criteria considered can be (i) any further addition (or deletion) of features produces superior subset and (ii) whether some evaluation function produces an optimal subset.

Validation

Validation does not form the core of the feature selection process. Rather, it validates the feature subset produced by feature selection by comparing it with previously established results on synthetic and real-world datasets.

Feature selection finds its applicability in diverse domains that deal with high dimension data. Some of use cases of feature selection are image recognition, bioinformatics, rule induction, and text categorization.

4.3 Fuzzy Rough Feature Selection

Feature attributes are crisp and real-valued, and herein traditional feature selectors encounter a problem particularly in terms of information loss. Although feature selection based on rough set theory can minimize the information loss and reveals the data interdependencies and reduces the dimensionality of the dataset immensely [33], yet it fails to ascertain the extent of similarity between two attribute values. For example, two close attribute values may differ due to noise, yet, rough set theory will consider these two values as completely different with different magnitudes. One possible way to tackle this problem is prior discretization of the dataset. However, this method remains inadequate, as the membership degrees of discretized values are not taken into account. It may be the case that two attribute values correspond to a “negative” label, but that one value is much more negative than the other, despite being significantly different. This certainly leads to information loss. A detailed discussion on this aspect can be found in [29].

It is therefore evident that there exists a requirement of some method that can accommodate data reduction of crisp real-valued datasets with measure of similarity extent among attribute values. The concept of fuzzy set theory [27] and the process of fuzzification provide a mechanism to assign membership values to the crisp real-valued attributes. The vagueness in the data can be modeled by assigning values of crisp attributes to multiple labels with variable membership degrees. This concept of reasoning under uncertainty can be explained by fuzzy concepts.

Fuzzy rough set feature selection combines the strength of rough set theory and fuzzy set theory. It can successfully handle the data dependencies and reduce data dimensionality without compromising the performance of classification and clustering (Chouchoulas and Shen, 2001). It can at the same time accommodate reasoning under uncertainty by assigning membership values to real-valued crisp attributes. Fuzzy rough set feature selection retains the most discriminative features while rest are removed without information loss. A fuzzy rough set consists of two fuzzy sets, a fuzzy lower and a fuzzy upper approximation, which are extensions of crisp rough set theory notions. Traditional rough set theory considers elements to be members with absolute certainty of lower approximation or are otherwise not members at all. In case of fuzzy rough set theory, elements have membership in the range [0, 1], thereby providing higher flexibility in handling uncertainty.

The following sub-section gives a detailed description of fuzzy rough set-based feature selection. Traditional fuzzy rough set approaches are supervised feature selection techniques; however, its unsupervised variants are also available in the literature [32].

4.3.1 *Supervised Fuzzy Rough Feature Selection*

Supervised fuzzy rough feature selection find reducts using evaluation metric that are guided by the decision labels of the concerned dataset. A detail account of fuzzy rough feature selection can be found in [15].

Let $I = (U', A_o)$ be an information system, where U' is a finite non-empty set of objects considered as the universe and A_o represents a finite non-empty attribute set such that $a : U' \to V_a$ for every $a \in A_o$. The attribute a may take its value from the set given by V_a. The equivalence relation associated with any $P \subseteq A_o$ is given by Eq. (4.1)

$$\text{IND}(P) = \{(x, y) \in U'^2 | \forall a \in P, \quad a(x) = a(y)\} \tag{4.1}$$

IND(P) generates partition denoted by $U'/\text{IND}(P)$ given by Eq. (4.2)

$$U'/\text{IND}(P) = \otimes\{U'/\text{IND}(\{a\})|a \in P\} \tag{4.2}$$

where the implication of $\otimes$ defined for the sets A_o and B_o is given by Eq. (4.3)

$$A_o \otimes B_o = \{X \cap Y | X \in A_o, Y \in B_o, X \cap Y \neq \emptyset\} \tag{4.3}$$

If $(x, y) \in \text{IND}(P)$, then it can be said that x, y are indiscernible by attributes from P. Let $X \subseteq U'$, the P-lower and P-upper approximations of X are given by Eqs. (4.4) and (4.5), respectively.

$$\underline{P}X = \left\{x \in U' | [x]_P \subseteq X\right\} \tag{4.4}$$

$$\overline{P}X = \left\{x \in U' | [x]_P \cap X \neq \emptyset\right\} \tag{4.5}$$

$\langle \underline{P}X, \overline{P}X \rangle$ forms a tuple and is called the rough set. $[x]_P$ denotes the equivalence classes of P indiscernibility relation. The positive region is given by Eq. (4.6)

$$\text{POS}_P(Q) = \bigcup_{x \in U'/Q} \underline{P}X \tag{4.6}$$

where P and Q represent set of attributes that induce equivalence relation over U'. All objects of U' lie in the positive region and are classified into U'/Q classes based on the information in P. For the set of attributes P and $Q \subset A_o$, the interdependency between them is given by Eq. (4.7)

$$k = \gamma_P(Q) = \frac{|\text{POS}_P(Q)|}{|U'|} \tag{4.7}$$

where $0 \leq k \leq 1$.

A minimal reduct R with respect to the initial set of attributes C_a, for a given attribute set D_a, is $\gamma_R(D_a) = \gamma_{C_a}(D)$. For $a \in R$, R is minimal if $\gamma_{R-\{a\}}(D_a) \neq \gamma_R(D_a)$.

A detailed description of fuzzy upper and lower approximation is found in [1].

A minimum subset of features that can preserve the degree of dependency for the whole dataset is defined as a fuzzy rough reduct which is calculated using the QuickReduct algorithm summarized in Algorithm 1 [32].

Algorithm 1 QuickReduct Algorithm

Input: C // conditional feature set
D// decision feature set
Output: R // an optimal reduct
1: begin
2: $R \leftarrow \emptyset$ // empty reduct set
3: $\gamma_{best} \leftarrow 0$ // best fuzzy rough dependency score set to 0.

4:$\gamma_{prev} \leftarrow 0$ // fuzzy rough dependency score of previous reduct set to 0
5: repeat
4:$T \leftarrow R$
5:$\gamma_{prev} \leftarrow \gamma_{best}$.
6: for each $x \in C - R$.
7: if $\gamma_{R\cup\{x\}}(D) > \gamma_T(D)$ then // evaluates fuzzy rough dependency score of the decision features of two pairs of attribute sets
8: $T \leftarrow R \cup \{x\}$ // appends attribute x to reduct R
9: $\gamma_{best}(D) > \gamma_T(D)$.
10: $R \leftarrow T$.
11: until $\gamma_{best} = \gamma_{prev}$ // fuzzy dependency score equals for the current reduct and
previous reduct
12: return $\boldsymbol{R}$.

4.3.2 Unsupervised Fuzzy Rough Feature Selection

Unsupervised fuzzy rough feature selection has similarity to fuzzy supervised approach except that the decision features of the supervised approach are replaced by any given set of features or group of features Q. Three different measures can be adopted to perform unsupervised feature selection based on fuzzy rough set theory. Fuzzy rough dependency measure, fuzzy rough boundary-based measure, and fuzzy rough discernibility measure may be employed for unsupervised fuzzy rough feature selection.

Given $P \subseteq A$ and Q is any given set of features or group of features, then dependency measure between P and Q

$$\gamma_P^{'}(Q) = \frac{\sum_{x\in U} \mu_{\mathrm{POS}_P(Q)}(x)}{|U|} \tag{4.8}$$

where $P \cap Q = \emptyset$ and $\mu_{\mathrm{POS}_{p(Q)}(x)}$ is fuzzy positive region.

$$\mu_{\mathrm{POS}_{R_P}(Q)}(x) = \sup \mu_{\underline{R_P} R_{QZ}}(x) \tag{4.9}$$

$$x \in U'$$

R_P represents the similarity relation induced by subset of features P and $\underline{R_P}$ is fuzzy lower approximation where a T transitive is used to approximate a fuzzy concept X,

R_{QZ} is the fuzzy tolerance class or equivalence class of object z.

Most fuzzy rough feature selections use lower approximations for feature selection. The lower approximations define the certainty regions of objects membership to a concept. Upper approximations give the uncertainty regions of an object. Hence, the upper approximation regions can help in discriminating between the subsets. The fuzzy rough boundary region for fuzzy tolerance class $R_{QZ}X$ is given by Eq. (4.10)

$$\mu_{\text{BND}_{(P)}(R_{QZ})}(x) = \mu_{\overline{R_P}R_{QZ}}(x) - \mu_{\underline{R_P}R_{QZ}}(x) \tag{4.10}$$

The upper approximation is given by Eq. (4.11).

$$\mu_{\overline{R_P}R_{QZ}}(x) = \sup T(\mu_{R_P}(x, y), \mu_{R_Q}(y, z) \tag{4.11}$$

$$y \in U'$$

$\text{BND}_P(x)$ is the boundary region of the object x. With the progress of search of optimal subset, the boundary region membership of the object decreases till a minimum is obtained. The total certainty of a feature P is given by Eq. (4.12)

$$\lambda_P(Q) = 1 - \frac{\sum_{z \in U} \sum_{x \in U} \mu_{\text{BND}_{R_P}(R_{Qz})}(x)}{|U|^2} \tag{4.12}$$

Classical discernibility relations can be extended to fuzzy tolerance relations such that it represents an objects' approximate equality. For a given combination of features P, a value is calculated to assess how good discernibility these features maintain relative to another subset of features Q among all objects and is given by Eq. (4.13)

$$f(P, Q) = \underbrace{T\big(c_{ij}(P, Q)\big)}_{1 \le i < j \le |U|} \tag{4.13}$$

with

$$c_{ij}(P, Q) = I(T\ \underbrace{(\mu_{R_a}(x_i, x_j))}), \mu_{R_Q}\big(x_i, x_j\big) \tag{4.14}$$

where $a \in P$.

4.3.3 Reduction Calculation

In unsupervised feature selection, the search is conducted in the space $P(C_a) \times P(C_a)$ so that a particular subset is compared with any other subset. To obtain the desired reduction, the algorithm initiates with all features with a linear backward search. The algorithm begins with all features in the dataset. Features are removed iteratively one by one to calculate the desired measure. If the measure is unaffected by removal of a feature, then it can be removed easily. The algorithm returns the full feature set in case of absence of interdependency in the dataset. The worst-case complexity of the search is $O(m)$, where m is the original number of features. Reduction is achieved for by using any of $\gamma'_T(\{x\})$, $\lambda_T(\{x\})$ or by fuzzy discernibility measure given by Eq. 4.14. The algorithm employed for reduction calculation is summarized in Algorithm 2 [32].

Algorithm 2. Unsupervised fuzzy rough feature selection

Input: X $(x_1, x_2, \ldots, x_m)$ // a dataset with m datapoints
Output: R // an optimal reduct
1: begin
2: $R \leftarrow C$ // assigns the entire conditional feature set C to the reduct R
3: for each $\boldsymbol{x} \in C$
4: $R \leftarrow R - \{x\}$ // each feature $\{\boldsymbol{x}\}$ is removed from R
5: if $CM(R, \{x\}) < 1$ // calculates the combined measure of evaluation
6: R $\leftarrow R \cup \{x\}$
7: return $\boldsymbol{R}$

4.4 Related Work

Research over the years has led to availability of extensive fuzzy rough feature selection techniques. In this section, the discussion is on some select fuzzy rough feature selection techniques across diverse application domains.

Kuncheva [26] conceptualized fuzzy rough set-based feature selection and used it in fuzzy pattern recognition. Subsequently, pioneering work is done by Shen and Jensen on fuzzy rough set-based feature selection methods [21–23, 36]. The concept of dependency functions of crisp rough sets is extended to their fuzzy counterparts. They developed the QuickReduct algorithm based on a fuzzy rough dependency function for feature selection. A minimal reduct is calculated using the QuickReduct algorithm. However, it has limitations as it could not exhaustively generate all the possible feature subsets. Despite limitations, experimental results with QuickReduct demonstrate satisfactory results, Zhao et al. [43] show that feature selection with QuickReduct has limitations in terms of mathematical foundation and theoretical analysis. To consolidate its mathematical formulation and theoretical analysis, Jensen and Shen [23, 24] propose a discernibility matrix using fuzzy for feature selection based on improved dependency function.

Hu et al. [19] point out that the QuickReduct algorithm has a computational complexity that exponentially increases with increase of input variables. Bhatt and Gopal [6, 7] show that stopping criteria of QuickReduct algorithm in certain cases led to non-convergence. Thus, they model new methods of feature selection based on fuzzy rough set theory that have improved time and computational complexity for feature selection in compact computational domain. Their methods have efficient termination criteria that achieves convergence on the considered datasets with improved reliability of selected set of features.

Selection of features based on fuzzy discernibility matrix [24] resulted in an over-reduct or under-reduct due to its stopping criteria [9, 41].

Jensen and Shen [24] in their paper highlight the efficiency of fuzzy rough set theory in dimensionality reduction. However, they observed that it also has limitations for large datasets. They have proposed three new approaches viz. fuzzy lower approximation-based feature selection, fuzzy boundary-based feature selection, and fuzzy discernibility-based feature selection on the basis of fuzzy similarity measures toward feature selection based on fuzzy rough set theory. Experimental results reveal that the methods can efficiently reduce dimensionality and simultaneously preserve accuracy of classification.

Wang et al. [39] argue that this model of fuzzy rough feature selection based on fuzzy rough dependency measure suffers a drawback. It is unable to maintain the maximal dependency function. They have put forward a new fuzzy rough set model that provides a good fit for the dataset under consideration. Moreover, it can tackle sample classification differences thereby minimizing misclassification. Their experimental results reveal that the proposed model is more efficient for datasets with numerous categories having large degree of overlap.

Anaraki and Eftekhari [4] employ the concept of fuzzy lower approximation-based feature selection (L-FRFS) to propose a new model of feature selection. It produces smaller reducts with better classification accuracy and execution time than L-FRFS, with respect to big datasets.

Qian et al. [34] propose a fuzzy rough feature selection accelerator called forward approximation, which can simultaneously perform dimensionality as well as sample reduction. The model is computationally efficient. The performance of the algorithm based on the accelerator is noteworthy particularly in case of large datasets.

Incremental methods of fuzzy rough feature selection using dynamic data is put forward by Yang et al. [41] using discernibility relation. Wang et al. [39] show that fuzzy rough set-based feature selection limited to batch processing is not very cost-effective and cannot be satisfactorily used with big datasets. Moreover, incremental fuzzy rough feature selections are advantageous as they have the flexibility to update knowledge from time to time.

Multi-label datasets have very high computational complexity. A fuzzy rough feature selection method for multi-label data is put forward by Qu et al. [35]. Association rules between labels are developed so as to collapse the combination labels into a sub-label set. Subsequently, the data that are relabeled are subjected to fuzzy rough feature selection. Although, the scale of labels is reduced and label overlapping avoided, yet it does not significantly improve the fuzzy rough feature selection.

Recently, Chen et al. [11] propose a graph theory-based fuzzy rough feature selection method. Their model considers to search a minimal transversal path from a hypergraph derivative. Thus, it avoids generation of hypergraph for the purpose of feature selection. The model efficiently minimizes time complexity in finding reducts, while dealing with large-scale datasets.

Ni et al. [30] propose a fuzzy rough set-based incremental feature selection algorithm. It reduces the data dimensionality without any domain knowledge. Experiments reveal that their model produces efficient results with minimal computation time and is found to be particularly effective in large-scale datasets. Table 4.1 summarizes the developments in the fuzzy rough feature selection domain over the past few decades.

Table 4.1 Developments in fuzzy rough feature selection

Authors	Developments
Jensen and Shen [21–23]	Dependency functions of crisp rough set is extended to their fuzzy counterpart
Bhatt and Gopal [6, 7]	QuickReduct algorithm in certain cases led to non-convergence arising due to stopping criteria
Bhatt and Gopal [6, 7]	Fuzzy rough set feature selection model that has improved time and computational complexity for feature selection in compact computational domain
Hu et al. [19]	With increase of input variables, computational complexity of QuickReduct algorithm increases exponentially
Zhao et al. [43]	QuickReduct has limitations in terms of its mathematical basis and theoretical analysis
Chen et al. [9] and Yang et al. [41]	Fuzzy discernibility matrix result in an over-reduct or under-reduct due to its stopping criteria
Anaraki and Eftekhari [4]	Concept of fuzzy lower approximation-based feature selection (L-FRFS) is used to propose a new model of feature selection
Qian et al. [34]	Fuzzy rough feature selection accelerator called forward approximation, which can simultaneously perform dimensionality as well as sample reduction
Wang et al. [39]	New fuzzy rough set model that provides a good fit for a considered dataset which can tackle sample classification differences thereby minimizing misclassification
Yang et al. [41]	Incremental methods of fuzzy rough feature selection using dynamic data using discernibility relation
Qu et al. [35]	Feature selection for multi-label data using association rules between labels using fuzzy rough set concepts
Chen et al. [11]	Graph theory-based fuzzy rough feature selection method
Ni et al. [30]	Fuzzy rough set-based incremental feature selection algorithm that reduces the data dimensionality without any domain knowledge

4.5 Potential of Fuzzy Rough Feature Selection in Text Categorization

The problem of assigning texts to predefined categories automatically is called text categorization [28, 31]. With the massive accumulation of online texts in digital libraries, text categorization is of great practical importance in extracting valuable information from the repositories. Text categorization is often confronted with the high dimensionality problem of the feature space. Even for a moderate collection, the feature space consists of hundreds of thousand terms. Data mining algorithms find it difficult to handle the prohibitively high data dimension, thereby rendering mining tasks intractable. Therefore, it becomes necessary to retain accuracy of categorization as well as to diminish the original feature space. Yang and Pederson [40] evaluate different feature selection techniques for text categorization problems in crisp domains. It is observed that the considered methods can effectively remove 50–90% of the terms while retaining the categorization accuracy.

Although, the potential of fuzzy rough set-based feature selection is evident from the reported literature in the period 2000–2020, yet feature selection in text categorization using fuzzy rough set-based models remain practically unexplored. All the reviewed literatures in Sect. 4.4 of the paper are based on diverse datasets other than text data. Jensen [20] reports text categorization using fuzzy rough feature selection (FRFS) with acceptable degree of dimensionality reduction and satisfactory classification. There exists immense scope of research in this domain utilizing the concept of fuzzy rough set models. Considering the breakthrough of feature selection using fuzzy rough sets in diverse domains, it can be anticipated that it will provide promising results in the area of text categorization. Therefore, there exists immense scope for researchers to make progress in the field of text categorization using fuzzy rough set models.

Fuzzy rough set feature selection is extremely useful in dimensionality reduction. However, for large-scale datasets fuzzy rough set-based techniques are time-consuming [11]. Most fuzzy rough set-based feature selection techniques require the entire dataset to be loaded in the memory, which may act as constraint in memory short environments when the dataset is too large like text datasets. One way to overcome the problem is to reduce the data dimension beforehand using manifold learning and subsequently the follow-up fuzzy rough set-based feature selection performed on an affinity matrix of much reduced dimension.

This paper proposes a hybrid feature selection technique called landmark-based fuzzy rough feature selection (LBFRFS) for data dimensionality reduction of large-scale datasets such as text datasets for memory short environments. The proposed technique (LBFRFS) hybridizes large-scale spectral clustering with landmark-based representation [10] with fuzzy rough set-based feature selection. The idea behind it is to construct an affinity matrix of much reduced dimension $p \times n$, where $p \ll n$ (n is the number of datapoints) using landmark-based spectral clustering. p represents the landmark points of the original dataset and the remaining points are linear combinations of these landmark points.

4.5.1 Proposed Hybrid Landmark-Based Fuzzy Rough Feature Selection (LBFRFS)

In this section, the landmark-based fuzzy rough feature selection (LBFRFS) is proposed based on the concepts of large-scale spectral clustering with landmark-based representation and fuzzy rough feature selection described in Sect. 4.3. Given a dataset $X = \{x_1, x_2, \ldots, x_n\}$, be a subset of R^m with n instances and m features, large-scale spectral clustering with landmark-based representation selects $p(\ll n)$ landmarks, the representative data points, and original data points are represented as linear combinations of the landmark points. Large-scale spectral clustering with landmark-based representation efficiently leads to construction of sparse affinity matrix of much reduced dimension $p \times n$, between the data points and landmark points where $p \ll n$ with property

$$W = \hat{Z}^T \hat{Z} \tag{4.15}$$

where W is the adjacency matrix and Z is the regression matrix that unfolds the underlying relationship between the data points X and landmark points.

A detailed description of large-scale spectral clustering with landmark-based representation can be found in [10]. The affinity matrix in Eq. 4.15 is further subjected to fuzzy rough set-based feature selection. The algorithm of the proposed LBFRFS is summarized in Algorithm 3.

Algorithm 3. Landmark-Based Fuzzy Rough Feature Selection

Input: N data points with M variables
u the number of landmark points
Output: R the optimal reduct
1. Select the cluster centers using k-means as landmark points.
2. Compute the affinity matrix $W = \hat{Z}^T \hat{Z}$
3. Assign W to reduct R.
4. For each $x \in W$,
5. Remove x from R.
6. Calculate the fuzzy rough dependency measure

$$\gamma'_P(Q) = \frac{\sum_{x \in U} \mu_{POS_P(Q)}(x)}{|U|}$$

7. If the fuzzy rough dependency measure is less than 1, then retain x in R else remove.
8. Return R

4.5.2 Experimental Results

In this section, the results of the experiments performed to demonstrate the effectiveness of the proposed LBFRFS are presented along with the benchmark datasets used and the performance evaluation metrics. The experimental results are evaluated for text categorization using k-means clustering.

4.5.2.1 Dataset

Three benchmark datasets used for the experimentation. The TDT2 dataset consists of 11,201 on-topic documents that have been classified into 96 categories. The experimental analysis is conducted on 9394 documents that belong to the largest 30 categories. Documents that appear in multiple categories are removed. Reuters 21,578 dataset consists of 21,578 documents distributed across 135 categories. Documents belonging to multiple categories are discarded leaving behind 8293 documents distributed across 65 categories which are considered for experiments. The 20 Newsgroups (NG) corpus is a collection of approximately 20,000 documents belonging to different newsgroups, classified into 20 categories. 1000 documents are taken from each group. There is a small fraction of documents that belong to more than one category which are discarded, leaving behind 18,846 documents.

4.5.2.2 Evaluation Metrics

The experimental results evaluated on standard metrics of accuracy and normalized mutual information.

Accuracy is defined as:

$$\text{Accuracy} = \frac{\text{TP} + \text{TN}}{\text{TP} + \text{TN} + \text{FP} + \text{FN}}$$

TP, TN, FP, and FN are true positives, true negatives, false positives, and false negatives, respectively.

Normalized mutual information (NMI) is defined as:

$$\overline{\text{MI}} = \frac{\text{MI}(C, C')}{\max(H(C), H(C'))}$$

where C represents set of cluster labels from the ground truth and C' represents the set of clusters labels after clustering, respectively. $H(C)$, $H(C')$ are entropies of C and C', respectively.

Their mutual information $\text{MI}(C, C')$ is defined as:

$$\mathrm{MI}(C, C') = \sum_{c_i \in C, C'_j \in C'} P(C_i, C'_j) \log_2 \frac{P\left(C_i, C'_j\right)}{P(C_i) \cdot P\left(C'_j\right)}$$

where $P(C_i)$ and $P\left(C'_j\right)$ denote probabilities that documents selected arbitrarily from the corpus belong to C_i and C'_j, respectively, and $P\left(C_i, C'_j\right)$ denote joint probabilities that the documents selected simultaneously belong to both the clusters. The range of $\overline{\mathrm{MI}}(C, C')$ is from 0 to 1. $\overline{\mathrm{MI}} = 0$ indicates that two clusters are independent while $\overline{\mathrm{MI}} = 1$, indicate that the clusters are identical.

4.5.2.3 Experiments and Analysis

Feature selection is carried out with the proposed LBFRFS over the considered datasets. Further, the results obtained with LBFRFS are compared with the unsupervised and supervised fuzzy rough feature selection techniques viz. unsupervised fuzzy rough lower approximation-based feature selection (UFRFS), unsupervised fuzzy boundary region-based feature selection (B-UFRFS) and unsupervised fuzzy discernibility-based feature selection (D-UFRFS), supervised fuzzy rough lower approximation-based feature selection (FRFS), supervised fuzzy boundary region-based feature selection (B-FRFS), and supervised fuzzy discernibility-based feature selection (D-FRFS). The datasets used in the experiment are labeled. However, before using the LBFRFS, UFRFS, B-UFRFS, and D-UFRFS, the decision variables are removed as they are unsupervised, and the approaches are applied on unlabelled data. In case of FRFS, B-FRFS, and D-FRFS the entire dataset with decision variables is used as they are supervised. The quality of the subset of features produced is evaluated in terms of *k*-means clustering. Two separate approaches are adopted when generating the reduct for considered datasets. Firstly, a near optimal reduct is selected from the dataset and secondly; a reduct is selected from each fold of tenfold cross validation. Subsequently, average for each fold is calculated. The reduct size produced by the considered feature selection techniques over the benchmark datasets is depicted in Table 4.2. Table 4.3 reports the results of the average reduct sizes produced by above mentioned techniques, respectively, using tenfold cross validation.

Table 4.2 Reduct size obtained using the considered feature selection techniques

Dataset	Features	Objects	FRFS	B-FRFS	D-FRFS	UFRFS	B-UFRFS	D-UFRFS	LBFRFS
TDT2	36,771	9394	1096	1246	1268	1437	1478	1506	1516
Reuters	18,933	8293	3160	3160	3158	4008	4015	4203	4199
20NG	26,214	18,846	3243	3242	3199	3578	3578	3596	3572

Table 4.3 Average reduct sizes obtained from tenfold cross validation using the feature selection techniques

Dataset	Features	Objects	FRFS	B-FRFS	D-FRFS	UFRFS	B-UFRFS	D-UFRFS	LBFRFS
TDT2	36,771	9394	1096.7	1246.6	1268.1	1437.5	1478.6	1506.8	1516.3
Reuters	18,933	8293	3160.4	3160.5	3158.2	4008.4	4015.6	4203.1	4199.1
20NG	26,214	18,846	3243.2	3242.4	3199.4	3578.2	3578.5	3596.3	3572.4

It is observed that LBFRFS produces reduct whose sizes are near comparable with those of the considered unsupervised fuzzy rough set approaches on TDT2 and Reuters datasets. With 20NG, LBFRFS produces remarkable performance in terms of reduct size, producing minimal reduct size compared to the unsupervised fuzzy rough set approaches. However, the supervised fuzzy rough set feature selection techniques have reduct size smaller than unsupervised approaches.

In order to assess the quality of reduct obtained, clustering is performed using k-means with the reduct. To randomize the experiments, different k values ($k = 5$, 10, 15, 20) are considered. For a given cluster k, 30 tests are performed randomly and their average results are noted with respect to the evaluation metric. The k-means clustering is performed 20 times with random initial points. The best result is noted in terms of k-means objective function. Clustering is done using the reduct obtained by the various techniques for $k = 5$, 10, 15, 20, and results are depicted in Table 4.4.

Table 4.4 Clustering accuracy of the feature selection techniques using k-means clustering

Dataset	$k = 5$						
	FRFS	B-FRFS	D-FRFS	UFRFS	B-UFRFS	D-UFRFS	LBFRFS
TDT2	79.36	7933	80.12	78.12	76.89	77.32	77.83
Reuters	81.22	80.97	81.07	81.70	81.74	80.32	**82.57**
20NG	80.66	81.09	80.79	81.54	82.09	82.06	**82.14**
$k = 10$							
TDT2	77.66	77.32	77.49	76.88	75.93	75.37	77.39
Reuters	75.18	74.94	75.17	75.69	75.17	75.19	75.21
20NG	72.84	72.81	72.87	71.39	71.08	71.50	71.59
$k = 15$							
TDT2	75.24	75.39	75.27	75.24	75.06	75.10	**76.28**
Reuters	71.36	71.38	70.19	72.03	73.05	71.29	**73.46**
20NG	77.33	78.53	79.29	78.18	79.44	79.61	**79.66**
$k = 20$							
TDT2	77.17	77.63	76.93	76.43	75.67	76.49	77.09
Reuters	74.11	75.29	75.18	74.17	75.16	73.35	72.74
20NG	76.55	76.28	77.16	77.27	77.29	78.37	78.64

Table 4.5 NMI of the feature selection techniques using k-means clustering

Dataset	$k = 5$						
	FRFS	B-FRFS	D-FRFS	UFRFS	B-UFRFS	D-UFRFS	LBFRFS
TDT2	72.67	73.38	73.62	75.62	75.88	74.39	74.88
Reuters	71.72	70.97	71.67	71.70	71.44	72.22	72.17
20NG	71.86	71.89	70.99	71.74	72.69	72.76	**72.79**
$k = 10$							
TDT2	70.44	70.42	70.46	70.08	70.09	70.37	**70.49**
Reuters	71.08	71.14	71.07	70.69	70.27	72.19	71.91
20NG	72.34	72.71	72.77	72.69	72.88	72.57	72.59
$k = 15$							
TDT2	69.14	70.09	70.27	69.24	70.06	70.27	69.18
Reuters	70.06	70.28	70.29	71.03	69.05	68.79	**71.46**
20NG	68.63	68.78	69.29	68.11	69.54	69.61	**69.65**
$k = 20$							
TDT2	67.76	67.63	67.32	67.83	66.79	66.97	67.09
Reuters	65.21	66.29	66.28	64.77	65.16	66.35	66.24
20NG	66.25	66.28	66.16	67.27	66.29	66.17	66.14

It is observed from Table 4.4 that LBFRFS produces best results for Reuters and 20 NG datasets for $k = 5$. Also, remarkable performance is obtained with the proposed technique on all the three considered datasets for $k = 15$. In the remaining cases, LBFRFS has comparable performance with the considered fuzzy rough feature selection techniques. It must be noted here that LBFRFS is an unsupervised technique yet produces better clustering accuracy than the supervised ones.

From Table 4.5, it is observed that the NMI measure of the clusters produces best results for 20 NG with $k = 5$, for TDT2 with $k = 10$, for Reuters, and 20 NG for $k = 15$. In all other cases, LBFRFS gives a close performance with the rest of the techniques considered.

4.6 Conclusions

Fuzzy rough sets are powerful mathematical tools which can model data dependency and uncertainty in data analysis. This concept has been widely applied by researchers for feature selection and dimensionality reduction. The paper reviews the major developments in fuzzy rough set-based feature selection over a period of twenty years. It has been observed that fuzzy rough set-based feature selection has successfully been applied to various application domains. However, its applicability in the field of text categorization remains practically unexplored. Considering its effectiveness and efficiency in diverse fields, fuzzy rough set-based feature selection may be expected to perform satisfactorily in the domain of text categorization. However, for large-scale datasets, fuzzy rough set-based techniques are time-consuming. Most fuzzy rough set-based feature selection techniques require the entire dataset to be loaded in the memory, which may be constrained in memory short environments when the dataset is too large like text datasets. To overcome this problem, the paper proposes a hybrid feature selection technique called landmark-based fuzzy rough feature selection (LBFRFS) for dimensionality reduction of large-scale datasets such as text datasets for memory short environments. It is observed from experimental results and analysis the effectiveness and efficiency of LBFRFS when compared with the state-of-the-art fuzzy rough feature selection techniques. As a future work, the efficiency of the proposed LBFRFS feature selection technique can be further evaluated with various classifiers to assess the predictive accuracy of the feature subset (reduct) produced over large benchmark text datasets.

References

1. Radzikowska, A.M., Kerre, E.E.: A comparative study of fuzzy rough sets. Fuzzy Sets Syst. **126**(2), 137–155 (2002)
2. Albrecht, A.A.: Stochastic local search for the feature set problem, with applications to microarray data. Appl. Math. Comput. **183**(2), 1148–1164 (2006)
3. Almuallim, H., Dietterich, T.G.: Learning with many irrelevant features. AAAI **91**, 547–552 (1991)
4. Anaraki, J.R., Eftekhari, M.: Improving fuzzy-rough quick reduct for feature selection. In: 2011 19th Iranian Conference on Electrical Engineering, pp. 1–6. IEEE (2011)
5. Belkin, M., Niyogi, P.: Laplacian eigenmaps and spectral techniques for embedding and clustering. In: Advances in Neural Information Processing Systems, pp. 585–591 (2002)
6. Bhatt, R.B., Gopal, M.: On fuzzy-rough sets approach to feature selection. Pattern Recogn. Lett. **26**(7), 965–975 (2005)
7. Bhatt, R.B., Gopal, M.: On the compact computational domain of fuzzy-rough sets. Pattern Recogn. Lett. **26**(11), 1632–1640 (2005)
8. Bishop, C.M.: Neural Networks for Pattern Recognition. Oxford University Press (1995)
9. Chen, D., Hu, Q., Yang, Y.: Parameterized attribute reduction with Gaussian kernel based fuzzy rough sets. Inf. Sci. **181**(23), 5169–5179 (2011)
10. Chen, X., Cai, D.: Large scale spectral clustering with landmark-based representation. In: Twenty-Fifth AAAI Conference on Artificial Intelligence (2011)
11. Chen, J., Mi, J., Lin, Y.: A graph approach for fuzzy-rough feature selection. Fuzzy Sets Syst. **391**, 96–116 (2020)
12. Dash, M., Liu, H.: Feature selection for classification. Intell. Data Anal. **1**(3), 131–156 (1997)
13. Dash, M., Liu, H.: Consistency-based search in feature selection. Artif. Intell. **151**(1–2), 155–176 (2003)
14. Devijver, P.A., Kittler, J.: Pattern Recognition: A Statistical Approach. Prentice Hall (1982)
15. Diao, R., Mac Parthaláin, N., Shen, Q.: Dynamic feature selection with fuzzy-rough sets. In: 2013 IEEE International Conference on Fuzzy Systems (FUZZ-IEEE), pp. 1–7. IEEE (2013)
16. Du¨ntsch, I., Gediga, G.: Rough Set Data Analysis: A Road to Non-invasive Knowledge Discovery. Methodos, Bangor (2000)

17. Fukunaga, K.: Introduction to Statistical Pattern Recognition, ser. Computer Science and Scientific Computing. Academic, Boston (1990)
18. Gennari, J.H., Langley, P., Fisher, D.: Models of incremental concept formation. Artif. Intell. **40**(1–3), 11–61 (1989)
19. Hu, Q., Xie, Z., Yu, D.: Hybrid attribute reduction based on a novel fuzzy-rough model and information granulation. Pattern Recogn. **40**(12), 3509–4352 (2007)
20. Jensen, R.: Combining Rough and Fuzzy Sets for Feature Selection. Doctoral dissertation, University of Edinburgh (2005)
21. Jensen, R., Shen, Q.: Semantics-preserving dimensionality reduction: rough and fuzzy-rough-based approaches. IEEE Trans. Knowl. Data Eng. **16**(12), 1457–1471 (2004)
22. Jensen, R., Shen, Q.: Fuzzy–rough attribute reduction with application to web categorization. Fuzzy Sets Syst. **141**(3), 469–485 (2004)
23. Jensen, R., Shen, Q.: Fuzzy-rough sets assisted attribute selection. IEEE Trans. Fuzzy Syst. **15**(1), 73–89 (2007)
24. Jensen, R., Shen, Q.: New approaches to fuzzy-rough feature selection. IEEE Trans. Fuzzy Syst. **17**(4), 824–883 (2008)
25. John, G.H., Kohavi, R., Pfleger, K.: Irrelevant features and the subset selection problem. In: Machine Learning Proceedings 1994, pp. 121–129. Morgan Kaufmann
26. Kuncheva, L.I.: Fuzzy rough sets: application to feature selection. Fuzzy Sets Syst. **51**(2), 147–153 (1992)
27. Zadeh, L.A.: Fuzzy sets. Inf. Control **8**, 338–353 (1965)
28. Leopold, E., Kindermann, J.: Text categorization with support vector machines. How to represent texts in input space? Mach. Learn. **46**(1–3), 423–444 (2002)
29. Beynon, M.J.: Stability of continuous value discretisation: an application within rough set theory. Int. J. Approx. Reas. **35**, 29–53 (2004)
30. Ni, P., Zhao, S., Wang, X., Chen, H., Li, C., Tsang, E.C.: Incremental feature selection based on fuzzy rough sets. Inf. Sci. **536**, 185–204 (2020)
31. Nigam, K., McCallum, A.K., Thrun, S., Mitchell, T.: Text classification from labeled and unlabeled documents using EM. Mach. Learn. **39**(2–3), 103–134 (2000)
32. Parthaláin, N.M., Jensen, R.: Measures for unsupervised fuzzy-rough feature selection. Int. J. Hybrid Intell. Syst. **7**(4), 249–259 (2010)
33. Pawlak, Z.: Rough sets. Int. J. Comput. Inform. Sci. **11**(5), 341–356 (1982)
34. Qian, Y., Wang, Q., Cheng, H., Liang, J., Dang, C.: Fuzzy-rough feature selection accelerator. Fuzzy Sets Syst. **258**, 61–78 (2015)
35. Qu, Y., Rong, Y., Deng, A., Yang, L.: Associated multi-label fuzzy-rough feature selection. In: 2017 Joint 17th World Congress of International Fuzzy Systems Association and 9th International Conference on Soft Computing and Intelligent Systems, pp. 1–6. IEEE, Otsu, Japan
36. Shen, Q., Jensen, R.: Selecting informative features with fuzzy-rough sets and its application for complex systems monitoring. Pattern Recogn. **3** (2004)
37. Tenenbaum, J.B., De Silva, V., Langford, J.C.: A global geometric framework for nonlinear dimensionality reduction. Science **290**(5500), 2319–2323 (2000)
38. Fayyad, U., Piatetsky-Shapiro, G., Smyth, P.: From data mining to knowledge discovery in databases. Artif. Intell. Mag. **17**(3), 37–54 (1996)
39. Wang, C., Qi, Y., Shao, M., Hu, Q., Chen, D., Qian, Y., Lin, Y.: A fitting model for feature selection with fuzzy rough sets. IEEE Trans. Fuzzy Syst. **25**(4), 741–753 (2016)
40. Yang, Y., Pedersen, J.O.: A comparative study on feature selection in text categorization. In: ICML, vol. 97, No. 412–420, p. 35 (1997)

41. Yang, Y., Chen, D., Wang, H., Tsang, E.C., Zhang, D.: Fuzzy rough set based incremental attribute reduction from dynamic data with sample arriving. Fuzzy Sets Syst. **312**, 66–86 (2017)
42. Yu, L., Liu, H.: Redundancy based feature selection for microarray data. In: Proceedings of the Tenth ACM SIGKDD International Conference on Knowledge Discovery and Data Mining, pp. 737–742 (2004)
43. Zhao, S., Tsang, E.C., Chen, D.: The model of fuzzy variable precision rough sets. IEEE Trans. Fuzzy Syst. **17**(2), 451–467 (2009)

Chapter 5
An Extensive Survey on Classification of Malaria Parasites in Patients Based on Fuzzy Approaches

Divya Srivastava, Samya Muhuri, Shashank Sheshar Singh, and Madhushi Verma

5.1 Introduction

Malaria is one of the world's most significant endemic diseases. According to the world health organization (WHO), 229 million malaria cases were estimated in 2019, killing another 409000 people. 94% of these cases and deaths are from the African region [1]. The disease is caused by a parasite called *Plasmodium malariae* which can be transported into the human bloodstream through a female anopheles mosquito [2]. Four varieties of these parasites (*P. malaria, P. vivax, P. falciparum*, and *P. ovale*) can infect humans. Each depends on the geographical and temperate condition, and another estimated 120 species affect other animals [3].

Over time, the most popular and effective way of detecting the presence of these parasites is to examine the human blood cells under a microscope by an expert. *Light microscopy* is one of the cost-effective approaches for detecting the presence of malaria parasites. In a simple description, the approach requires a microscopic expert to take the blood sample of a patient, add some staining solutions to expose the visibility of the malaria parasite, examine them under the light microscope, and finally, interpret the results of the examination for diagnosis. However, one of the

Samya Muhuri, Shashank Sheshar Singh, Madhushi Verma: These authors contributed equally to this work.

D. Srivastava (✉) · M. Verma
School of Computer Science Engineering and Technology, Bennett University, Greater Noida 201310, Uttar Pradesh, India
e-mail: divya.srivastava@bennett.edu.in

M. Verma
e-mail: madhushi.verma@bennett.edu.in

S. Muhuri · S. S. Singh
Department of Computer Science and Engineering, Thapar Institute of Engineering & Technology, Patiala 147004, Punjab, India
e-mail: samya.Muhuri@bennett.edu.in

S. S. Singh
e-mail: shashank.sheshar@gmail.com

T. Som et al. (eds.), *Fuzzy, Rough and Intuitionistic Fuzzy Set Approaches for Data Handling*, Forum for Interdisciplinary Mathematics,
https://doi.org/10.1007/978-981-19-8566-9_5

limitations of this approach is that it is time-consuming to comprehend malaria parasites' behavior fully, hence costly to train an expert for a vast country like India [4]. Also, the success of this approach depends on the level of experience of the expert to achieve good efficiency [5, 6].

A new approach called Rapid Diagnostic Tests (RDT) was introduced to address these limitations. This approach tests for a malaria parasite antigen. WHO has recommended RDTs as an effective alternative approach for detecting malaria parasites [7, 8]. As RDT does not require high expert knowledge, it is very effective in rural areas in the unavailability of a microscopic expert.

In 2017, RDT was responsible for the classification of malaria parasites in approximately 47% of the malaria-endemic countries across the world [9]. The simplicity of RDTs comes with a cost. Since four malaria parasites infect humans, four antigens for each species will be tested and incur additional costs. RDTs are more expensive as compared to light microscopy examination. Furthermore, RDTs do not provide quantification of its results [10]. According to a comparative study by [11], from the total of 1724 blood samples tested, 128 (13.3%) were false negatives. Therefore, RDT results are not very accurate.

Polymerase chain reaction (PCR) is another approach that is considered the most accurate compared to all the approaches discussed thus far. It is capable of detecting the presence of malaria parasites and their species at a molecular level. However, despite the high accuracy of this approach, its complexity has made it infeasible to replace existing classification methods because it is time-consuming and requires skilled experts [12].

According to [13], the requirements for any malaria-detecting approach should consider the following: the time and cost per test, level of experience required to process, provide quantification description of parasite stages, resistance to drugs, and other interacting indicators. Several alternative approaches have been developed to achieve these requirements. The automated classification approach has gained much interest in recent times. Due to the advancement in computational systems, different machine learning techniques have been adopted to locate parasites in an image of blood smears [14–18]. In recent times, deep learning is considered more often in the automated classification of malaria parasites. With the success of Alexnet [19] in 2012, convolutional neural network (CNN) has gained interest in various image recognition areas. Harnessing the properties of CNN to recognize malaria parasites was first considered by [4] to differentiate infected and uninfected cells from images. Since then, several studies have considered CNN for feature extraction in malaria parasite recognition, like in [3, 4, 20–26]. Meanwhile, training a deep learning model requires an enormous amount of data to fit properly, which is not readily available in the medical application area. Other automated classification approaches involve different image processing approaches to segment infected parasites from the red blood cells (RBC). These approaches include clustering [27–30], morphological approach [31–34], contour segmentation [35–37], fuzzy systems [38, 39] , and many more.

5.1.1 *Fuzzy Logic*

Fuzzy logic takes after the human dynamic procedure and manages ambiguous and loose data. The word fuzzy alludes to things which are not satisfactory or obscure. Any occasion, interaction, or capacity that is changing ceaselessly cannot generally be characterized as one or the other valid or bogus, which implies that we want to characterize such exercises in a Fuzzy way.

5.1.1.1 Preliminaries

Fuzzy Logic: Fuzzy logic takes after the human powerful system and oversees vague and free information. The word fuzzy insinuates things which are not agreeable or are dark. Any event, cooperation, or limit that is changing interminably cannot by and large be described as either legitimate or sham, which suggests that we need to portray such practices in a Fuzzy manner.

Fuzzy Set: Fuzzy Set: A fuzzy set is a class of things with a continuum of grades of interest. Such a set is depicted by an interest (brand name) work which allots to each fight a grade of enlistment heading off to some place in the range of nothing and one. The ideas of consideration, association, convergence, supplement, connection, convexity, and so forth are reached out to such sets.

Fuzzy Numbers: A fuzzy number is a hypothesis of a standard, certifiable number as in it does not suggest one single worth yet rather to a related arrangement of likely characteristics, where each possible worth has its own heap some place in the scope of 0 and 1. This weight is known as the participation work. A fuzzy number is along these lines an uncommon instance of a curved, standardized fuzzy arrangement of the genuine line.

Fuzzification and De-fuzzification: Fuzzification is a stage to decide how much information has a place with every one of the proper fuzzy sets through the participation capacities. De-fuzzification is the method involved with getting a solitary number from the yield of the collected fuzzy set. It is utilized to move fuzzy derivation results into a fresh yield. All in all, de-fuzzification is acknowledged by a dynamic calculation that chooses the best fresh worth dependent on a fuzzy set.

5.1.2 *Fuzzy Logic Application on Disease Diagnosis*

In the recent era, medical diagnostic procedures are aided with technologies and became independent of human intervention. Thus, fuzzy-based logic is mainly utilized in medical diagnosis to enhance accuracy and precision. In disease diagnosis, fuzzy systems can help us in four broad domains: physical symptoms, medical history, pathological tests, and histological investigations of the patient. Fuzzy set theory and fuzzy logic are highly applicable for developing knowledge-based medical diagnostic systems due to their precise performance in uncertainty. In [40], a fuzzy-based decision support system is designed for diagnosing coronary artery. In another work [41], a fuzzy-based methodology is presented for the classification of arrhythmic and ischemic beat. A five-layer fuzzy ontology is also designed for decision support application of diabetes [42]. The researchers have shown fuzzy systems for locating abnormalities in bone scintigraphy [43]. In [44], the authors have shown the method of dealing with the uncertainty of medical data using several features of fuzzy logic.

5.2 Fuzzy Logic and Fuzzy-Based Malaria Diagnosis Framework

The fuzzy systems approach is highly considered for inspecting malaria parasites based on any microscopic image of the RBC. This section discusses existing literature that is related to the classification of malaria from blood images. Fuzzification and de-fuzzification are the two key steps for recognizing malaria parasites from the blood samples. On the contrary, a fuzzy rule-based system can be designed to examine malaria from the patients' symptoms. The section will demonstrate different steps of fuzzy rule-based diagnosis framework based on conventional and unconventional methods.

5.2.1 Fuzzy Logic

Fuzzy logic takes after the human unique method and oversees vague and free information. The word fuzzy insinuates things which are not palatable or are dark. Any event, association, or limit that is changing interminably cannot for the most part be described as either legitimate or counterfeit, which suggests that we need to portray such practices in a Fuzzy manner.

5.2.2 Pre-processing the Data

Malaria can be recognized utilizing two types of data, which are conventional and non-conventional. The conventional method includes examination of blood samples, whereas non-conventional methods purely depend on observing symptoms of the patient. As all types of medical data involve complexity and vagueness, the diagnosis tools need some level of decision support. Though data gathered from images and symptoms are treated differently, data pre-processing is necessary for accurate analysis. Pre-processing steps involve data normalization, denoising, interpolation, and registration. The processed data then passes through the fuzzy framework for disease detection. Whereas blood image samples go through fuzzification, de-fuzzification, and clustering steps, several fuzzy-based rules can be developed for deciding symptoms of malaria.

5.2.3 Fuzzification and De-fuzzification

Fuzzification is an approach of transferring a crisp input value to a fuzzy value based on the available information in the knowledge base. Three popular membership functions that have been employed in the data fuzzification process are Gaussian, triangular, and trapezoidal. In the de-fuzzification approach, best crisp value is selected based on a fuzzy set for the decision-making algorithm. Next, the appropriate features should be considered from the blood samples to detect the trace of malaria parasites.

5.2.4 Feature Selection and Clustering

There are several features in the images that can differentiate malaria parasites from the blood samples. As shown in [45], image exposure compensation and edge enhancement have increased the efficiency of the approach. Afterward, fuzzy c-means algorithm is utilized for distinguishing malaria parasites. Ghosh et al. [46] have shown the fuzzy diversion method over entropy of the blood images to segment the parasite. Histogram thresholding is introduced in [47] to locate infected cells. We have shown the frequently used stages for detecting malaria parasites from images in Fig. 5.1.

5.2.5 Malaria Parasite Classification from Blood Images

Chayadevi et al. [48] have proposed a fuzzy logic system to automatically segment the color of the malaria parasite in digital images of a blood film. Their approach has been used for pre-processing the image to highlight specific features like colors and fractal feature. Some other approaches consist of classification of the segmented malaria parasites (using either Adaptive Resonance Theory Neural Network (ARTNN), Back Propagation Network (BPN), or SVM). The fuzzy segmentation approach is described in different steps as follows: (a) using a suitable clustering algorithm to perform pixel classification for a specified number of regions, (b) selecting the center of each region, defining their membership function, and finding their key weight and a threshold value, and (c) iterating through all the n pixels and calculating each region's membership function for all the unclassified pixels. Finally, the fuzzy rules have been implemented to classify each pixel and assign them to corresponding regions.

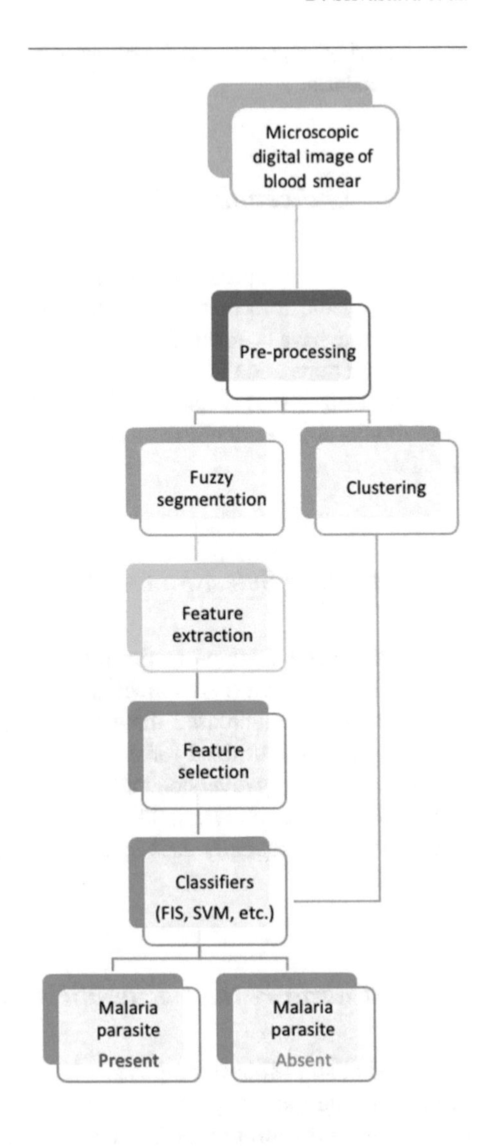

Fig. 5.1 Frequently used stages in surveyed literature for detecting malaria parasite in microscopic digital image of blood smear

Madhu [49] proposed a fuzzy set method to enhance edges in a microscopic digital image. The approach was sectioned into four phases; (a) grayscale conversion, which transforms the original given image into a grayscale image, and a min-max normalization is performed, (b) Gaussian membership functions of type II fuzzy sets are computed, and with the help of a threshold value, the boundary of the membership values are measured, (c) Hamacher t-conorm is introduced to compute additional membership values based on lower and upper membership values, and (d) a median filter algorithm is applied to obtain edge-enhanced microscopic digital images, which enhances the classification of these malaria parasites. In [50], the same author exploits the Einstein t-conorm function to compute the membership

Table 5.1 Fuzzy-based rules for detecting malaria disease

Rule	Symptoms							Inference
	Fever	Headache	Vomiting	Dizziness	Nausea	Liver fattiness	Weakness	
1	Gentle	Gentle	Gentle	Gentle	Gentle	Gentle	Gentle	Gentle
2	Gentle	Gentle	Gentle	Average	Average	Average	Acute	Gentle
3	Average	Gentle	Average	Average	Average	Acute	Acute	Average
4	Average	Gentle	Average	Average	Acute	Acute	Acute	Average
5	Average	Average	Average	Average	Acute	Acute	Acute	Acute
6	Acute	Average	Average	Acute	Acute	Acute	Very acute	Acute
7	Acute	Acute	Acute	Average	Average	Very acute	Very acute	Acute
8	Acute	Acute	Acute	Acute	Acute	Very acute	Very acute	Acute
9	Very acute	Very acute	Very acute	Very acute	Acute	Acute	Acute	Very acute
10	Very acute	Very acute	Very acute	Very acute	Very acute	Very acute	Very acute	Very acute

function for the segmentation-infected regions in the RBC. Further implements the inverse of the Gaussian gradient function for its final segmentation of the malaria parasites.

Ghosh et al. [51] proposed an extension of fuzzy divergence with Yager's measure to segment malaria parasites in a microscopic digital image. Their membership values were computed using Cauchy's membership function to measure the threshold for the segmentation of malaria parasites in the RBC.

Adaptive Neuro-Fuzzy Inference System (ANFIS) was introduced in 1993 [52]. Rather than manually selecting a fixed membership function, it is designed to permit adaptability and adjustment of the membership functions. Both fuzzy logic and neural network framework have been used to achieve more accuracy [53].

Appiah [54] proposed a malaria diagnostic system using the ANFIS approach, which uses clinical data to diagnose a patient's malaria status. Tsegay and Anusuya [55] also proposed a similar diagnostic system utilizing both Artificial Neural Network (ANN) and fuzzy logic. All these abovementioned papers are mostly focused on detecting the presence of malaria parasites in microscopic digital images of blood smears using fuzzy logic (Table 5.1).

The exactness of our forecast calculation with the marked test information is calculated by using precision and recall. A disarray lattice is displayed in Table 5.2 that permits perception of the exhibition of an expectation calculation. True Positive, and Negative, False Positive, and Negative are addressed as Tr_p, TR_n, F_p, and F_n, respectively. Reliable analytical metrics like precision, recall, F-measure, accuracy, and specificity are formulated in Eqs. 5.1–5.4.

Table 5.2 Confusion matrix representation

Values	Real positive (+)	Real negative (−)
Predicted true (+)	Tr_p	Tr_n
Predicted false (−)	F_p	T_n

$$\text{Precision}\,(P_r) = \frac{\mathrm{Tr}_p}{\mathrm{Tr}_p + F_p} \tag{5.1}$$

$$\text{Recall}\,(R_c) = \frac{\mathrm{Tr}_p}{\mathrm{Tr}_p + F_n} \tag{5.2}$$

$$F\text{-measure} = \frac{2 * p_r * R_c}{P_r + R_c} \tag{5.3}$$

$$\text{Accuracy} = \frac{\mathrm{Tr}_p + \mathrm{Tr}_n}{\mathrm{Tr}_p + \mathrm{Tr}_n + F_p + F_n} \tag{5.4}$$

$$\text{Specificity} = \frac{\mathrm{Tr}_n}{\mathrm{Tr}_n + F_p} \tag{5.5}$$

5.3 Classification of Malaria from Non-conventional Method

Other than the traditional blood test, several off-bit procedures have been also utilized to identify malaria. The popular methods are quantitative buffy coat (QBC) method [56], rapid diagnostic tests [57], ParaScreen [58], SD Bioline [59], and polymerase chain reaction (PCR) [60]. In [61], authors have shown that both saliva and urine can be utilized as the noninvasive sources of DNA for molecular classification of both *P. falciparum* and *P. vivax*. Though till date, these non-conventional methods are not fully utilized due to lack of infrastructure or less experimental data.

Medical practitioners have also emphasized the symptoms of the patients to anticipate the presence of malaria. The common symptoms include fever, headache, vomiting, dizziness, nausea, liver fattiness, and weakness. Based only on the symptoms, it would become very difficult to judge whether a patient is suffering from malaria or not. As the solution, some fuzzy rules can be formulated on the symptoms to provide a benchmark to the decision support system as shown in [62]. In Table 5.1, we have shown some fuzzy rules based on the symptoms of malaria. The degree of truth on the rules can be determined by calculating minimum nonzero input values based on operators and 'root sum square' method. The membership function would describe the probability of severity of a patient to be affected by malaria.

Table 5.3 Comparison of different approaches

Method	Approach	Sensitivity (%)	Specificity (%)	Accuracy (%)	Comment
Oladele et al. [63]	Combination of neural network and fuzzy logic	89.26	83.21	90.02	Performing better than only fuzzy-based systems
Bias et al. [39]	Histogram analysis based on fuzzy logic	90.53	84.93	90.12	Performing better than conventional edge detection methods
Preethi et al. [64]	ANN and DCNN	94.85	87.14	95.10	Perform better in noise removal
Muda et al. [65]	Hybrid K-means and median-cut	95.12	89.62	95.32	Perform better for object segmentation
Ghosh et al. [46]	Thresholding fuzzy divergence	95.63	90.08	95.84	Performing better only for some specific stages of malaria parasites

Here, we have discussed several methods for diagnosing malaria parasite from conventional and non-conventional methods based on fuzzy system. As shown in Fig. 5.2, a knowledge base can be developed by utilizing both patients' pathological history and symptoms. Comparison among different approaches is tabulated in Table 5.3. It is seen that most of the fuzzy-based frameworks have achieved more than 90% accuracy. It is also observed that the proper image segmentation algorithms always lead toward better accuracy in malaria parasite detection by the pathologists in peripheral blood smears.

5.4 Detection of Malaria from Soft-Computing Methods

In today's world, deep learning-based models have enhanced the efficiency of classification algorithms. In [66], a customized convolutional neural network model has been proposed for malaria parasite detection from blood smears. A computer-aided diagnosis framework is also developed for finding a trace of parasites from blood images [67]. Recently, some prominent works have been proposed by utilizing the power of neural network [68, 69]. A deep learning-based feature extraction procedure is proposed in [70]. The performance of several convolutional neural networks is depicted in [25]. For detecting malaria parasites, a data-driven approach is proposed in [71]. A deep transfer learning framework is suggested in [72]. All these frame-

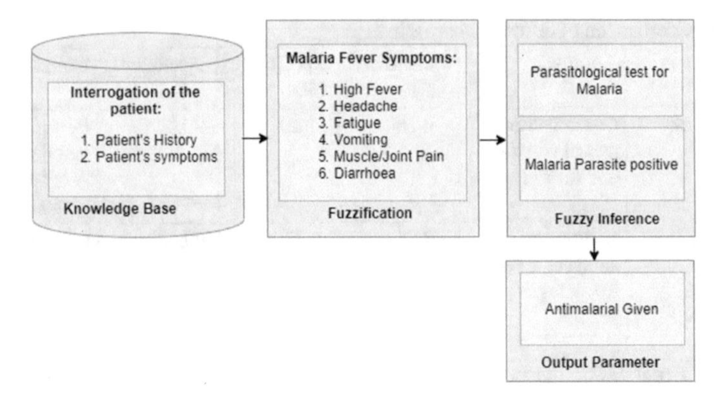

Fig. 5.2 Fuzzy-based malaria detection framework

works have established a strong base for detecting malaria parasites mostly from blood images mostly in unsupervised nature. Though they have not directly used the fuzzy-based logic, the efficiency of these methods is considered as the baseline for future research.

5.5 Conclusion and Future Research Scope

In this current manuscript, we have discussed several approaches for detection of malaria parasites. The methods include blood sample segmentation and realizing symptoms of the patients. We have considered different fuzzy-based decision-making systems that can remove human interpretation and assist the medical practitioner. The scope of machine learning algorithms has also been communicated that can enhance the faithfulness of the automated medical diagnosis. Our work not only opens up the path in medical investigation but also encourages the scientists of interdisciplinary research areas.

The current topic has wide future perspective as fuzzy-based systems have more capabilities than the conventional methods to determine malaria parasites. In the future, conventional and non-conventional approaches can be blended for acquiring better accuracy and precision. A knowledge base can be developed by utilizing both patients' pathological history and symptoms. Fuzzy rules can be developed from the symptoms, whereas the fuzzy interface can be fabricated for blood image segmentation. The result of both interfaces can be fused to make an automated decision. Also, several machine learning algorithms can be maneuvered for better fidelity of our approach.

References

1. Malaria. https://www.who.int/health-topics/malaria#. Accessed 11 Aug 2021
2. Collins, W.E., Jeffery, G.M.: Plasmodium malariae: parasite and disease. Clin. Microbiol. Rev. **20**(4), 579–592 (2007)
3. Arrow, K.J., Panosian, C., Gelband, H., et al.: The parasite, the mosquito, and the disease. In: Saving Lives, Buying Time: Economics of Malaria Drugs in an Age of Resistance. National Academies Press (US) (2004)
4. Liang, Z., Powell, A., Ersoy, I., Poostchi, M., Silamut, K., Palaniappan, K., Guo, P., Hossain, M.A., Sameer, A., Maude, R.J., et al.: CNN-based image analysis for malaria diagnosis. In: 2016 IEEE International Conference on Bioinformatics and Biomedicine (BIBM), pp. 493–496 (2016). IEEE
5. Wang, L., Wang, X.-k., Peng, J.-j., Wang, J.-q.: The differences in hotel selection among various types of travellers: a comparative analysis with a useful bounded rationality behavioural decision support model. Tour. Manag. **76**, 103961 (2020)
6. Yang, Y., Hu, J., Liu, Y., Chen, X.: A multiperiod hybrid decision support model for medical diagnosis and treatment based on similarities and three-way decision theory. Expert Syst. **36**(3), 12377 (2019)
7. Kim, S.H., Nam, M.-H., Roh, K.H., Park, H.C., Nam, D.H., Park, G.H., Han, E.T., Klein, T.A., Lim, C.S.: Evaluation of a rapid diagnostic test specific for plasmodium vivax. Trop. Med. Int. Health **13**(12), 1495–1500 (2008)
8. Ugah, U.I., Alo, M.N., Owolabi, J.O., Okata-Nwali, O.D., Ekejindu, I.M., Ibeh, N., Elom, M.O.: Evaluation of the utility value of three diagnostic methods in the detection of malaria parasites in endemic area. Malaria J. **16**(1), 1–8 (2017)
9. WHO, U.: World Malaria Report. World Health Organization, Geneva (2017)
10. Poostchi, M., Silamut, K., Maude, R.J., Jaeger, S., Thoma, G.: Image analysis and machine learning for detecting malaria. Transl. Res. **194**, 36–55 (2018)
11. Berzosa, P., de Lucio, A., Romay-Barja, M., Herrador, Z., González, V., García, L., Fernández-Martínez, A., Santana-Morales, M., Ncogo, P., Valladares, B., et al.: Comparison of three diagnostic methods (microscopy, RDT, and PCR) for the detection of malaria parasites in representative samples from equatorial guinea. Malaria J. **17**(1), 1–12 (2018)
12. Tangpukdee, N., Duangdee, C., Wilairatana, P., Krudsood, S.: Malaria diagnosis: a brief review. Korean J. Parasitol. **47**(2), 93 (2009)
13. Vink, J., Laubscher, M., Vlutters, R., Silamut, K., Maude, R., Hasan, M., De Haan, G.: An automatic vision-based malaria diagnosis system. J. Microsc. **250**(3), 166–178 (2013)
14. Go, T., Kim, J.H., Byeon, H., Lee, S.J.: Machine learning-based in-line holographic sensing of unstained malaria-infected red blood cells. J. Biophotonics **11**(9), 201800101 (2018)
15. Das, D.K., Ghosh, M., Pal, M., Maiti, A.K., Chakraborty, C.: Machine learning approach for automated screening of malaria parasite using light microscopic images. Micron **45**, 97–106 (2013)
16. Park, H.S., Rinehart, M.T., Walzer, K.A., Chi, J.-T.A., Wax, A.: Automated detection of *P. falciparum* using machine learning algorithms with quantitative phase images of unstained cells. PLoS ONE **11**(9), 0163045 (2016)
17. Kunwar, S.: Malaria detection using image processing and machine learning (2018)
18. Manning, K., Zhai, X., Yu, W.: Image analysis and machine learning based malaria assessment system. Digit. Commun. Netw. (2021)
19. Krizhevsky, A., Sutskever, I., Hinton, G.E.: Imagenet classification with deep convolutional neural networks. Adv. Neural Inf. Process. Syst. **25**, 1097–1105 (2012)
20. Mehanian, C., Jaiswal, M., Delahunt, C., Thompson, C., Horning, M., Hu, L., Ostbye, T., McGuire, S., Mehanian, M., Champlin, C., et al.: Computer-automated malaria diagnosis and quantitation using convolutional neural networks. In: Proceedings of the IEEE International Conference on Computer Vision Workshops, pp. 116–125 (2017)

21. Hung, J., Carpenter, A.: Applying faster R-CNN for object detection on malaria images. In: Proceedings of the IEEE Conference on Computer Vision and Pattern Recognition Workshops, pp. 56–61 (2017)
22. Masud, M., Alhumyani, H., Alshamrani, S.S., Cheikhrouhou, O., Ibrahim, S., Muhammad, G., Hossain, M.S., Shorfuzzaman, M.: Leveraging deep learning techniques for malaria parasite detection using mobile application. Wirel. Commun. Mob. Comput. **2020** (2020)
23. Umer, M., Sadiq, S., Ahmad, M., Ullah, S., Choi, G.S., Mehmood, A.: A novel stacked CNN for malarial parasite detection in thin blood smear images. IEEE Access **8**, 93782–93792 (2020)
24. Kumar, A., Sarkar, S., Pradhan, C.: Malaria disease detection using CNN technique with SGD, RMSprop and ADAM optimizers. In: Deep Learning Techniques for Biomedical and Health Informatics, pp. 211–230. Springer (2020)
25. Militante, S.V.: Malaria disease recognition through adaptive deep learning models of convolutional neural network. In: 2019 IEEE 6th International Conference on Engineering Technologies and Applied Sciences (ICETAS), pp. 1–6 (2019). IEEE
26. Loh, D.R., Yong, W.X., Yapeter, J., Subburaj, K., Chandramohanadas, R.: A deep learning approach to the screening of malaria infection: automated and rapid cell counting, object detection and instance segmentation using mask R-CNN. Comput. Med. Imaging Graph. **88**, 101845 (2021)
27. Aimi Salihah, A.-N., Yusoff, M., Zeehaida, M.: Colour image segmentation approach for detection of malaria parasites using various colour models and k-means clustering (2013)
28. Nasir, A.A., Mashor, M., Mohamed, Z.: Segmentation based approach for detection of malaria parasites using moving k-means clustering. In: 2012 IEEE-EMBS Conference on Biomedical Engineering and Sciences, pp. 653–658 (2012). IEEE
29. Abbas, N., Mohamad, D., et al.: Microscopic RGB color images enhancement for blood cells segmentation in YCBCR color space for k-means clustering. J. Theor. Appl. Inf. Technol. **55**(1), 117–125 (2013)
30. Abdul-Nasir, A.S., Mashor, M.Y., Halim, N.H.A., Mohamed, Z.: The cascaded moving k-means and fuzzy c-means clustering algorithms for unsupervised segmentation of malaria images. In: AIP Conference Proceedings, vol. 1660, p. 070111 (2015). AIP Publishing LLC
31. Mustafa, W.A., Abdul-Nasir, A.S., Mohamed, Z., Yazid, H.: Segmentation based on morphological approach for enhanced malaria parasites detection. J. Telecommun. Electron. Comput. Eng. (JTEC) **10**(1–16), 15–20 (2018)
32. Savkare, S., Narote, S., et al.: Automatic detection of malaria parasites for estimating parasitemia. Int. J. Comput. Sci. Secur. (IJCSS) **5**(3), 310 (2011)
33. Di Ruberto, C., Dempster, A., Khan, S., Jarra, B.: Morphological image processing for evaluating malaria disease. In: International Workshop on Visual Form, pp. 739–748 (2001). Springer
34. Loddo, A., Di Ruberto, C., Kocher, M.: Recent advances of malaria parasites detection systems based on mathematical morphology. Sensors **18**(2), 513 (2018)
35. Abidin, S.R., Salamah, U., Nugroho, A.S.: Segmentation of malaria parasite candidates from thick blood smear microphotographs image using active contour without edge. In: 2016 1st International Conference on Biomedical Engineering (IBIOMED), pp. 1–6 (2016). IEEE
36. Kumarasamy, S.K., Ong, S., Tan, K.S.: Robust contour reconstruction of red blood cells and parasites in the automated identification of the stages of malarial infection. Mach. Vis. Appl. **22**(3), 461–469 (2011)
37. Imran Razzak, M., Naz, S.: Microscopic blood smear segmentation and classification using deep contour aware CNN and extreme machine learning. In: Proceedings of the IEEE Conference on Computer Vision and Pattern Recognition Workshops, pp. 49–55 (2017)
38. Sharma, M., Mittal, R., Choudhury, T., Satapathy, S.C., Kumar, P.: Malaria detection using improved fuzzy algorithm. In: Intelligent Computing and Information and Communication, pp. 653–665. Springer (2018)
39. Bias, S., Reni, S.K., Kale, I.: A novel fuzzy logic inspired edge detection technique for analysis of malaria infected microscopic thin blood images. In: 2017 IEEE Life Sciences Conference (LSC), pp. 262–265 (2017). IEEE

40. Tsipouras, M.G., Exarchos, T.P., Fotiadis, D.I., Kotsia, A.P., Vakalis, K.V., Naka, K.K., Michalis, L.K.: Automated diagnosis of coronary artery disease based on data mining and fuzzy modeling. IEEE Trans. Inf. Technol. Biomed. **12**(4), 447–458 (2008)
41. Tsipouras, M.G., Voglis, C., Fotiadis, D.I.: A framework for fuzzy expert system creation—application to cardiovascular diseases. IEEE Trans. Biomed. Eng. **54**(11), 2089–2105 (2007)
42. Lee, C.-S., Wang, M.-H.: A fuzzy expert system for diabetes decision support application. IEEE Trans. Syst. Man Cybern. Part B (Cybern.) **41**(1), 139–153 (2010)
43. Yin, T.-K., Chiu, N.-T.: A computer-aided diagnosis for locating abnormalities in bone scintigraphy by a fuzzy system with a three-step minimization approach. IEEE Trans. Med. Imaging **23**(5), 639–654 (2004)
44. John, R.I., Innocent, P.R.: Modeling uncertainty in clinical diagnosis using fuzzy logic. IEEE Trans. Syst. Man Cybern. Part B (Cybern.) **35**(6), 1340–1350 (2005)
45. Hendrawan, Y.F., Angkoso, C.V., Wahyuningrum, R.T.: Colour image segmentation for malaria parasites detection using cascading method. In: 2017 International Conference on Sustainable Information Engineering and Technology (SIET), pp. 83–87 (2017). IEEE
46. Ghosh, M., Das, D., Chakraborty, C., Ray, A.K.: Plasmodium vivax segmentation using modified fuzzy divergence. In: 2011 International Conference on Image Information Processing, pp. 1–5 (2011). IEEE
47. Hussain, M., Bora, D.J.: An analytical study on different image segmentation techniques for malaria parasite detection. In: 2018 International Conference on Research in Intelligent and Computing in Engineering (RICE), pp. 1–7 (2018). IEEE
48. Chayadevi, M., Raju, G.: Automated colour segmentation of malaria parasite with fuzzy and fractal methods. In: Computational Intelligence in Data Mining, vol. 3, pp. 53–63. Springer (2015)
49. Madhu, G.: Gaussian membership function and type ii fuzzy sets based approach for edge enhancement of malaria parasites in microscopic blood images. In: International Conference on ISMAC in Computational Vision and Bio-Engineering, pp. 651–664 (2018). Springer
50. Golla, M.: A novel algorithm for segmentation of parasites in thin blood smears from microscopy using type ii fuzzy sets and inverse Gaussian gradient. Int. J. Comput. Vis. Image Process. (IJCVIP) **9**(3), 1–22 (2019)
51. Ghosh, M., Chakraborty, C., Ray, A.K.: Yager's measure based fuzzy divergence for microscopic color image segmentation. In: 2013 Indian Conference on Medical Informatics and Telemedicine (ICMIT), pp. 13–16 (2013). IEEE
52. Jang, J.-S.: ANFIS: adaptive-network-based fuzzy inference system. IEEE Trans. Syst. Man Cybern. **23**(3), 665–685 (1993)
53. Piero, P.: Adaptive neural fuzzy inference systems (ANFIS): analysis and applications. Lect. Not. 1–41 (2000)
54. Appiah, R.: Implementation of Adaptive Neuro Fuzzy Inference System for Malaria Diagnosis (A Case Study at Kwesimintsim Polyclinic), PhD thesis (2016)
55. Tsegay, G., Anusuya, R.: Decision support diagnosis system using artificial neural network and fuzzy logic modeling in case of malaria. Artif. Intell. Syst. Mach. Learn. **9**(1), 16–20 (2017)
56. Kochareka, M., Sarkar, S., Dasgupta, D., Aigal, U.: A preliminary comparative report of quantitative buffy coat and modified quantitative buffy coat with peripheral blood smear in malaria diagnosis. Pathog. Glob. Health **106**(6), 335–339 (2012)
57. Moody, A.: Rapid diagnostic tests for malaria parasites. Clin. Microbiol. Rev. **15**(1), 66–78 (2002)
58. Nigussie, D., Legesse, M., Animut, A., Mulu, A., et al.: Evaluation of paracheck pf o and parascreen pan/pf o tests for the diagnosis of malaria in an endemic area, south Ethiopia. Ethiop. Med. J. **46**(4), 375–381 (2008)
59. Kashosi, T.M., Mutuga, J.M., Byadunia, D.S., Mutendela, J.K., Mulenda, B., Mubagwa, K.: Performance of SD bioline malaria Ag Pf/Pan rapid test in the diagnosis of malaria in South-Kivu, DR Congo. Pan Afr. Med. J. **27** (2017)
60. Postigo, M., Mendoza-León, A., Pérez, H.A.: Malaria diagnosis by the polymerase chain reaction: a field study in southeastern Venezuela. Trans. Roy. Soc. Trop. Med. Hyg. **92**(5), 509–511 (1998)

61. Buppan, P., Putaporntip, C., Pattanawong, U., Seethamchai, S., Jongwutiwes, S.: Comparative detection of plasmodium vivax and plasmodium falciparum DNA in saliva and urine samples from symptomatic malaria patients in a low endemic area. Malaria J. **9**(1), 1–7 (2010)
62. Djam, X., Wajiga, G., Kimbi, Y., Blamah, N.: A fuzzy expert system for the management of malaria (2011)
63. Oladele, T.O., Ogundokun, R.O., Misra, S., Adeniyi, J.K., Jaglan, V.: A joint neuro-fuzzy malaria diagnosis system. J. Phys.: Conf. Ser. **1767**, 012038 (2021)
64. Preethi, S., Arunadevi, B., Prasannadevi, V.: Malaria Parasite Enumeration and Classification Using Convolutional Neural Networking, pp. 225–245 (2021)
65. Muda, T.Z.T., Salam, R.A.: Blood cell image segmentation using hybrid k-means and median-cut algorithms. In: 2011 IEEE International Conference on Control System, Computing and Engineering, pp. 237–243 (2011). IEEE
66. Yang, F., Poostchi, M., Yu, H., Zhou, Z., Silamut, K., Yu, J., Maude, R.J., Jaeger, S., Antani, S.: Deep learning for smartphone-based malaria parasite detection in thick blood smears. IEEE J. Biomed. Health Inform. **24**(5), 1427–1438 (2019)
67. Pattanaik, P.A., Mittal, M., Khan, M.Z.: Unsupervised deep learning cad scheme for the detection of malaria in blood smear microscopic images. IEEE Access **8**, 94936–94946 (2020)
68. Prakash, S.S., Kovoor, B.C., Visakha, K.: Convolutional neural network based malaria parasite infection detection using thin microscopic blood smear samples. In: 2020 Second International Conference on Inventive Research in Computing Applications (ICIRCA), pp. 308–313 (2020). IEEE
69. Mitrović, K., Milošević, D.: Classification of malaria-infected cells using convolutional neural networks. In: 2021 IEEE 15th International Symposium on Applied Computational Intelligence and Informatics (SACI), pp. 000323–000328 (2021). IEEE
70. Raj, M., Sharma, R., Sain, D.: A deep convolutional neural network for detection of malaria parasite in thin blood smear images. In: 2021 10th IEEE International Conference on Communication Systems and Network Technologies (CSNT), pp. 510–514 (2021). IEEE
71. Qin, B., Wu, Y., Wang, Z., Zheng, H.: Malaria cell detection using evolutionary convolutional deep networks. In: 2019 Computing, Communications and IoT Applications (ComComAp), pp. 333–336 (2019). IEEE
72. Var, E., Tek, F.B.: Malaria parasite detection with deep transfer learning. In: 2018 3rd International Conference on Computer Science and Engineering (UBMK), pp. 298–302 (2018). IEEE

Chapter 6
Application of Feature Extraction and Feature Selection Followed by SMOTE to Improve the Prediction of DNA-Binding Proteins

Anoop Kumar Tiwari, Shivam Shreevastava, Neelam Kumari, Arti Malik, and Tanmoy Som

6.1 Introduction

DNA-binding proteins (DNA-BPs) (Fig. 6.1) are defined as the proteins that perform interaction and binding with DNA. DNA-BP acts as a key factor in the structural composition of the DNA as well as in gene regulations [1–3]. The main functions of the DNA-BPs can be categorized into two parts. Firstly, the DNA is organized and constructed. Secondly, various cellular processes include transcription, DNA recombination, replication, modification, and repair. So, they can be employed as potential element in developing drugs for curing cancers and genetic diseases [4–6]. In recent years, DNA-binding proteins have drawn attentions of the researchers due to their important role in different bimolecular functions [7, 8]. DNA-binding proteins can be identified through several biological experimental techniques, such as filter binding assays, genetic analysis, chromatin immunoprecipitation on microarrays, X-ray crystallography, and gene analysis [9]. Detecting DNA-BPs by using experimental approaches is always expensive and time-consuming as protein sequence

A. K. Tiwari
Department of Computer Science and Information Technology, Central University of Haryana, Mahendergarh, Haryana 123031, India

S. Shreevastava (✉)
Department of Mathematics, School of Basic & Applied Sciences, Harcourt Butler Technical University, Kanpur, Uttar Pradesh 208002, India
e-mail: shivam.s@hbtu.ac.in

T. Som
Department of Mathematical Sciences, IIT (BHU), Varanasi, Uttar Pradesh 221005, India

N. Kumari · A. Malik
Division of Mathematics, School of Basic & Applied Sciences, Galgotias University, Greater Noida, Uttar Pradesh 201310, India

T. Som et al. (eds.), *Fuzzy, Rough and Intuitionistic Fuzzy Set Approaches for Data Handling*, Forum for Interdisciplinary Mathematics,
https://doi.org/10.1007/978-981-19-8566-9_6

data is increasing expeditiously due to advancement of next-generation technologies for high-throughput DNA sequencing. Researchers have approximated that the transcription factors number solitary can be either about 3000 or 10% out of available protein-coding genes in the human genome. Therefore, there is a continuous requirement of a quick and reliable computational tool that can handle this large volume of protein sequence data to identify DNA-BPs. Here, machine learning-assisted computational methods are always preferable as these are more quick, less expensive, and reliable when compared to available conventional methods. However, performance assessment of a machine learning algorithm is usually degraded due to scarcity of appropriate pre-processing steps including extraction of instructive features, selection of relevant and/or non-redundant features, resampling for imbalanced dataset, and choosing adequate learning algorithms.

In the case of peptide sequences with variety of lengths, feature extraction can play a key role in designing well-performed predictors. Feature extraction process produces a constant length of feature vectors from the different lengths of peptide sequences that reflect the necessary correlation with the target to establish a potential

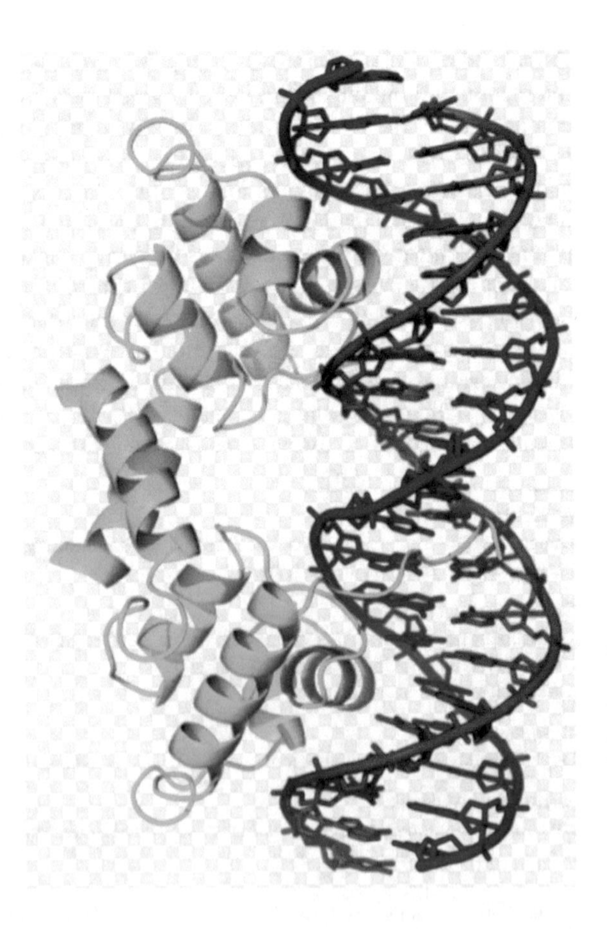

Fig. 6.1 DNA-binding protein structure [10]

classifier. Feature extraction techniques can provide various characteristics of the data points to the machine learning algorithms as it produces different interesting representative features, which leads to improve the average performances of the learned models.

The computational implementation of pattern recognition and other machine learning tasks are hampered by high dimensionality of datasets. In many real-world applications, data is constantly generated and expanded through various sources. The most important task in the field of knowledge discovery in database (KDD) is mining-required knowledge from real-valued datasets. A dataset consists of many attributes but some of them are not useful for learning tasks because they minimize the actual performance of algorithms and increase the complexity and hence training and testing times. Feature selection improves classification accuracy and prediction ability of proposed algorithms by removing inconsistent and redundant attributes [11–13]. The method of choosing the most informative attributes of a given dataset in order to minimize classification time, complexity, and cost is known as feature selection. Feature selection attains the predetermined goal while giving the maximum classification accuracy for test data. Areas like document classification, computer vision, and object recognition are based upon several applications of feature selection as it focuses on the interpretation of selected features rather than extracted features.

Pawlak gave the idea of rough set approximations that acquires information from the information system itself [14]. Rough set theory (RST) does not rely on prior model conventions and handles the vagueness available in the information systems. Its wide range of applications in the areas of decision support, data mining, information discovery, and pattern recognition attracts researcher from various fields. To deal with real-valued datasets, several discretization techniques are needed in order to apply RST before feature selection which results in information loss.

To overcome this issue, Dubois and Prade combined rough set with fuzzy set (proposed by Zadeh [15]) and proposed fuzzy rough set (FRS) concept which handles both vagueness and uncertainty in an information system [16]. Researchers implemented FRS theory on many real-valued datasets for feature selection and classification as this method does not require discretization [17–19]. Jensen and Shen presented tolerance-based FRS technique for feature selection and applied their model on various datasets for classification accuracies [20–23].

Another important issue is class imbalance, which occurs due to huge difference among total objects related to positive and negative classes. Class imbalance has received a lot of attention in the literature because it has a direct impact on machine learning algorithms necessary to solve prediction problems in bioinformatics datasets [24]. In data mining, machine learning, and pattern recognition tasks, this class imbalance problem is almost universal. Class imbalanced data usually leads to performance loss but some treatments like cost-sensitive learning, ensemble learning, and sampling are capable to enhance prediction performance [25, 26].

In this paper, we present a new methodology to improve the prediction performance of different learning algorithms for discriminating DNA-binding proteins and non-DNA-binding proteins. Firstly, same informative features are extracted from training and testing sets. Then, redundant and irrelevant features are removed from training sets by using fuzzy rough feature selection with harmony search [27]. Next, same features are removed from testing sets. Training and testing sets are converted into optimally balanced datasets using SMOTE. Now, numerous machine learning algorithms are applied on both training and testing sets and performances are recorded using percentage split of 70:30. Moreover, a schematic framework is also presented for proposed methodology for better understanding. The results of the performed experiment show that proposed methodology outperforms when compared to the previously reported results [28]. Furthermore, we used the fuzzy rough attribute evaluator technique to rate input features. Finally, ROC curves are presented to visualize the performances of different classifiers in a suitable way. A schematic representation can be seen in Fig. 6.2.

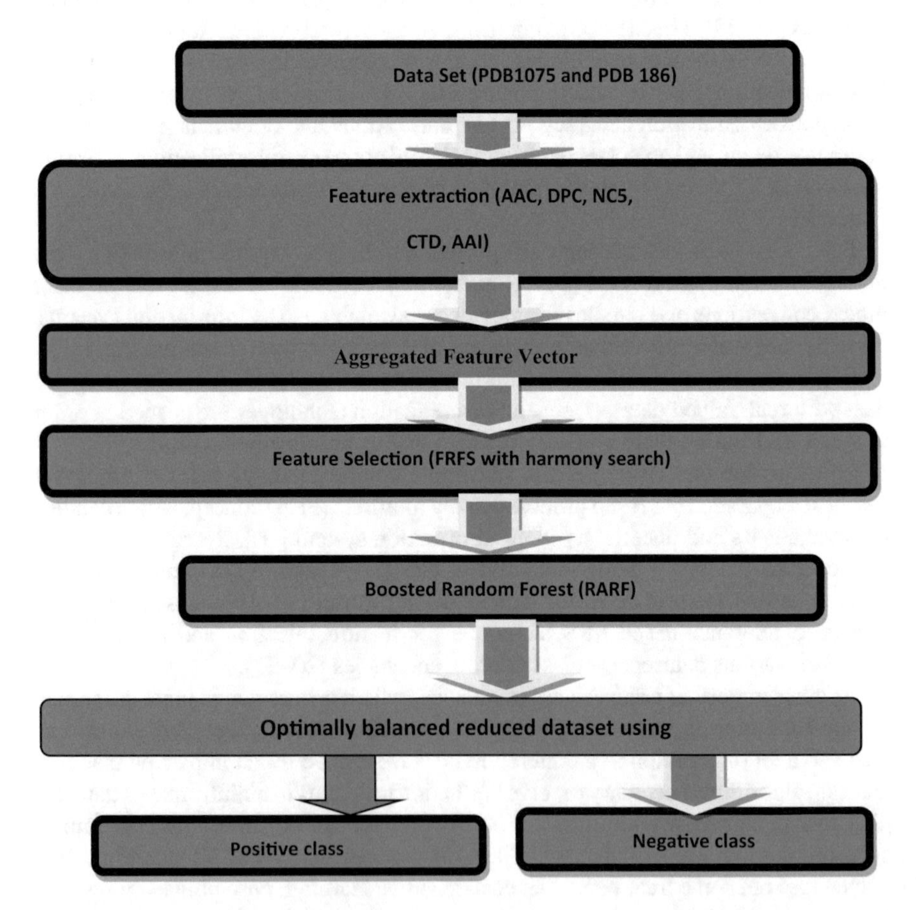

Fig. 6.2 Schematic framework for proposed methodology

Table 6.1 Characteristics of the datasets and size of the reduct sets

Dataset	Instances	Attributes	Reduct size	
			FRFS	ReliefF
PDB 1075	1075	459	128	168
PDB 186	186	459	128	168

6.2 Materials and Methods

6.2.1 Dataset

In the current study, datasets PDB1075 and PDB186 are used for the experimental analysis which has a wide application in the prediction of DbPs. The sequence of proteins can be found in the international protein database: PDB (https://www.rcsb.org/). In this paper, we consider the dataset PDB1075 (created by Liu et al. [29]) as training set and the dataset PDB186 (created by Lou et al. [30]) as independent testing set. Table 6.1 depicts the basic details of the above two datasets.

6.2.2 Input Features

Positive and negative samples can be differentiated by different sequences of amino acids. Various factors like composition, physicochemical properties, and selections and arrangements of amino acids are responsible for the creation of different sequences of amino acids. The following sequence-based features are extracted by using iFeature web server [31] for experimental analysis: AAC, DPC, NC5, CTD, and AAI.

(i) Amino acid composition (AAC)

AAC represents the occurrences of amino acids in a given peptide normalized by the length of the sequence. Its application can be seen widely in the field of Bioinformatics [32]. Here, it is of fixed length with 20 features in total. Its mathematical formulation is given as follows:

$$P = [fv_1, fv_2, \ldots, fv_{20}] \tag{6.1}$$

where $fv_i = R_i/L$ for all $i = 1, 2, 3, \ldots, 20$ is the normalized frequency of the ith amino acid in a given peptide. R_i is the quantity of type i amino acid, and L is the length of the sequence observed in a peptide.

(ii) Dipeptide composition (DPC)

In a given peptide, DPC represents the composition of a residue pair which further describes the fraction of amino acids and their order. It consists of 400 vectors and can be formulated as follows:

$$P = \left[fv_1,\ fv_2, \ldots,\ fv_j, \ldots,\ fv_{400}\right] \tag{6.2}$$

where frequency of jth amino acid pair in {AA, AC, AD, ..., YY} is represented by fv_j.

(iii) Binary profile (NC5)

In this feature, each amino acid is encoded as a 0/1 vector of dimension 20. N- or C-terminus of m-Amino acids (Amino Acids of length m) can be translated for a given peptide as follows:

$$\mathrm{BPF}(m) = [b(P_1), b(P_2), b(P_3), \ldots, b(P_m)] \tag{6.3}$$

where the dimension of BPF(m) is $20 \times m$. After taking $m = 5$ at both termini, we obtain BPFN5 and BPFC5. A 200-dimensional feature vector is formed by combining these two termini.

(iv) Composition-Transition-Distribution (CTD)

In this feature, C stands for composition, which signifies the composition of amino acids, T (transition) indicates the percentage of those amino acid residues which have certain characteristics that are followed by other types of amino acids, and distribution (D) calculates the length of a sequence within which 1%, 25%, 50%, 75%, and 100% of the amino acid residues with certain characteristics are located. In CTD feature, composition, transition, and distribution are 21, 21, and 105-dimensional feature vector, respectively.

(v) Amino acid index (AAI)

Previously, out of 566 AAIs in the AAindex database, eight high-quality AAIs (accessions LIFS790101, TSAJ990101, MAXF760101, BIOV880101, CEDJ970104, BLAM930101, MIYS990104, and NAKH920108) were identified from 566 total AAIs in the AAindex database by applying a clustering technique. AAI generates a 160 (=20 amino acids $\times$ 8 properties) dimensional vector, which has been widely applied in numerous sequence-based prediction tasks [33].

6.2.3 Classification Protocol. RF Boosted

Our experiments are carried out independently using eight extensively used machine learning algorithms that are commonly used for classification and prediction tasks on biological datasets. We applied boosted random forest (BRF) [34] algorithm for our experiments and found that it is better performing algorithm as compared to other.

A BRF algorithm consists of two parts: AdaBoost and the random forest classifier algorithm which itself consists of multiple decision trees. A decision tree builds models similar to an actual tree. Firstly, the algorithm divides data into smaller subsets while adding branches to the tree. The outcome of this algorithm is a tree with decision nodes and leaf nodes. A decision node consists of two or more branches which represent the value of each feature tested while the leaf node holds the result value on the patient's prospective condition (target value). Thus, the random forest provides the final result by averaging the results obtained from multiple trees.

6.2.4 Optimal Balancing Protocol

Estimation parameter like overall precision favors the majority class for those datasets, which consist of imbalanced negative and positive class instances. Which further leads to less sensitivity and higher specificity in prediction of minority class instances. To deal with this problem, Synthetic Minority Over-sampling Technique (SMOTE) is used in order to balance the reduced testing collection as follows. In recent years, SMOTE has been actively studied to address class imbalance issued [35, 36].

SMOTE: It creates synthetic instances by selecting a minority sample and its nearest neighbors randomly as a form of over-sampling. Then, it creates an artificial minority class instance by inserting one of the nearest neighboring minority class instance. SMOTE samples can be classified as linear combinations of two samples of same kind that are connected to the minority class (p and p^k) and are given as follows:

$$s = p + i * \left(p^k - p\right) \tag{6.4}$$

where i belongs to [0, 1] and p^k is chosen at random from the five minority class instances or samples closest to p. The default value of nearest neighbors for SMOTE in WEKA [37] is 5.

6.2.5 Feature Selection Protocol

Similar and overlapped attributes in bioinformatics datasets affect the classification task and increase complexity. Overlapping of interclass functions and the presence of similar attributes are main reasons for ambiguity and indistinctness. To handle such issues, Pawlak proposed rough set approximation based on indiscernibility relation. Zadeh's fuzzy set tackles ambiguity available in a real-valued dataset. Dubois and Prade combined these two theories and presented a fuzzy rough set theory which can handle vagueness and indiscernibility both. It also overcomes the shortcomings of rough set for not being applicable to real-valued datasets directly without discretization, which further leads to information loss. Fuzzy rough set theory has vast applications in decision-making and Classification problems [38]. In this paper, we used fuzzy rough feature selection (FRFS) technique to improve the prediction of binding proteins. Pseudocode for fuzzy rough feature selection can be given as below:

Algorithm 1 The algorithm for FRFS is as follows:

1. Input fuzzy information system
2. Find the decision classes $U'/\underline{Q} = \left\{\underline{Q}_1, \underline{Q}_2, \ldots, \underline{Q}_k\right\}$
3. Initialize $\underline{C'} \leftarrow \{P'_1, P'_2, \ldots, P'_m\}$, $R' \leftarrow \varnothing$
4. For every $P'_i \in \underline{C'} - R'$,
5. For $i = 1, 2, \ldots, m$, compute lower approximation $X \downarrow_t P'(\mathrm{x})$
6. Compute positive region

$$\mathrm{Pos}_{R\cup\{P_i\}}(x)$$

$$\text{Calculate degree of dependency} = \frac{\sum_{x\in U}|\mathrm{Pos}_{R\cup\{P_i\}}(x)|}{|U|}, \text{ for each } P'_i \in \underline{C'} - R'.$$

7. Compute significance $\mathrm{Sig}_{P_i}^{R\cup\{P_i\}}$
8. Find attribute P_t with greatest $\mathrm{Sig}_{P_i}^{R\cup\{P_i\}}$
9. While $\mathrm{Sig}_{P_i}^{R\cup\{P_i\}} > \delta$, $\underline{C} \leftarrow \underline{C} - R$ and $R \leftarrow R \cup \{P_t\}$
10. Return $Reduct$.

6.2.6 *Performance Evaluation Metrics*

The eight machine learning algorithms' relative prediction output is estimated using threshold-dependent and threshold-independent parameters. These parameters are derived from the confusion matrix values, specifically true positives (TP), which is the number of correctly predicted binding proteins, false negatives (FN), which is the number of incorrectly predicted binding proteins, true negatives (TN), which is the number of correctly predicted non-binding proteins, and false positives (FP), which is the number of incorrectly predicted non-binding proteins. Confusion matrix diagram is given in Fig. 6.3.

Sensitivity: It can be calculated by using ratio of true-positive (TP) value to sum of true-positive and false-negative (FN) values and denotes the percentage of correctly predicted binding proteins as follows:

$$\text{Sensitivity} = \frac{\mathrm{TP}}{(\mathrm{TP} + \mathrm{FN})} \times 100 \tag{6.5}$$

Fig. 6.3 Confusion matrix [39]

		Predicted condition	
	Total population = P + N	Positive (PP)	Negative (PN)
Actual condition	Positive (P)	True positive (TP)	False negative (FN)
	Negative (N)	False positive (FP)	True negative (TN)

Specificity: It expresses the percentage of correctly predicted non-binding proteins and can be determined using the following formula:

$$\text{Specificity} = \frac{\text{TN}}{(\text{TN} + \text{FP})} \times 100 \tag{6.6}$$

Accuracy: This parameter denotes the percentage of correctly predicted binding and non-binding proteins as follows:

$$\text{Accuracy} = \frac{(\text{TP} + \text{TN})}{(\text{TP} + \text{FP} + \text{TN} + \text{FN})} \times 100 \tag{6.7}$$

AUC: This evaluation metric is robust to imbalanced nature of the proteomics datasets. It highlights the area under curve (AUC) of a receiver operating characteristics curve (ROC). It is better binding protein predictor if its value is close to 1 and worst if its value is 0. In random ranking process, it takes a value of 0.5.

Mathew's correlation coefficient (MCC): It is widely applicable as a performance parameter in the field of binary classification. MCC value can be calculated as follows:

$$\text{MCC} = \frac{(\text{TP} \times \text{TN} - \text{FP} \times \text{FN})}{\sqrt{(\text{TP} + \text{FP})(\text{TP} + \text{FN})(\text{TN} + \text{FP})(\text{TN} + \text{FN})}} \tag{6.8}$$

If MCC value is 1, then it is considered as best for non-binding protein predictor. Furthermore, we carried out all our experiments using the open-source java-based machine learning platform WEKA.

6.3 Experimental Analysis

In the current paper, entire experiments are conducted with PDB 1075 and PDB 186 datasets containing DNA-binding proteins, where PDB 1075 is training dataset whereas PDB 186 is testing dataset. Firstly, 459 features are extracted by using iFeature web server. Then, we apply fuzzy rough feature selection with harmony search on PDB 1075 dataset to eliminate irrelevant and redundant features. Further, the same features are eliminated from PDB 186 dataset as well (Table 6.1). Next, the reduced datasets (PDB 1075 and PDB 186) are converted into balanced datasets by using SMOTE. Now, we use eight well-known and widely used classifiers, namely sequential minimization optimization (SMO), IBK, PART, JRIP, Naïve Bayes, random forest, rotation forest, and RealAdaBoost random forest to evaluate the performances of the classifiers for original and reduced datasets, respectively. Tables 6.2 and 6.3 represent the performances of the learning algorithms based on various evaluation parameters for original datasets. Performances of the machine learning algorithms over reduced datasets produced by fuzzy rough feature selection with harmony search and ReliefF algorithm are recorded in Tables 6.4, 6.5, 6.6, and 6.7. From Table 6.1, we can observe that fuzzy rough feature selection with harmony search is producing more reduced datasets when compared to ReliefF algorithm as FRFS and ReliefF have produced 128 and 168 features, respectively. By observing the experimental results (Tables 6.2, 6.3, 6.4, 6.5, 6.6, and 6.7), we can conclude that feature selection based on fuzzy rough set-based concept is more effective for the prediction of DNA-binding protein when compared to crisp approaches such as ReliefF. For training (PDB1075) and testing (PDB 186) datasets, boosted random forest (RARF) generates the best results when compared to other learning techniques for both original as well as reduced datasets produced by crisp and fuzzy assisted approaches. For PDB 1075, the best results are reported by RARF with sensitivity, specificity, accuracy, AUC, and MCC of 90.5, 91.2, 92.1, 0.899, and 0.877, respectively, as mentioned in Table 6.4. For PDB 186, the best results are yet again produced by RARF with sensitivity, specificity, accuracy, AUC, and MCC of 84.3, 82.8, 84.6, 0.569, and 0.811, respectively, as recorded in Table 6.4. Experimental results clearly demonstrate the supremacy of the presented methodology for the AUC of DNA-binding proteins as the previously available literatures have reported a maximum accuracy of 86.05% for PDB 1075 and 75.30 for PDB 186 while our approach has provided a maximum accuracy of 92.1 for PDB 1075 and 84.6% for PDB 186. ROC is a convenient way to visualize the performance of all eight learning algorithms. Figures 6.4 and 6.5 are a plot of ROC for reduced PDB 1075 and PDB 186 datasets.

All the experiments are performed using percentage split of 70:30 and tenfold cross validation. Extraction of features was initially done with iFeature server, while other experiments, such as feature selection, classification, and visualization, were performed in Weka 3.8 on hardware platform with Intel(R) Core(TM) i5-8265U CPU @ 1.60 GHz, 1.80 GHz with 8.00 GB RAM.

Table 6.2 Performance evaluation metrics of learning algorithms with original PDB1075 dataset using percentage split of 70:30

Learning algorithm	Sensitivity	Specificity	Accuracy	MCC	AUC
IBK	58.6	54.7	56.5	0.132	0.566
Naïve Bayes	55.9	74.7	65.8	0.312	0.722
JRip	69.1	66.5	67.7	0.355	0.685
Rotation forest	71.7	74.1	73.0	0.459	0.791
Random forest	75.0	70.6	72.7	0.455	0.797
SMO	74.3	62.4	68.0	0.368	0.683
RARF	79.8	78.6	79.5	0.502	0.811
PART	59.9	60.0	59.9	0.198	0.622

Table 6.3 Performance evaluation metrics of learning algorithms with original PDB186 dataset using percentage split of 70:30

Learning algorithm	Sensitivity	Specificity	Accuracy	MCC	AUC
IBK	76.7	23.1	51.8	0.003	0.499
Naïve Bayes	63.3	65.4	64.3	0.286	0.630
JRip	50.0	65.4	57.1	0.155	0.558
Rotation forest	70.0	30.8	51.8	0.008	0.613
Random forest	70.0	69.2	69.6	0.392	0.772
SMO	60.0	69.2	64.3	0.292	0.646
RARF	73.3	72.5	75.6	0.488	0.763
PART	53.3	65.4	58.9	0.188	0.588

Table 6.4 Performance evaluation metrics of learning algorithms with optimally balanced reduced PDB1075 dataset produced by FRFS using percentage split of 70:30

Learning algorithm	Sensitivity	Specificity	Accuracy	MCC	AUC
IBK	65.6	64.4	69.5	0.228	0.566
Naïve Bayes	68.9	74.7	72.7	0.445	0.722
JRip	74.7	69.8	72.7	0.446	0.685
Rotation forest	78.8	79.5	80.4	0.566	0.822
Random forest	86.8	84.8	88.2	0.622	0.819
SMO	82.9	83.7	78.9	0.466	0.721
RARF	90.5	91.2	92.1	0.877	0.899
PART	69.5	68.2	69.2	0.211	0.655

Table 6.5 Performance evaluation metrics of learning algorithms with optimally balanced reduced PDB186 dataset produced by FRFS using percentage split of 70:30

IBK	Sensitivity	Specificity	Accuracy	MCC	AUC
IBK	70.3	56.5	68.9	0.211	0.511
Naïve Bayes	69.8	68.6	66.9	0.286	0.630
JRip	62.1	68.9	71.1	0.298	0.558
Rotation forest	70.0	62.9	72.1	0.312	0.615
Random forest	79.8	78.8	81.9	0.511	0.792
SMO	72.1	73.5	74.7	0.421	0.646
RARF	84.3	82.8	84.6	0.569	0.811
PART	65.9	68.2	70.8	0.455	0.615

Table 6.6 Performance evaluation metrics of learning algorithms with optimally balanced reduced PDB1075 dataset produced by ReliefF using percentage split of 70:30

Learning algorithm	Sensitivity	Specificity	Accuracy	MCC	AUC
IBK	60.6	59.7	64.5	0.298	0.598
Naïve Bayes	68.9	69.5	70.3	0.511	0.698
JRip	72.1	70.8	74.5	0.556	0.721
Rotation forest	77.4	78.9	79.5	0.611	0.798
Random forest	83.4	81.2	82.6	0.599	0.711
SMO	78.8	79.2	75.5	0.511	0.723
RARF	86.2	88.7	86.8	0.777	0.866
PART	72.4	70.33	74.5	0.321	0.711

Table 6.7 Performance evaluation metrics of learning algorithms with optimally balanced reduced PDB186 dataset produced by ReliefF [28] using percentage split of 70:30

IBK	Sensitivity	Specificity	Accuracy	MCC	AUC
IBK	72.1	64.8	71.2	0.345	0.611
Naïve Bayes	72.9	71.8	69.8	0.412	0.711
JRip	65.5	70.2	72.4	0.398	0.655
Rotation forest	68.9	66.8	69.5	0.299	0.566
Random forest	75.4	77.8	78.2	0.612	0.792
SMO	73.2	70.8	70.9	0.511	0.698
RARF	80.3	81.2	81.9	0.566	0.829
PART	70.3	69.5	70.2	0.459	0.609

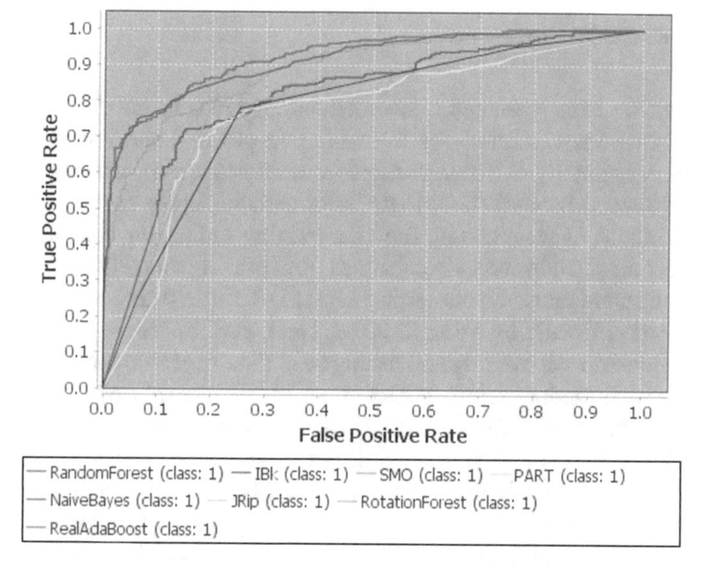

Fig. 6.4 AUC of different machine learning algorithms for reduced PDB1075 dataset produced FRFS

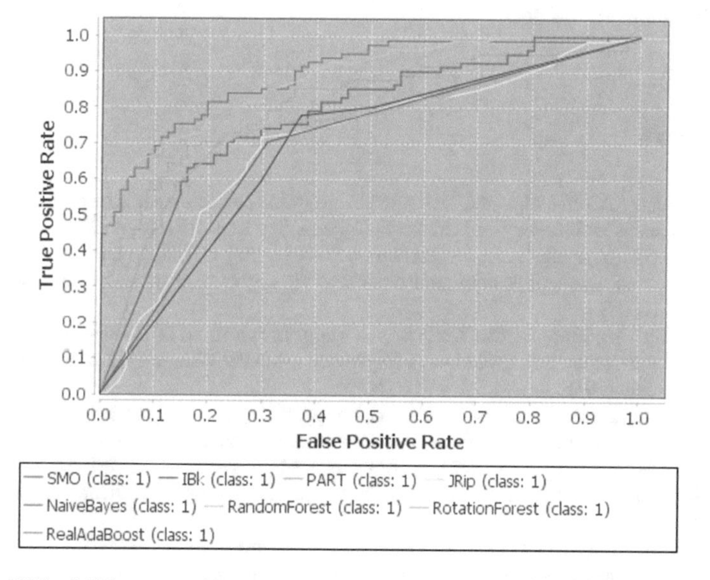

Fig. 6.5 AUC of different machine learning algorithms for reduced PDB186 dataset produced FRFS

6.4 Conclusion

In this paper, we have presented a new methodology to enhance the discriminating ability of numerous classifiers for DNA-binding proteins and non-DNA-binding proteins from optimally balanced reduced training and testing sets based on FRFS technique followed by SMOTE. Firstly, informative features such as AAC, DPC, NC5, CTD, and AAI were chosen from the training and testing sets. Secondly, we selected relevant and non-redundant features from training sets (PDB 1075). Thirdly, same features were chosen from testing sets (PDB 186). Training and testing sets were balanced optimally by using SMOTE. Next, performance of different classifiers was explored over training and testing sets. Finally, we obtain the best results by using boosted random forest (RARF) with percentage split of 70:30 validation over optimally balanced reduced PDB 1075 dataset with sensitivity, specificity, accuracy, MCC, and AUC of 90.5, 91.2, 92.1, 0.877, and 0.899, respectively. Comprehensive experiments clarify that our proposed methodology is better than previous approaches. From the entire results, we notice that average performances of artificial intelligence models in differentiating DNA-binding proteins from non-DNA-binding proteins can be enhanced by using feature extraction, feature selection with fuzzy rough set-based technique with harmony search followed by SMOTE, and selection of suitable learning model.

References

1. Chowdhury, S.Y., Shatabda, S., Dehzangi, A.: iDNAProt-ES: identification of DNA-binding proteins using evolutionary and structural features. Sci. Rep. **7**(1), 1–14 (2017)
2. Fang, Y., Guo, Y., Feng, Y., Li, M.: Predicting DNA-binding proteins: approached from Chou's pseudo amino acid composition and other specific sequence features. Amino Acids **34**(1), 103–109 (2008)
3. Fu, X., Zhu, W., Liao, B., Cai, L., Peng, L., Yang, J.: Improved DNA-binding protein identification by incorporating evolutionary information into the Chou's PseAAC. IEEE Access **6**, 66545–66556 (2018)
4. Lu, W., Song, Z., Ding, Y., Wu, H., Cao, Y., Zhang, Y., Li, H.: Use Chou's 5-step rule to predict DNA-binding proteins with evolutionary information. BioMed Res. Int. **2020** (2020)
5. Rahman, M.S., Shatabda, S., Saha, S., Kaykobad, M., Rahman, M.S.: DPP-PseAAC: a DNA-binding protein prediction model using Chou's general PseAAC. J. Theor. Biol. **452**, 22–34 (2018)
6. Singhvi, N., Singh, Y., Shukla, P.: Computational approaches in epitope design using DNA binding proteins as vaccine candidate in Mycobacterium tuberculosis. Infect. Genet. Evol. **83**, 104357 (2020)
7. Talanian, R.V., McKnight, C.J., Rutkowski, R., Kim, P.S.: Minimum length of a sequence-specific DNA binding peptide. Biochemistry **31**(30), 6871–6875 (1992)
8. Wang, Y., Ding, Y., Guo, F., Wei, L., Tang, J.: Improved detection of DNA-binding proteins via compression technology on PSSM information. PLoS One **12**(9), e0185587 (2017)
9. Yang, W., Deng, L.: PreDBA: a heterogeneous ensemble approach for predicting protein-DNA binding affinity. Sci. Rep. **10**(1), 1–11 (2020)
10. https://en.wikipedia.org/wiki/DNA-binding_protein

11. Chandrashekar, G., Sahin, F.: A survey on feature selection methods. Comput. Electr. Eng. **40**(1), 16–28 (2014)
12. Dash, M., Liu, H.: Feature selection for classification. Intell. Data Anal. **1**(1–4), 131–156 (1997)
13. Guyon, I., Elisseeff, A.: An introduction to variable and feature selection. J. Mach. Learn. Res. **3**, 1157–1182 (2003)
14. Pawlak, Z.: Rough Sets: Theoretical Aspects of Reasoning About Data. Springer Science & Business Media, Berlin (2012)
15. Zadeh, L.A.: Fuzzy Sets, Fuzzy Logic, and Fuzzy Systems: Selected Papers, pp. 394–432. World Scientific (1996)
16. Dubois, D., Prade, H.: Putting rough sets and fuzzy sets together. In: Intelligent Decision Support, pp. 203–232. Springer, Berlin (1992)
17. Sheeja, T., Kuriakose, A.S.: A novel feature selection method using fuzzy rough sets. Comput. Ind. **97**, 111–116 (2018)
18. Wang, C., Huang, Y., Ding, W., Cao, Z.: Attribute reduction with fuzzy rough self-information measures. Inf. Sci. **549**, 68–86 (2021)
19. Wang, C., Huang, Y., Shao, M., Fan, X.: Fuzzy rough set-based attribute reduction using distance measures. Knowl.-Based Syst. **164**, 205–212 (2019)
20. Jensen, R., Shen, Q.: Fuzzy–rough attribute reduction with application to web categorization. Fuzzy Sets Syst. **141**(3), 469–485 (2004)
21. Jensen, R., Shen, Q.: Semantics-preserving dimensionality reduction: rough and fuzzy-rough-based approaches. IEEE Trans. Knowl. Data Eng. **16**(12), 1457–1471 (2004)
22. Jensen, R., Shen, Q.: Fuzzy-rough sets assisted attribute selection. IEEE Trans. Fuzzy Syst. **15**(1), 73–89 (2007)
23. Jensen, R., Shen, Q.: New approaches to fuzzy-rough feature selection. IEEE Trans. Fuzzy Syst. **17**(4), 824–838 (2008)
24. Japkowicz, N., Stephen, S.: The class imbalance problem: a systematic study. Intell. Data Anal. **6**(5), 429–449 (2002)
25. Pirizadeh, M., Alemohammad, N., Manthouri, M., Pirizadeh, M.: A new machine learning ensemble model for class imbalance problem of screening enhanced oil recovery methods. J. Petrol. Sci. Eng. **198**, 108214 (2021)
26. Usmani, S.S., Bhalla, S., Raghava, G.P.: Prediction of antitubercular peptides from sequence information using ensemble classifier and hybrid features. Front. Pharmacol. **9**, 954 (2018)
27. Dubey, M., et al.: A systematic review on harmony search algorithm: theory, literature, and applications. Math. Probl. Eng. **2021** (2021)
28. Spolaôr, N., et al.: ReliefF for multi-label feature selection. In: 2013 Brazilian Conference on Intelligent Systems. IEEE (2013)
29. Liu, B., Xu, J., Lan, X., et al.: iDNA-Prot|dis: identifying DNA-binding proteins by incorporating amino acid distance-pairs and reduced alphabet profile into the general pseudo aminoacid composition. PLoS One **9**(9), article e106691 (2014)
30. Lou, W., Wang, X., Chen, F., Chen, Y., Jiang, B., Zhang, H.: Sequence based prediction of DNA-binding proteins based on hybrid feature selection using random forest and Gaussian naive Bayes. PloS one. 9(1):e86703 (2014)
31. Chen Z., et al.: iFeature: a python package and web server for features extraction and selection from protein and peptide sequences. Bioinformatics **34**(14), 2499–2502 (2018)
32. Jain, P., Tiwari, A.K., Som, T.: Enhanced prediction of anti-tubercular peptides from sequence information using divergence measure-based intuitionistic fuzzy-rough feature selection. Soft. Comput. **25**(4), 3065–3086 (2021)
33. Kawashima, S., Kanehisa, M.: AAindex: amino acid index database. Nucleic Acids Res. **28**(1), 374–374 (2000)
34. Breiman, L.: Random forests. Mach Learn. **45**, 5–32 (2001). https://doi.org/10.1023/A:1010933404324
35. Chawla, N.V., Bowyer, K.W., Hall, L.O., Kegelmeyer, W.P.: SMOTE: synthetic minority over-sampling technique. J. Artif. Intell. Res. **16**, 321–357 (2002)

36. Mukherjee, M., Khushi, M.: SMOTE-ENC: a novel SMOTE-based method to generate synthetic data for nominal and continuous features. Appl. Syst. Innov. **4**(1), 18 (2021)
37. Hall, M., Frank, E., Holmes, G., Pfahringer, B., Reutemann, P., Witten, I.H.: The WEKA data mining software: an update. ACM SIGKDD Explor. Newsl. **11**(1), 10–18 (2009)
38. Jensen, R., Shen, Q.: Computational intelligence and feature selection: rough and fuzzy approaches (2008)
39. https://en.wikipedia.org/wiki/Confusion_matrix

Chapter 7
Perspectives of Soft Computing in Multiscale Modeling for Fluid Flow Systems

Debabrata Datta and Tushar Kanti Pal

7.1 Introduction

In a modeling scheme, when several models structured at more than one scales are required to describe a complex system wherein physics also changes as per models, we call that kind of modeling as multiscale modeling [1]. Multiscale modeling in physical and biological sciences is focused toward the computation of material properties on one level based on multilevel information. Particular approaches are used to describe the system at each level. Multiscale modeling plays an important role in computational materials engineering. Scales which generally considered as multiscale are microscale, mesoscale and macroscale. In general, there exists a substantial amount of nonlinearity in fluid dynamics and due to this, nonlinearity complex interactions exist among multiscales. Multiscale transport phenomena take birth from the nonlinearity of the fluid mechanical problems.

In order to carry out the modeling of a multiscale systems with its uncertain parameters, decision-makers take the route of soft computing where fuzzy set is used to address the uncertainty of the system. It is worth to remember that in multiscale modeling, both the spatial and temporal scales are taken into account [1]. It is one kind of fusion of decision theory and mathematics at multiscale giving birth to a field known as multiscale decision-making. In the field of meteorology, multiscale modeling refers to the interaction between weather systems of different spatial and temporal scales. Lattice Boltzmann method (LBM) is one kind of multiscale modeling applied to investigate flow of fluid through any microchannel where macroscale Navier–Stokes equation fails [2].

D. Datta (✉)
Department of Information Technology, Heritage Institute of Technology, Kolkata, India
e-mail: debabrata.datta@heritageit.edu

T. K. Pal
Technology Development Division, Bhabha Atomic Research Centre, Mumbai, India

T. Som et al. (eds.), *Fuzzy, Rough and Intuitionistic Fuzzy Set Approaches for Data Handling*, Forum for Interdisciplinary Mathematics,
https://doi.org/10.1007/978-981-19-8566-9_7

Computational fluid dynamics (CFD) play a major role in multiscale modeling. Soft computing approach of handling uncertainty of parameters of a multiscale system using fuzzy set theory to turn the multiscale system by using fuzzy set theory converts multiscale modeling as intelligent system, because soft computing provides a machine learning technique within multiscale system while addressing the uncertainty of the system. Modeling turbulence by CFD in multiscale is a challenging task. The challenges are valid for both numerical and physical. Literature reviews of multiscale modeling pertaining to industrial applications or domain of physics using soft computing approach can be found elsewhere in [1–5]. One of the key challenges is the simulation of physical processes across a range of scales from the macro to microscales. Navier–Stokes equations describe macroscale-based physics problems and thus accessible to conventional CFD simulation [2]. However, in microscale similar problems, the continuum approximation no longer holds. Problems pertaining to microscale, we need to solve corresponding kinetic equations. Therefore, many systems use a holistic approach of aggregation of different simulation across a range of scales, which is known as multiscale approach.

It is also necessary to investigate the parameters associated with the governing equations of the problem. Very often, governing equations are numerically solved using conventional CFD approach with deterministic value of those parameters which address the characteristics of the flow of fluid. However, deterministic values of the representative parameters of the governing system of equations are biased and very often, they are uncertain due to insufficiency of the experimental determinations. In order to face this difficulty of this deterministic values, it is required to address the uncertainty scheme of those parameters which can be probabilistic as well as possibilistic [3]. Possibilistic approach of addressing uncertainty of physical parameters of fluid flow problem is preferred compared to probabilistic approach because of less sample size of relevant experiments. Possibilistic approach may also call as soft computing. This chapter will explore the soft computing approach of multiscale modeling for fluid mechanical system. Fuzzy set theoretical method will address the uncertainty of the relevant parameters in the form of an alpha cut which is basically an interval. So, the said computational strategy will be applied to define the uncertainty of the system in terms of a range [4]. The basic purpose of this chapter is to demonstrate the multiscale modeling embedded with soft computing to address the uncertainty of fluid flow. Remaining part of this chapter is designed into this way. An overview of soft computing is described in Sect. 7.2. The concept of multiscale modeling is described in Sect. 7.3. Multiscale modeling is carried out using lattice Boltzmann method (LBM). Hence, Sect. 7.4 presents the mathematical details of LBM. Section 7.5 presents a fusion of soft computing and multiscale modeling presenting that a new scheme of computation, where LBM is fused with fuzzy set theoretic approach of soft computing. Results of few case studies of the fusion of soft computing and multiscale modeling are also presented in Sect. 7.5. Conclusion of the chapter is described in Sect. 7.6, wherein we have highlighted the lessons learned by soft computing approach in modeling multiscale problems.

7.2 Overview of Soft Computing

Mathematical modeling can be done for relatively simple systems. Complex systems that arise in flow of fluid through microstructures (e.g., flow of blood in human body), in biology, in medicine and many other management systems cannot be addressed by conventional mathematical models and analytical methods [5]. Soft computing on one hand facilitates to quantify epistemic uncertainty associated with an engineering system (e.g., uncertainty of hydrological modeling, uncertainty of any civil structures, etc.) and on the other hand, soft computing provides machine learning that can automate the system to built as smart system (e.g., fuzzy logic control-based washing machine, fuzzy logic-based air conditioned machine, etc.). Machine learning (ML) is rapidly progressing in the field of bioscience including medical science. ML explores the methodology of learning from data by computer like a human. As such, it is a multidisciplinary subject that combines statistics, physics, computer science and mathematics [5].

The interpretation of data appropriately provides the success rate of ML approach to solve a problem. For example, a recent success story of ML in cancer diagnostic has shown that it is possible to classify skin cancer into either malignant or benign using photographic images from smartphones.

7.2.1 Why We Do Soft Computing?

Mathematical model and analysis can be done for relatively simple systems. Complex systems that arise in flow of fluid through microstructures (e.g., flow of blood in human body), in biology, in medicine and many other management systems cannot be addressed by conventional mathematical models and analytical methods [5]. Soft computing on one hand facilitates to quantify epistemic uncertainty associated with an engineering system (e.g., uncertainty of hydrological modeling, uncertainty of any civil structures, etc.) and on the other hand, soft computing provides machine learning that can automate the system to built as smart system (e.g., fuzzy logic control-based washing machine, fuzzy logic-based air conditioned machine, etc.). Machine learning (ML) is rapidly progressing in the field of bioscience including medical science. ML explores the methodology of learning from data by computer like a human. As such, it is a multidisciplinary subject that combines statistics, physics, computer science and mathematics [5]. The interpretation of data appropriately provides the success rate of ML approach to solve a problem. For example, a recent success story of ML in cancer diagnostic has shown that it is possible to classify skin cancer into either malignant or benign using photographic images from smartphones.

7.3 Multiscale Modeling

Traditional approaches of mathematical or numerical modeling focus only on one scale. The scale of a problem here we means micro, meso and macro. In general, engineering problems are solved traditionally at macroscale. But, very often, problems are required to investigate by atomistic scale because of the composite materials involved in the problem. Modeling of a problem using atomistic scale is known as microscale modeling. For example, molecular dynamics-based problem and quantum mechanics-based problem need the construction of the Hamiltonian (total energy) of the system, which is based on microscale. In this context, Dirac had stated that it is very difficult to implement quantum mechanical-based problem in practical sense. In order to remove this difficulty as well as to retain the macroscale property of the material properties, one needs one more scale known as mesoscale which will bridge the gap between microscale and macroscale. Therefore, multiscale modeling is defined as that modeling in which all these scales are involved to some extent. In multiscale modeling, governing constitutive equations are dependent on micro, meso and macrophysics of the material properties of the system. In general, multiscale models are solved numerically. Hence, numerical methods for capturing macro characteristics of the complex systems with microscopic models are to be redefined. Multiscale modeling is categorized into two classes, viz. (a) sequential multiscale modeling and (b) concurrent multiscale modeling. In the case of sequential multiscale modeling, precomputed details of the constitutive relations using microscale models are used in macroscale models [5]. In concurrent multiscale modeling, as an example, molecular dynamics approach (microscale method) is coupled to extended finite element method (XFEM), which is a macroscale modeling. This implies that concurrent multiscale modeling is a suitable modeling for solving three-dimensional crack problems in which dislocations are computed at atomistic levels and overall model is numerically solved using XFEM.

We need a new mathematical/numerical approach for embedding soft computing in multiscale modeling through machine learning algorithms. By machine learning algorithm, we can deal multiscale data for use in multiscale modeling. Multiscale data refers to complex signal that may come from functions, curves and images and we require to decompose multiscale data using either Fourier or wavelet decomposition. Medical images always contain various textures (e.g., edge of the images) that require multiscale methods; however, data pertaining to these textures at different scales are ambiguous (not detectable perfectly) and hence they are treated using fuzzy set theory, wherein soft computing plays major role in this type of embedded multiscale modeling. Machine learning incorporated with physics-based simulation can immediately benefit from multiscale modeling toward the generation of synthetic data which can further use to learn the scale of data [6].

Multiscale modeling with multiscale data (change of physics) can be also found in differential equations:

(a) Multiscale modeling with multiscale data in differential equation

Propagation of wave packet as the initial solution $u(x, 0) = \text{Const}(x)\text{e}^{i\zeta(x)/\tau}$ of the wave equation $\partial_t^2 u = \Delta u$, is a multiscale problem because solution depends on the two scales, such as envelope of the wave packet and scale of the wavelength, τ. In the field of mechanics of composite materials, as per elasticity theory, modeling of material property, stress tensor, $\nabla.\tau = 0$ and $\lambda(\nabla.u)I + \mu\left(\nabla u + (\nabla u)^{\text{T}}\right)$ depends on the heterogeneity of lame's constants that implies two scale rather two phase problem.

(b) Multiscale modeling with multiphysics solution

The flow of non-Newtonian fluid, for example, blood in the cardiovascular system possesses multiphysics solution because the cardiovascular system changes continually to fulfill the demands of the organisms in the diseased and healthy states. Researchers have also investigated that the elastodynamic contraction of the heart is also an example of multiscale multiphysics solution due to the variability of spatial patterns of inherent heterogeneity. The governing equations describing the flow of highly viscous non-Newtonian fluid are nonlinear and highly coupled, and based on multiphysics that is ruled by laws of thermodynamics and fluidity of the system. A fluid can be either Newtonian or non-Newtonian and accordingly laws of physics change justifying that the system possess multiscale multiphysics solution. Multiscale modeling can forecast the behavior of a medical system at different length and temporal scale that involves chemical, physiological, biological and biomedical systems.

Traditional method of modeling such system is inefficient because we know that such kind of problem is formulated mathematically by Navier–Stokes' equation which addresses the macroscale property of the fluid. But the fluid property of blood depends on the microscale so without intervening microscale property into Navier–Stokes' equation, it is not possible to obtain any biologically meaningful solution. The difficulty of this situation lies on the complexity of mathematics. As per Paul Dirac, we can quote "the underlying physical laws necessary for the mathematical theory of a large part of physics and the whole of chemistry are thus completely known, and the difficulty is only that the exact application of these laws leads to equations much too complicated to be soluble." With a view to the deficiency of traditional modeling of many real-life problem such as fluid flow through microporous system for modeling medical images, neutron transport through diffusion for modeling dynamics of nuclear reactor is solved by implementing lattice Boltzmann method (LBM) which is known as multiscale modeling and based on the mesoscopic scale. Therefore, it is mandatory to describe the mathematical and numerical details of classical (traditional) LBM.

7.4 Mathematical Structure of Classical LBM

The lattice Boltzmann is a mesoscopic scale modeling that bridges the gap between microscopic scale and macroscopic scale by taking into consideration of an ensemble of particles as a unit with the behavior of each particle. The system is as shown in Fig. 7.1. The ensemble of particles in the mesoscopic scale is evolved by a distribution function which acts as a representative for collection of particles.

In LBM, we enjoy the advantage of both macroscopic and microscopic characters of the system, with manageable computer resources [6]. LBM is based on Boltzmann transport equation, and in this section, we present its mathematical formulation.

7.4.1 *Formulation of Boltzmann Transport Equation*

Let $f(r, c, t)$ represents the number of molecules (particles) at time t, having position coordinates ranged between r and $r + \mathrm{d}r$ with a velocity lying between c and $c + \mathrm{d}c$. Let F represents the force which is applied on the system represented by $f(r, c, t)$ and the application of force F will disturb the order of the molecules. Basically, the molecules (particles) of the system will collide with each other. This situation can be described as a balanced system between number of molecules after collision and

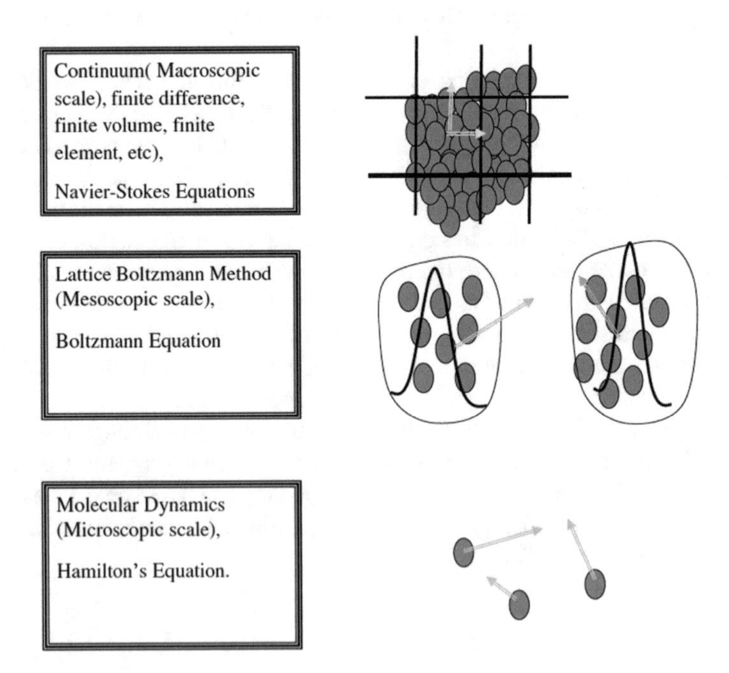

Fig. 7.1 Techniques of simulation

the same before collision. Mathematically, we can write this balanced situation as

$$\begin{aligned}&\text{No. of molecules after collision} - \text{no. of molecules before collision}\\&\quad= \text{net no of molecules}\end{aligned}$$

$$f(r + c\mathrm{d}t, c + F\mathrm{d}t, t + \mathrm{d}t)\mathrm{d}r\mathrm{d}c - f(r, c, t)\mathrm{d}r\mathrm{d}c = \Omega(f)\mathrm{d}r\mathrm{d}c\mathrm{d}t \tag{7.1}$$

Dividing both sides of Eq. (7.1) by dt, we have

$$\frac{\mathrm{d}f}{\mathrm{d}t} = \Omega(f) \tag{7.2}$$

Since $f = f(r, c, t)$, we can write

$$\mathrm{d}f = \frac{\partial f}{\partial r}\mathrm{d}r + \frac{\partial f}{\partial c}\mathrm{d}c + \frac{\partial f}{\partial t}\mathrm{d}t \tag{7.3}$$

$$\frac{\mathrm{d}f}{\mathrm{d}t} = \frac{\partial f}{\partial r}c + \frac{F}{m}\frac{\partial f}{\partial c} + \frac{\partial f}{\partial t} = \Omega(f) \tag{7.4}$$

In absence of external force F, we can write Eq. (7.4) as

$$\frac{\mathrm{D}f}{\mathrm{d}t} = c\frac{\partial f}{\partial r} + \frac{\partial f}{\partial t} = \Omega(f) \tag{7.5}$$

Now, finally, in general (in three-dimension form), we can write Eq. (7.5) as

$$\frac{\partial f}{\partial t} + c\nabla f = \Omega(f) \tag{7.6}$$

Equation (7.6) is known as Boltzmann transport equation without external force F. In Eq. (7.6), c and ∇f are vectors. The main issue of Eq. (7.6) is that it is an integro-differential equation and difficult to solve. The macroscale properties of a fluid mechanical system are governed by fluid density ρ, velocity vector of the fluid, $\vec{u}$ and internal energy, e. Hence, using macroscopic scale, we can write these properties as

$$\rho(r, t) = \int mf(r, c, t)\mathrm{d}c \tag{7.7a}$$

$$\rho(r, t)u(r, t) = \int mcf(r, c, t)\mathrm{d}c \tag{7.7b}$$

$$\rho(r, t)e(r, t) = \frac{1}{2}\int mu_a^2 f(r, c, t)\mathrm{d}c \tag{7.7c}$$

where m signifies the molecular mass and u_α represents the velocity of the particle in the direction of α, velocity relative to the fluid velocity as $u_\alpha = c - u$. By using laws of the kinetic theory of gas, we can write the energy as, $e = \frac{3}{2m} k_B T$.

Now, in Eq. (7.6), $\Omega(f)$ is called as collision operator, and using Bhatnagar-Gross-Krook (BGK) formalism [6], collision operator can be written as

$$\Omega(f) = \omega(f^{\text{eq}} - f) = \frac{1}{\tau}(f^{\text{eq}} - f) \tag{7.8}$$

where ω is known as collision frequency and τ is known as relaxation factor. The local equilibrium distribution function, f^{eq}, is obtained from Maxwell–Boltzmann distribution. Substitution of collision operator, Ω, from Eq. (7.8) into Eq. (7.6), we can write the final form of Boltzmann transport equation as

$$\frac{\partial f}{\partial t} + c\nabla f = \frac{1}{\tau}(f^{\text{eq}} - f) \tag{7.9}$$

Equation (7.9) is finally discretized using a numerical scheme into a specific direction (lattice point) to transform the Boltzmann transport equation into its lattice mode, named as lattice Boltzmann equation and this is written as

$$\frac{\partial f_i}{\partial t} + c_i \nabla f_i = \frac{1}{\tau}\left(f_i^{\text{eq}} - f_i\right) \tag{7.10}$$

We replace Navier–Stokes equation used in computational fluid dynamics (CFD) simulation by Eq. (7.10), which is defined as the engine of the LBM.

7.4.2 Arrangements of Lattice Structure

The generalized structure of lattice is expressed as $D_n Q_m$, where n represents the dimension of the problem in hands ($n = 1$ for 1D, $n = 2$ for 2D, $n = 3$ for 3D) and m refers to the speed model, number of linkages. The lattice structures taken into account for any problem are shown in Fig. 7.2.

The lattice structure signifies that the central node is connected with its neighboring node. Hence, we define lattice speed as a specified speed due to which the linkage of the neighboring nodes takes place with the stream of fictitious particles

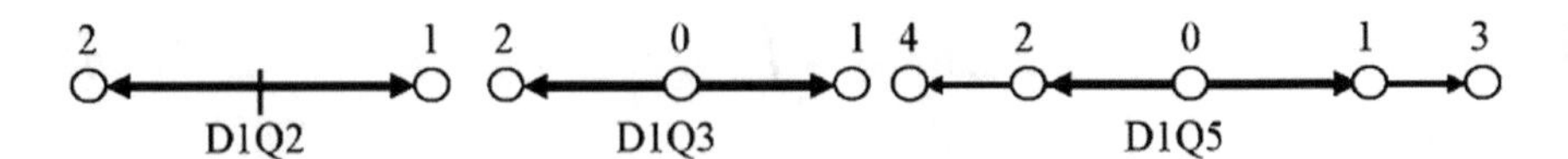

Fig. 7.2 Lattice structure

evolved from the central node. A few of the lattice structures generally used in the computation of LBM is presented in Sect. 7.4.2.1.

7.4.2.1 Lattice Structure D1Q3 and D1Q2

The computation schema of LBM method depends on the lattice structure. For example, the lattice structure D1Q3 represents one dimension problem and three velocity vectors (c_0, c_1 and c_2) for f_0, f_1 and f_2 which equal to 0, 1 and -1, respectively. For the convenience of computation, we assume spatial incremental step dx is equal to the temporal increment step dt. Otherwise, we can write $c_1 = \Delta x/\Delta t$ and $c_2 = \Delta x/\Delta t$; where Δx and Δt are the length linkage and corresponding time step, respectively. In this arrangement, physics guides us that the total number of fictitious particles at any instant of time will be always within the number of particles used in the structure (here, in D1Q3, it is three particles). The central particle is the fixed particle which has zero velocity. During streaming, the other two particles move either to the left or to the right node. The weighting factors ω_i in D1Q3 structure have values of 4/6, 1/6 and 1/6 for $f_0, f_1$and f_2, respectively. The speed of sound, c_s, in lattice units for D1Q3 is $1/\sqrt{3}$. In a similar way, the lattice structure D1Q2 signifies one dimension system and two lattice points. The weighting factors ω_i have values of 1/2 and 1/2 for f_1 and f_2, respectively. The speed of sound in this structure is $1/\sqrt{2}$. In the computation, we can consider more than one dimension as well as more lattice points such as, D1Q5, D2Q9 and D3Q15.

7.4.2.2 Weighting Factors of Lattice Structure D1Q5

In this lattice structure, as per the hypothesis, the total number of fictitious particles at any instant of time cannot exceed five particles. The weighting factors, ω_i, are 6/12, 2/12, 2/12, 1/12 and 1/12 for f_0, f_1, f_2, f_3 and f_4, respectively. The speed of sound in lattice units is $1/\sqrt{3}$.

7.4.2.3 Weighting Factors of Two-Dimensional Lattice Structure D2Q9

The LBM model with the lattice structure D2Q9 has high velocity vectors, with the central particle speed being zero, 2.5. The speeds are $c(0, 0)$, $c(1, 0)$, $c(0, 1)$, $c(1, 0)$, $c(0, 1)$, $c(1, 1)$, $c(1, 1)$, $c(1, 1)$ and $c(1, 1)$ for $f_0, f_1, f_2, f_3, f_4, f_5, f_6, f_7$ and f_8, respectively. The weighting factors for corresponding distribution functions are 4/9, 1/9, 1/9, 1/9, 1/9, 1/36, 1/36, 1/36 and 1/36. A lattice structure of D2Q9 is as shown in Fig. 7.3.

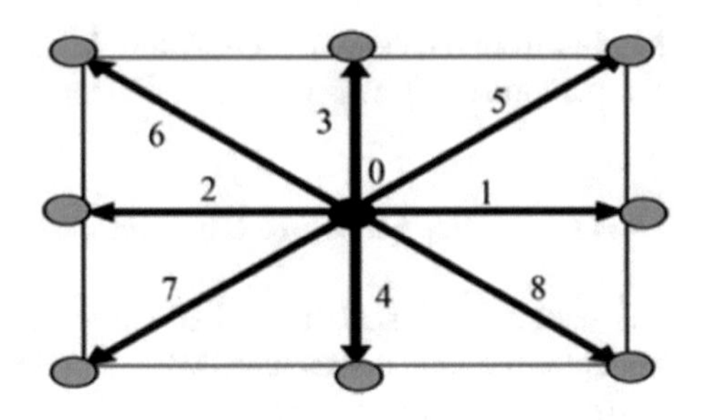

Fig. 7.3 D2Q9 lattice structure

7.4.2.4 Multiscale Modeling with LBM—A Case Study

Lattice Boltzmann model of groundwater flow as a multiscale modeling is presented here. Parent equation to describe the physical process is known as advection–diffusion. Governing equation for this study is written as

$$\frac{\partial C}{\partial t} + \frac{\partial (u_i C)}{\partial x_i} = \frac{\partial}{\partial x_i}\left(D_i \frac{\partial C}{\partial x_i}\right) \tag{7.11}$$

where c represents the concentration, t signifies the time, D_i signifies the dispersion coefficient in ith direction, u_i represents the fluid velocity and x_i is the Cartesian coordinate in the ith direction. It is required to solve the model governed by the equation as shown, and to solve the above equation at pore scale by lattice Boltzmann method, we treat the concentration of species as a distribution function which obeys the Boltzmann equation given below

$$f_\alpha(x + e_\alpha \Delta t, t + \Delta t) - f_\alpha(x, t) = -\frac{1}{\tau}\left(f_\alpha - f_\alpha^{\mathrm{eq}}\right) \tag{7.12}$$

where f_α presents the distribution function of particles, f_α^{eq} signifies the local equilibrium distribution function, Δt as the time step, e_α as the vector of the particle speed, τ as the single relaxation time and x presents the space vector for the used Cartesian coordinate system. We can write the expression of concentration C as

$$C(x, t) = \sum_\alpha f_\alpha(x, t) \tag{7.13}$$

The equilibrium distribution function is evaluated with the following constraints on f_α^{eq} as

$$\sum_\alpha f_\alpha^{\mathrm{eq}} = C \tag{7.14a}$$

$$\sum_\alpha e_{\alpha i} f_\alpha^{\mathrm{eq}} = u_i C \tag{7.14b}$$

$$\sum_{\alpha} e_{\alpha i} e_{\alpha j} f_{\alpha}^{\mathrm{eq}} = \lambda_i e_x e_y C \delta_{ij} = \begin{cases} \lambda_x e_x e_y C, \ i = j = x \\ \lambda_y e_x e_y C, \ i = j = y \end{cases} \tag{7.14c}$$

We can evaluate f_{α}^{eq} based on the constraints (Eqs. 7.14a, 7.14b and 7.14c). Chapman–Enskog expansion of distribution function can be applied to recover advection–diffusion equation. By doing this algebraic operation, we get parameters of equilibrium distribution function (λs) in terms of macroscopic parameter (diffusion coefficient, D_i) which can be written as

$$\Delta t < \frac{\left(\tau - \frac{1}{2}\right) \Delta x^2 \Delta y^2}{D_x \Delta y^2 + D_y \Delta x^2} \tag{7.15}$$

It is known that the advection–diffusion process is a process where both advection and diffusion take place simultaneously. Let us consider a problem of concentration of pollutant in an aquatic medium (river or lake) or a drop of ink in a porous medium such as parchment paper. The concentration of pollutant or ink drop diffuses with the corresponding stream moves away from the source. The physics involved in this case is known as advection–diffusion and the defined problem is said to be a transport problem.

Let us consider one-dimensional advection–diffusion equation in Cartesian coordinate system as,

$$\frac{\partial \theta}{\partial t} + u \frac{\partial \theta}{\partial x} = \alpha \frac{\partial^2 \theta}{\partial x^2} \tag{7.16}$$

where the symbols have usual significances (u signifies the velocity of flow and α signifies the diffusion coefficient). The LBM form of advection–diffusion problem can be written as

$$f_k(x + \Delta x, t + \Delta t) = f_k(x, t)[1 - \omega] + \omega f_k^{\mathrm{eq}}(x, t) \tag{7.17}$$

$$f_k^{\mathrm{eq}} = w_k \theta(x, t) \left[1 + \frac{c_k . \vec{u}}{c_s^2} \right] \tag{7.18}$$

$$c_k = \frac{\Delta x}{\Delta t} i + \frac{\Delta y}{\Delta t} j \tag{7.19}$$

The speed of sound c_s for lattice structure D1Q2, D2Q4 and D3Q6 for $c_k = 1$ is estimated as

$$C_s = \frac{1}{\sqrt{2}}$$

and the same for lattice structure D1Q3, D2Q5, D2Q9 and D3Q15 is

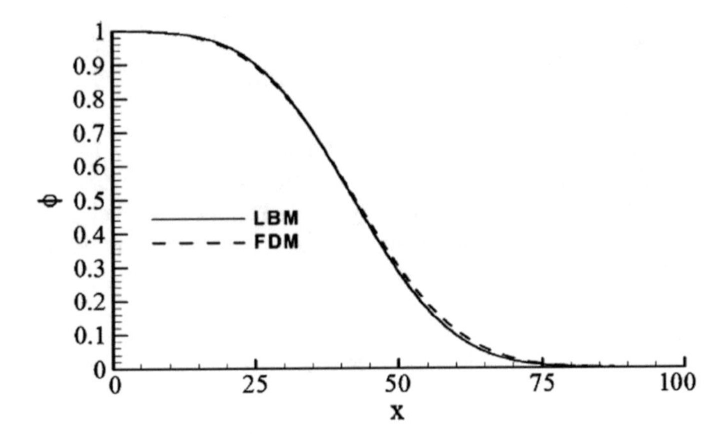

Fig. 7.4 Comparison of LBM and finite difference outcome of ADE

$$C_s = \frac{1}{\sqrt{2}}$$

The outcome of the lattice Boltzmann method-based multiscale modeling of the advection–diffusion problem is shown in Fig. 7.4. The result is compared with the finite difference solution of the same.

7.4.3 *Soft Computing in Multiscale Modeling*

Multiscale models contain parameters which play role in scaling of the system, but those parameters are basically uncertain due to their imprecision or ambiguity. Uncertainty of multiscale models can be quantified by implementing the uncertainty of those parameters of multiscale models. Soft computing plays the role for quantifying uncertainty of the governing parameters associated with multiscale model. Fuzzy set theoretic approach is used to address the uncertainty (epistemic uncertainty) and here in this case, we have used triangular fuzzy number for expressing the membership function of the uncertain fuzzy parameters of the multiscale model under investigation. Very often, one can apply machine learning algorithm to learn the temporal uncertainty of multiscale model with soft computing. For example, we can have fuzzy autoregressive integrated moving average (f-ARIMA), fuzzy deep learning, etc. Nowadays, multiscale simulations in the field of biology, medical science and other branches of engineering sciences, search to infer the uncertain dynamical behavior of the system. Machine learning can apply for soft computing and multiscale modeling can proceed with the outcome of the machine learning guided soft computing. We can also implement soft computing by Dempster–Shafer evidence theory (possibilistic computation) where we can choose the most feasible uncertain parameters among many on the basis of rank and that kind of method

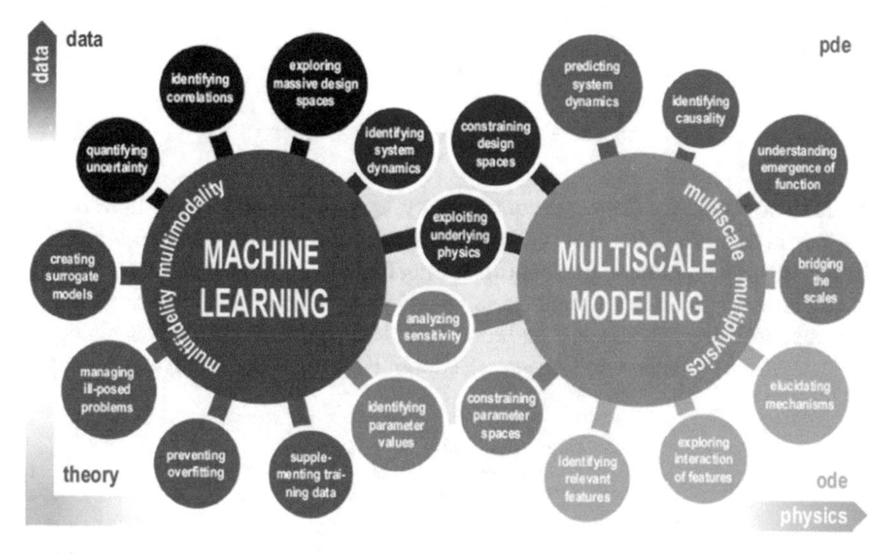

Fig. 7.5 Soft computing embedded multiscale modeling and machine learning

is known as multi criteria decision-making method (MCDM). Robustness, efficiency, sensitivity, specificity of multiscale modeling with soft computing depends on machine learning tools. It is known that integration of fuzzy data from different resources can be made using Bayesian learning to build predictive simulation tools in multiscale modeling. Figure 7.5 presents the integration of machine learning and soft computing embedded multiscale modeling. Details of multiscale modeling with soft computing can be found elsewhere in [7].

7.5 Fusion of Soft Computing and Multiscale Modeling

Multiscale modeling of fluid flow analysis is carried out by lattice Boltzmann method, in which soft computing plays the role of quantification of uncertainty of the fluid flow model such as solute transport through geological medium. Uncertainty of the parameters of the fluid flow model is categorized as epistemic uncertainty due to their imprecision (less sample size for their experimental determination) and hence the modeling is modified using fuzzy set theoretic concept of soft computing. We have proposed a perspective of fuzzy set theory-based lattice Boltzmann method and our approach generates a new dimension of lattice Boltzmann known as "Fuzzy Lattice Boltzmann" (FLB). We have solved solute transport through geological media using FLB which on one hand exhibits a multiscale modeling and on the other hand estimates the uncertainty (epistemic or knowledge uncertainty) associated with the system. In fact, parameters of the governing equation of solute transport problem are assessed using alpha cut value of fuzzy parameters of the governing equation

(Eq. 7.12)

$$\frac{\partial(\theta C)}{\partial t} = \frac{\partial}{\partial x_j}\left(\vartheta D_a \frac{\partial C}{\partial x_j}\right) - \frac{\partial\left(u_{aj} C\right)}{\partial x_j} \tag{7.20}$$

where $j = x, y$ and z for Cartesian coordinate system and Einstein summation convention rule is utilized, C represents concentration of solute in groundwater, ϑ signifies the porosity, D_a signifies apparent diffusion coefficient, u_{aj} represents Darcy velocity of groundwater along the direction j. On the basis of valid assumptions for a homogeneous formation of rock mass and for a suitable time frame shorter than geological time scale (order of a few million years), simplified form of Eq. (7.20) can be written as

$$\frac{\partial C}{\partial t} = D_a \frac{\partial^2 C}{\partial x_j^2} - u_j \frac{\partial C}{\partial x_j} \tag{7.21}$$

where $u_j = u_{aj}/\vartheta$ is pore water velocity. Finally, governing equation of solute transport in FLB mode can be shaped as

$$\frac{\partial \widetilde{C}}{\partial t} = \widetilde{D}_x \frac{\partial^2 \widetilde{C}}{\partial x^2} - \widetilde{u}_x \frac{\partial \widetilde{C}}{\partial x} \tag{7.22}$$

where symbols have usual significances. Definition and corresponding fuzzy arithmetic are skipped here as they can be found elsewhere in [7]. However, for the sake of its completeness, a short description pertaining to its membership function and corresponding α-cut representation with some algebraic properties as required for fuzzy LBM scheme is described in short.

7.5.1 Definition of Fuzzy Set

A fuzzy set is defined as a pair of two numbers, such as

$$\tilde{A} = \left\{\left(x, \mu_{\tilde{A}}(x)\right) | x \in R, \mu_{\tilde{A}}(x) \in [0, 1]\right\}$$

where $\mu_{\tilde{A}}(x)$ represents the membership function of the crisp value x. The shape of the membership function can be a continuous function such as triangular, trapezoidal, Gaussian and sigmoidal. If the membership function is convex, bounded and normal, then the corresponding fuzzy set is labeled as fuzzy number such as triangular fuzzy number, trapezoidal fuzzy number, Gaussian fuzzy number and sigmoidal fuzzy number. The detail of the arithmetic operation of two such fuzzy numbers is found elsewhere in [8]. Basically, alpha cut of a fuzzy number is defined as an interval on real number scale and one can use that interval for further operation as required in multiscale modeling through soft computing. In that sense, it is required to define alpha cut of a fuzzy number and we have defined that in Sect. 7.5.2.

7.5.2 *Alpha (α)-Cut and Algebraic Properties of Fuzzy Number*

Alpha (α)-cut of a fuzzy set is defined as the set of crisp values whose membership values are greater than or equal to alpha. The shape of the triangular membership function of a fuzzy set is given by

$$\mu(x) = \begin{cases} \frac{x_L - a}{b-a}, & a \le x_L \le b \\ \frac{c - x_R}{c-b}, & b \le x_R \le c \\ 0 & \text{otherwise} \end{cases} \tag{7.23}$$

Now, using Eq. (7.23), alpha cut representation of a fuzzy number with respect to its triangular membership function $\mu(x)$ can be written as

$$A_\alpha = [x_L^\alpha, x_R^\alpha] = [a + (b-a)\alpha,\ c - (c-b)\alpha] \tag{7.24}$$

7.5.3 *Mathematical Structure of Fuzzy Lattice Boltzmann Scheme*

Governing equation of solute transport (advection–diffusion equation) in the framework of alpha cut representation of fuzzy parameters (diffusion coefficient $\widetilde{D_x}$, flow velocity $\widetilde{u_x}$) can be written as

$$\frac{\partial C_\alpha}{\partial t} = D_{x\alpha}\frac{\partial^2 C_\alpha}{\partial x^2} - u_{x\alpha}\frac{\partial C_\alpha}{\partial x} \tag{7.25}$$

We have solved Eq. (7.25) numerically using standard single relaxation time (SRT) of LB scheme. Accordingly, standard SRT mode of discrete velocity LB equation can be written as

$$f_i\left(\vec{r} + \vec{e_i}\,\Delta t, t + \Delta t\right) = f_i(\vec{r}, t) + \Omega_i^{\mathrm{BGK}}(\vec{r}, t) \tag{7.26}$$

$$\Omega_i^{\mathrm{BGK}}(\vec{r}, t) = \frac{1}{\tau}\left[f_i^{\mathrm{eq}}(\vec{r}, t) - f_i(\vec{r}, t)\right] \tag{7.27}$$

where $f_i(\vec{r}, t)$ is particle distribution function at spatiotemporal coordinate $(\vec{r}, t)$ along ith direction, $\overrightarrow{e_i}$ represents particle velocity along ith direction, $\Omega_i^{\mathrm{BGK}}(\vec{r}, t)$ is BGK collision operator [9] along ith direction at same spatiotemporal coordinate, Δt is time step, α is relaxation coefficient and $f_i^{\mathrm{eq}}(\vec{r}, t)$ is particle equilibrium distribution function along ith direction. α-cut representation of Eqs. (7.18) and (7.19) can be written as

$$f_{i\alpha}\left(\vec{r} + \overrightarrow{e_i}\,\Delta t, t + \Delta t\right) = f_{i\alpha}(\vec{r}, t) + \Omega_{i\alpha}^{\mathrm{BGK}}(\vec{r}, t) \tag{7.28}$$

$$\Omega_{i\alpha}^{\mathrm{BGK}}(\vec{r}, t) = \frac{1}{\tau}\left[f_{i\alpha}^{\mathrm{eq}}(\vec{r}, t) - f_{i\alpha}(\vec{r}, t)\right] \tag{7.29}$$

where the particle equilibrium distribution function for an ADE can be written as

$$f_{i\alpha}^{\mathrm{eq}}(\vec{r}, t) = w_i C_\alpha(\vec{r}, t)\left(1 - \frac{\overrightarrow{e_i}\,.\overrightarrow{u_{x\alpha}}}{e_s^2}\right) \tag{7.30}$$

Here, in Eq. (7.30), w_i are the weights for particle's distribution function along ith direction and e_s is "pseudo-sound speed" [9]. Lattice structures such as D1Q2 and D1Q3 for 1D, D2Q4 and D2Q5 are for 2D and D3Q15 for 3D are used to solve Eqs. (7.28), (7.29) and (7.30). Figure 7.6 presents the lattice structure D3Q15.

Owing to the macroscopic particle density as the zero order velocity moment of distribution function, concentration of solute in terms of discrete particle distribution function can be expressed as

$$C_\alpha(\vec{r}, t) = \sum_i f_{i\alpha}(\vec{r}, t) \tag{7.31}$$

Equilibrium distribution function and weight factors defined in Eq. (7.23) satisfy the following properties

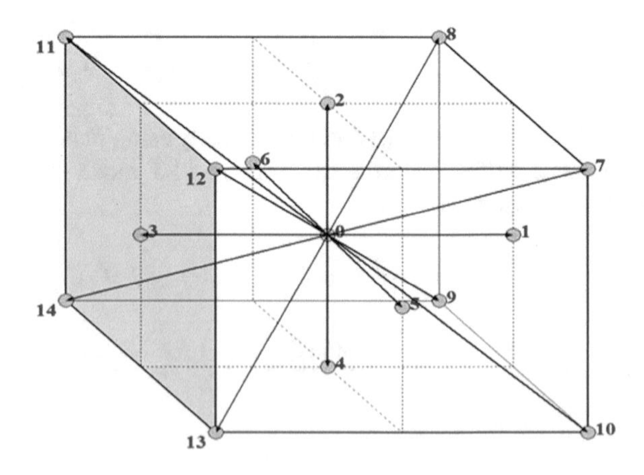

Fig. 7.6 Lattice structure D3Q15

$$\sum_i w_i = 1 \tag{7.32}$$

$$\sum_i f_{i\alpha}^{\text{eq}}(\vec{r}, t) = C_\alpha(\vec{r}, t) \tag{7.33}$$

$$\sum_i e_{ix} f_{i\alpha}^{\text{eq}}(\vec{r}, t) = u_{x\alpha} C_\alpha(\vec{r}, t) \tag{7.34}$$

$$\sum_i e_{ix} e_{iy} f_{i\alpha}^{\text{eq}}(\vec{r}, t) = e_s^2 C_\alpha(\vec{r}, t) \delta_{xy} \tag{7.35}$$

where δ_{xy} is the Dirac delta function, which is equal to 1 when $x = y$ and equal to 0 when $x \neq y$. Assuming that time step (Δt) is small and equal to ε (i.e., $\Delta t = \varepsilon$), and using multiscale Chapman–Enskog expansion [9] technique Eqs. (7.28) and (7.29) can be rewritten as

$$f_{i\alpha}\left(\vec{r} + \vec{e_i}\,\varepsilon, t + \varepsilon\right) = f_{i\alpha}(\vec{r}, t) + \frac{1}{\tau}\left[f_{i\alpha}^{\text{eq}}(\vec{r}, t) - f_{i\alpha}(\vec{r}, t)\right] \tag{7.36}$$

$$f_{i\alpha} = f_{i\alpha}^{(0)} + \sum_{n=1}^{\infty} \varepsilon^n f_{i\alpha}^{(n)} = f_{i\alpha}^{(0)} + \varepsilon f_{i\alpha}^{(1)} + \varepsilon^2 f_{i\alpha}^{(2)} + \vartheta\left(\varepsilon^3\right) \tag{7.37}$$

Taylor series expansion of the left-hand side of Eq. (7.29) with respect to time and space around point $(\vec{r}, t)$ can be generalized as

$$\sum_{n=1}^{\infty} \frac{\varepsilon^n}{n!}\left(\frac{\partial}{\partial t} + e_{ij}\frac{\partial}{\partial x_j}\right)^n f_{i\alpha}(\vec{r}, t) = \frac{1}{\tau}\left[f_{i\alpha}^{\text{eq}}(\vec{r}, t) - f_{i\alpha}(\vec{r}, t)\right] \tag{7.38}$$

$$\begin{aligned} \text{or, } & \sum_{n=1}^{\infty} \frac{\varepsilon^n}{n!}\left(\frac{\partial}{\partial t} + e_{ij}\frac{\partial}{\partial x_j}\right)^n \left(f_{i\alpha}^{(0)} + \varepsilon f_{i\alpha}^{(1)} + \varepsilon^2 f_{i\alpha}^{(2)} + \vartheta\left(\varepsilon^3\right)\right) \\ & = \frac{1}{\tau}\left[f_{i\alpha}^{\text{eq}} - \left(f_{i\alpha}^{(0)} + \varepsilon f_{i\alpha}^{(1)} + \varepsilon^2 f_{i\alpha}^{(2)} + \vartheta\left(\varepsilon^3\right)\right)\right] \end{aligned} \tag{7.39}$$

Grouping terms of the same order in ε yield the following successive approximations [9]

$$O\left(\varepsilon^0\right) : f_{i\alpha}^{\text{eq}} = f_{i\alpha}^{(0)} \tag{7.40}$$

$$O\left(\varepsilon^1\right) : \left(\frac{\partial}{\partial t} + e_{ij}\frac{\partial}{\partial x_j}\right) f_{i\alpha}^{(0)} = -\frac{1}{\tau} f_{i\alpha}^{(1)} \tag{7.41}$$

$$O\left(\varepsilon^2\right) : \left(\frac{\partial}{\partial t} + e_{ij}\frac{\partial}{\partial x_j}\right) f_{i\alpha}^{(1)} + \frac{1}{2}\left(\frac{\partial}{\partial t} + e_{ij}\frac{\partial}{\partial x_j}\right)^2 f_{i\alpha}^{(0)} = -\frac{1}{\tau} f_{i\alpha}^{(2)} \tag{7.42}$$

we have now

$$\left(1-\frac{1}{2\tau}\right)\left(\frac{\partial}{\partial t}+e_{ij}\frac{\partial}{\partial x_j}\right)f_{i\alpha}^{(1)}=-\frac{1}{\tau}f_{i\alpha}^{(2)} \tag{7.43}$$

Adding Eq. (7.33) with $\varepsilon \times$ Eq. (7.35), we have

$$\left(\frac{\partial}{\partial t}+e_{ij}\frac{\partial}{\partial x_j}\right)f_{i\alpha}^{(0)}+\varepsilon\left(1-\frac{1}{2\tau}\right)\left(\frac{\partial}{\partial t}+e_{ij}\frac{\partial}{\partial x_j}\right)f_{i\alpha}^{(1)}=-\frac{1}{\tau}\left(f_{i\alpha}^{(1)}+\varepsilon f_{i\alpha}^{(2)}\right) \tag{7.44}$$

Using the properties of distribution function, following constraints on fluctuating parts of the distribution function can be imposed and accordingly we obtain

$$\sum_i f_{i\alpha}^{(k)}(\vec{r},t)=0,\quad k=1,2,3,\ldots \tag{7.45}$$

And

$$\frac{\partial}{\partial t}\sum_i f_{i\alpha}^{(k)}=0,\quad k=1,2,3,\ldots \tag{7.46}$$

From Eq. (7.38), we obtain

$$\begin{aligned}\frac{\partial}{\partial t}\sum_i f_{i\alpha}^{(0)}+\frac{\partial}{\partial x_j}\sum_i e_{ij}f_{i\alpha}^{(0)}-\tau\left(1-\frac{1}{2\tau}\right)\frac{\partial}{\partial x_j}\sum_i e_{ij}\frac{\partial f_{i\alpha}^{(0)}}{\partial t}\\-\tau\left(1-\frac{1}{2\tau}\right)\frac{\partial}{\partial x_j}\sum_i e_{ij}e_{ik}\frac{\partial f_{i\alpha}^{(0)}}{\partial x_k}=0\end{aligned} \tag{7.47}$$

The term containing time derivative of equilibrium distribution function is smaller compared to the other three terms and hence, it can be omitted and can be treated as error term. Using the properties of equilibrium distribution function, the above equation (Eq. 7.39) can be written as

$$\frac{\partial C_\alpha}{\partial t}+\frac{\partial\left(u_{j\alpha}C_\alpha\right)}{\partial x_j}-\left(\tau-\frac{1}{2}\right)\frac{\partial}{\partial x_j}\left(e_s^2\frac{\partial C_\alpha}{\partial x_j}\right)=0 \tag{7.48}$$

Equation (7.40) can be rearranged as

$$\frac{\partial C_\alpha}{\partial t}=\frac{\partial}{\partial x_j}\left(D\frac{\partial C_\alpha}{\partial x_j}\right)-\frac{\partial\left(u_{j\alpha}C_\alpha\right)}{\partial x_j} \tag{7.49}$$

where

$$D = e_s^2\left(\tau - \frac{1}{2}\right) \tag{7.50}$$

Since LB solutions are more stable in lattice unit compared to physical unit, we have solved LBE in lattice unit. In lattice unit, Eq. (7.42) can be written as

$$D^* = e_s^2\left(\tau - \frac{1}{2}\right) \tag{7.51}$$

where D_x^* is the lattice diffusion coefficient. For unit lattice time step $\Delta t^* = 1$ and unit lattice spatial step = 1, lattice velocity $e = dx^*/\Delta t^* = 1$ and $e_s = e/\sqrt{3}$ (valid for D1Q3, D2Q5, D2Q9). Substituting these values in Eq. (7.43), we obtain the value of lattice diffusion coefficient as

$$D^* = \frac{1}{3}\left(\tau - \frac{1}{2}\right) \tag{7.52}$$

The relation between lattice diffusion coefficient and physical diffusion coefficient is

$$D_x^* = \frac{D_x}{\frac{\Delta x^2}{\Delta t}} \tag{7.53}$$

We further obtain the expression for physical time step as

$$\Delta t = \frac{\Delta x^2}{3D_x}\left(\tau - \frac{1}{2}\right) \tag{7.54}$$

For $\tau = 1$, Eq. (7.46) can be written as

$$\Delta t = \frac{\Delta x^2}{6D_x} \tag{7.55}$$

Equation (7.47) shows that corresponding to each α-cut value of diffusion coefficient, we have different time step value (Δt) and hence, Eq. (7.47) can be written as

$$\Delta t_\alpha = \frac{\Delta x^2}{6D_{x\alpha}} \tag{7.56}$$

By using alpha (α)-cut representative values (except α-cut = 1) of system parameters, we have four combinations of diffusion coefficient and groundwater velocity, such as, $[\mathrm{Dx_{Low}}, \mathrm{ux_{Low}}]_\alpha$, $\left[\mathrm{Dx_{Low}}, \mathrm{ux_{High}}\right]_\alpha$, $\left[\mathrm{Dx_{High}}, \mathrm{ux_{Low}}\right]_\alpha$, $\left[\mathrm{Dx_{High}}, \mathrm{ux_{High}}\right]_\alpha$. LBM in fuzzy mode is solved using each combination of input parameters. Basically, various combinations are used as per fuzzy vertex theory [8]. Collision and streaming operations of the system are depicted as:

Collision process [10]

In collision process, particles distribution function relaxes toward local equilibrium distribution function and it can be described by the following equation

$$f_{i\alpha}^{*}(\vec{r}, t + \Delta t) = f_{i\alpha}(\vec{r}, t) + \Omega_{i\alpha}^{\mathrm{BGK}}(\vec{r}, t) \tag{7.57}$$

where symbols have usual significances.

Streaming process [10]

In this process, particles move from one lattice point to nearest lattice point along the direction of the lattice velocity. Computationally, this process is just memory swapping and algorithmically, it can be written as

$$f_{i\alpha}\left(\vec{r} + \vec{e_i}\,\Delta t, t + \Delta t\right) = f_{i\alpha}^{*}(\vec{r}, t + \Delta t) \tag{7.58}$$

Additional bounce-back boundary conditions [11–13] are imposed at obstacle sites and along boundary walls at which particles reverse its direction after collision with obstacles or boundary walls. Mathematically, bounce-back algorithm [11–13] for a D1Q3 lattice can be written as

$$f_{(i\pm1)\alpha}(\vec{r} + e_i\Delta t, t + \Delta t) = f_{i\alpha}(\vec{r} + e_i\Delta t, t + \Delta t) \tag{7.59}$$

where $+$ sign for $i = 1$ and $-$ sign for $i = 2$.

7.5.4 Uncertainty Analysis

Uncertainty of concentration of solute is mainly due to the imprecision in measurements of the solute diffusion coefficient and groundwater flow velocity and accordingly, the α-cut representation of the solute concentration has been computed using single relaxation time LB scheme in presence of the fuzziness of the solute diffusion coefficient and groundwater flow velocity [14–16]. The membership function of the solute concentration results as a generalized triangular fuzzy number. Analytical solution of the governing solute transport equation with following initial and boundary conditions

$$\widetilde{C}(x, t) = 0, \quad x = \pm\infty \tag{7.60}$$

can be written as

$$\widetilde{C} = \frac{C_0}{\sqrt{4\pi\widetilde{D}_x t}}\exp\left(-\frac{(x - \widetilde{u}_x t)^2}{4\widetilde{D}_x t}\right) \tag{7.61}$$

The uncertainty of solute concentration is compared that of outcome from Author's fuzzy LB numerical scheme. Uncertainty output in both the cases is expressed in terms of the α-cut representation as lower and upper bound for any specified α-cut. The degree of uncertainty [17] of solute concentration is given by

$$\Delta C|_{\alpha} = \left[\frac{C^{\mathrm{U}} - C^{\mathrm{L}}}{C^{\mathrm{U}} + C^{\mathrm{L}}}\right]_{\alpha} \tag{7.62}$$

where C^{L} and C^{U} are the alpha cut representation of lower and upper bound of the solute concentration.

7.5.5 Results and Discussion

Input parameters for testing the soft computing method (Fuzzy LBM) of multiscale modeling of fluid flow analysis are as follows:

Initial concentration of effluent of a particular chemical is taken into account as 1 kg/m^3 and is discharged as a point source at downstream distance, $x = 10$ m in a channel of length $L = 400$ m. The fuzziness of the model parameters are expressed as triangular fuzzy number because the imprecision result is around the most likely value (experimental measured mean).

Accordingly, the triangular fuzzy number of the solute diffusion coefficient, $\widetilde{D_x}$ (m^2/s) $= \langle 0.010, 0.014, 0.017\rangle$, and velocity of water $\widetilde{u_x}$ (m/s) $= \langle 0.01, 0.012, 0.014\rangle$.

The problem with deterministic values of diffusion coefficient and water velocity has been solved using LBM by Zhou [8]. Membership function of diffusion coefficient and velocity for various α-cut values are given in tabular from in Table 7.1 and in graphical form in Fig. 7.7a, b, respectively.

Uncertainty in solute concentration is expressed in terms of a closed interval bounded by lower and upper value of solute concentration. The simulation is carried out using lattice unit, i.e., $\mathrm{d}x_{\mathrm{l}} = \mathrm{d}y_{\mathrm{l}} = 1$ lbu, and $\mathrm{d}t_{\mathrm{l}} = 1$ lbu to achieve numerical stability. Corresponding spatial step lengths are taken as, $\mathrm{d}x = \mathrm{d}y = 1$ and time step for each α-cut is calculated using Eq. (7.47). In order to test the stability of the LB scheme [18–20] with the step lengths dx, dy and dt as mentioned, numerical calculation of spatial profile of solute concentration with most likely value of the model parameters has been carried out and the results are compared with corresponding analytical results. The comparison of numerical and analytical solutions is shown graphically in Fig. 7.8 and tabulated in Table 7.2.

The numerical calculation of the upper and lower bound of the solute concentration for a specific time and at each length of the domain results the fuzziness of the solute concentration. The fuzziness of the solute concentration [21, 22] at lengths, $x =$ 125 m, 140 m, and 150 m for specific time ($t = 10{,}000$ s) is represented in terms

Table 7.1 Member values of diffusion coefficient and groundwater velocity at various α-cut font sizes of headings

α-cut value	Diffusion coefficient (m^2/s)		Groundwater velocity (m/s)	
	Lover value	Upper value	Lover value	Upper value
0	0.01	0.017	0.01	0.014
0.1	0.0104	0.0167	0.0102	0.0138
0.2	0.0108	0.0164	0.0104	0.0136
0.3	0.0112	0.0161	0.0106	0.0134
0.4	0.0116	0.0158	0.0108	0.0132
0.5	0.012	0.0155	0.011	0.013
0.6	0.0124	0.0152	0.0112	0.0128
0.7	0.0128	0.0149	0.0114	0.0126
0.8	0.0132	0.0146	0.0116	0.0124
0.9	0.0136	0.0143	0.0118	0.0122
1.0	0.014		0.0120	

of the membership function of solute concentration and the results are shown in Fig. 7.9a–c.

Finally, spatial profiles of solute concentration using four different combinations of fuzzy input parameters at any α-cut of 0.8 and 0.5 are computed for a total time of simulation (t = 10,000 s) and corresponding results are presented in Figs. 7.10a, b and 7.11, respectively. Lower and upper concentrations of solute at each spatial point for the same α-cut value and total simulation time are extracted from the four different spatial profiles. Spatial profiles of lower and upper bound of solute concentration are shown in Fig. 7.12, for α-cut value of 0.5, respectively. It can be observed from these figures that transport of solute results an uncertainty of solute concentration due to the fuzziness of the model parameters. The uncertainty increases with decrease of α-cut value which is an obvious fact because lower the α-cut value higher is the fuzziness in the input parameters. Comparison between analytical model and Fuzzy LBM model-based uncertainty [22] is shown graphically in Fig. 7.13 and a good agreement between the two results is established.

7.6 Conclusions

Multiscale modeling of fluid flow (solute transport) is explained using lattice Boltzmann method. Only single relaxation time (SRT) has been used to solve transport equation that is advection–dispersion equation (ADE) using fuzziness of the governing parameters such as parameters representing diffusion coefficient and ground water velocity. Fuzzy set theory of soft computing plays an important role to present the uncertainty of the model parameters of the system. Fuzziness of the

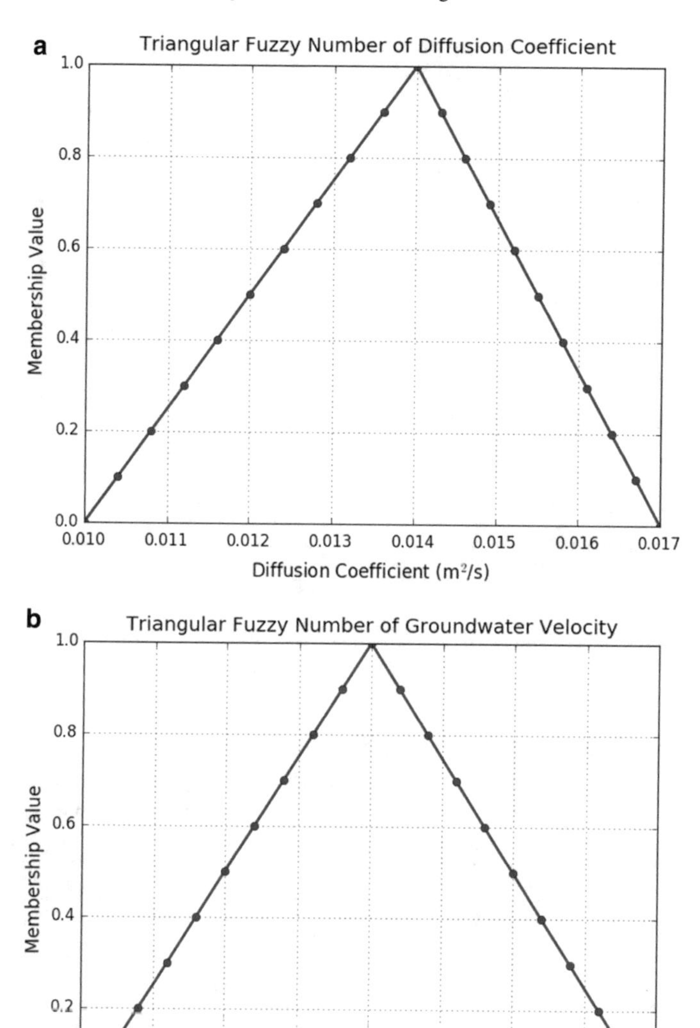

Fig. 7.7 **a** Triangular membership function of diffusion coefficient. **b** Triangular membership function of groundwater velocity

governing parameters is described by triangular fuzzy number because the imprecise measurement is interpreted as around mean. Explicit representation of that imprecise measurement is written as [mean $-c$, mean $+c$] where c signifies as tolerance limits (95% confidence level). Uncertainty of the problem under consideration using soft computing approach is categorized as epistemic uncertainty. Lattice Boltzmann

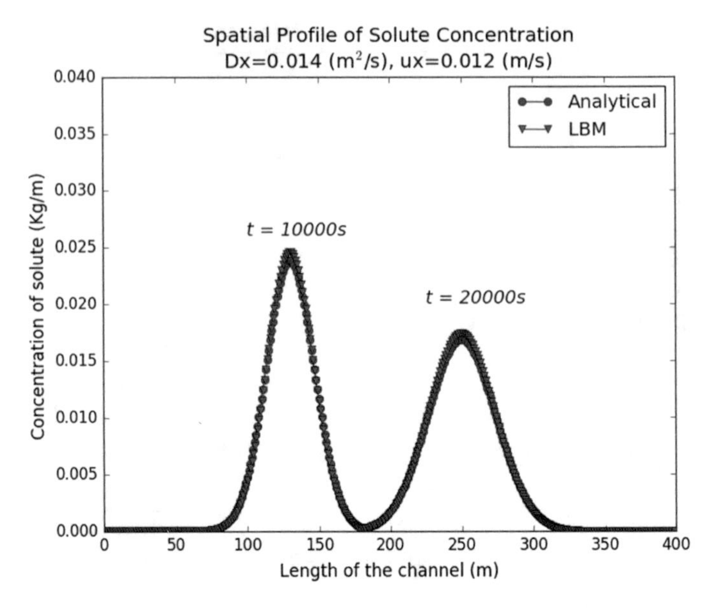

Fig. 7.8 Spatial concentration profile (numerical and analytical) of solute with most likely values of model parameters

Table 7.2 Member values of concentration at distance 150 m and time 10,000 s

Membership value	LBM solution of concentration at 100 m		Analytical Solution of concentration at 100 m	
	Lover value	Upper value	Lover value	Upper value
0	0.004757	0.015940	0.030839	0.030839
0.1	0.006748	0.018532	0.029597	0.029597
0.2	0.008994	0.020848	0.027902	0.027902
0.3	0.011389	0.022781	0.025906	0.025906
0.4	0.013810	0.024258	0.023739	0.023739
0.5	0.015889	0.025243	0.021505	0.021505
0.6	0.017561	0.025732	0.019284	0.019284
0.7	0.019210	0.025746	0.017138	0.017138
0.8	0.020754	0.025334	0.015108	0.015108
0.9	0.022224	0.024552	0.013313	0.013313
1.0	0.023469		0.004488	

method of solving fuzzy solute transport is achieved in multiscale modeling. It can be stated that Fuzzy LBM is effective and accurate for not only solving the advection–dispersion equation with uncertainty in terms of the fuzziness of the governing parameters but also provides a tool for developing a fuzzy inference system using Mamdani implication [8].

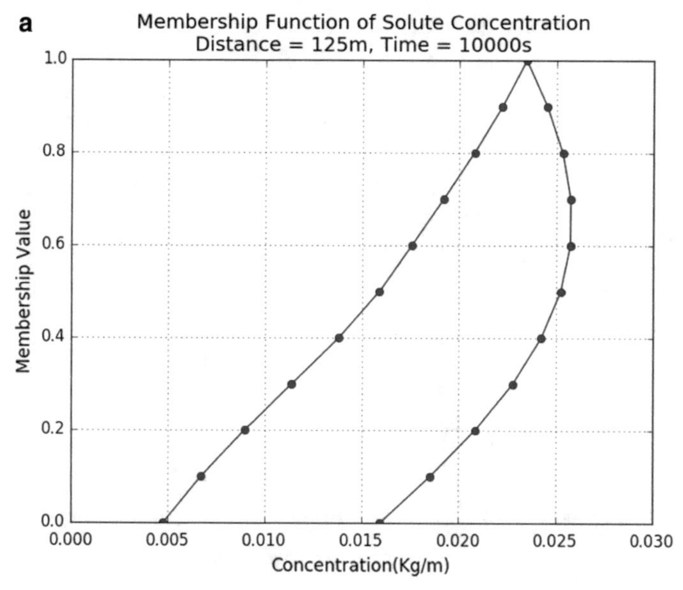

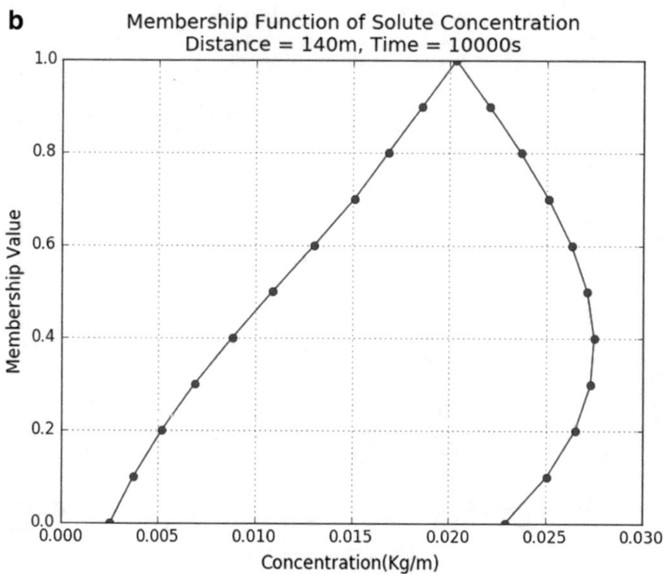

Fig. 7.9 **a** Membership function of solute concentration at length 125 m. **b** Membership function of solute concentration at length 140 m

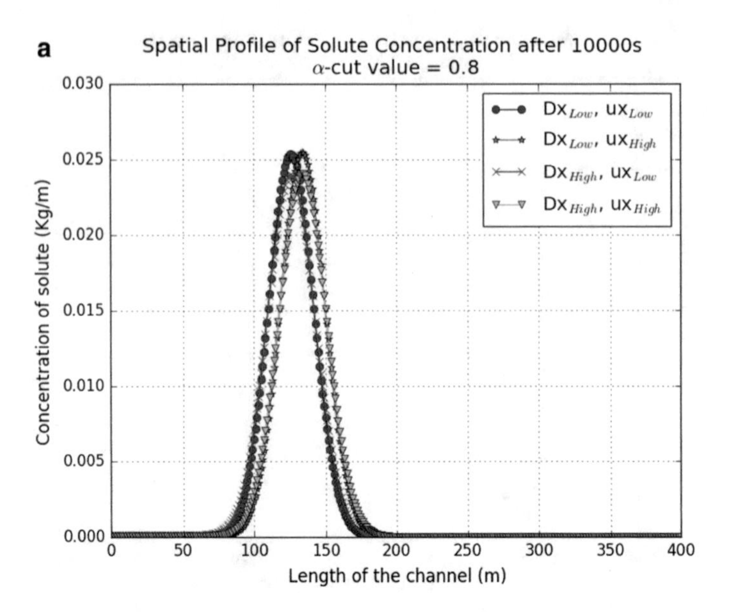

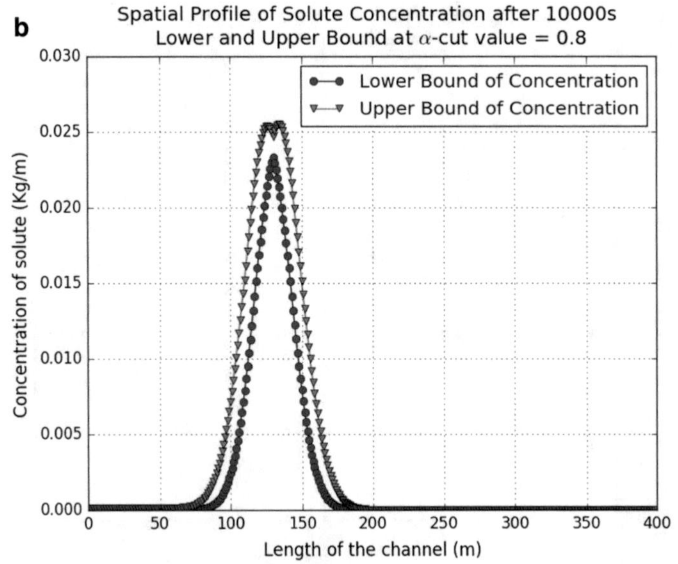

Fig. 7.10 **a** Membership function of solute concentration versus length of channel. **b** Membership function of solute concentration at various length

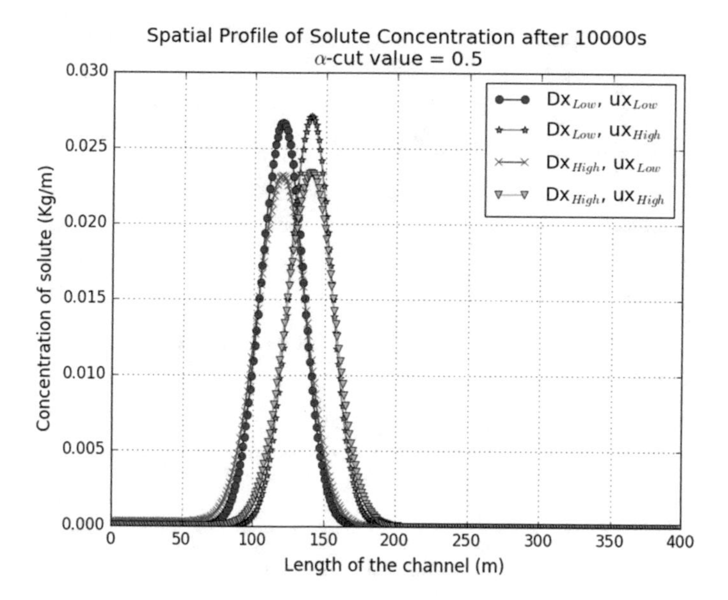

Fig. 7.11 Length profile of membership function of concentration of solute

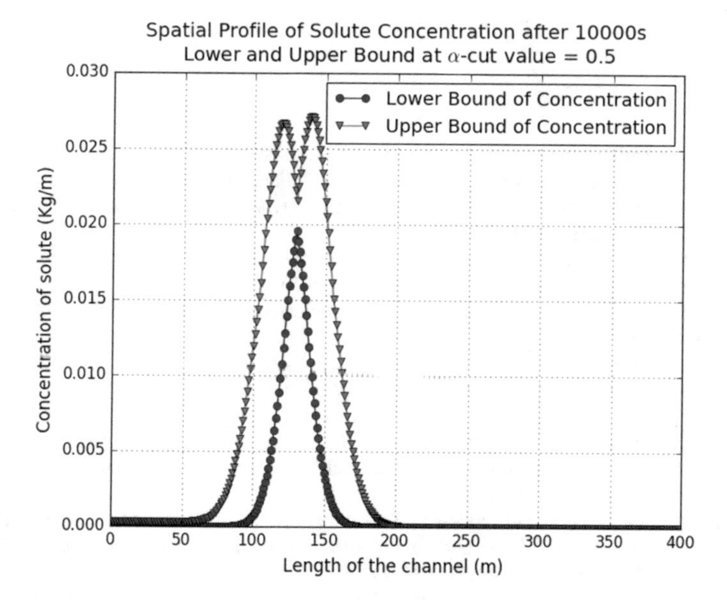

Fig. 7.12 Length-wise variation of membership function of concentration of solute

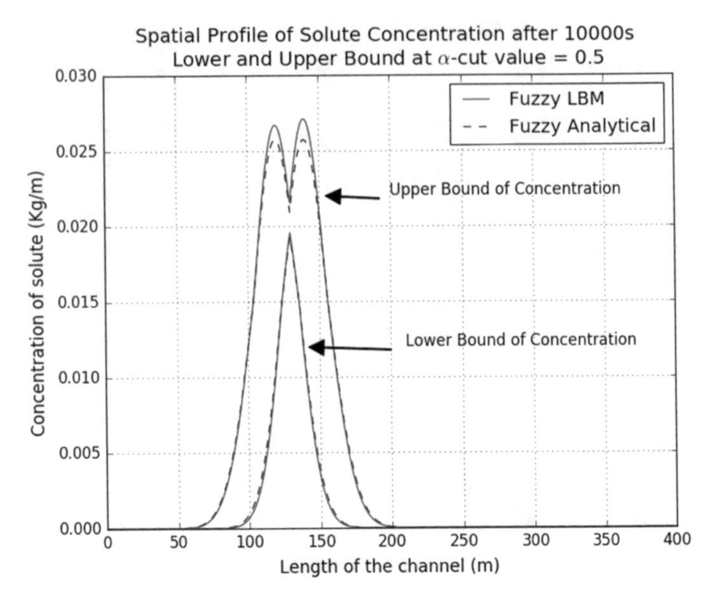

Fig. 7.13 Analytical and lattice Boltzmann method-based membership function of concentration of solute

References

1. Brandt, A.: Multiscale scientific computation: review 2001. In: Barth, T.J., et al. (eds.) Multiscale and Multiresolution Methods: Theory and Applications. Yosemite Educational Symposium Conference Proceedings, 2000. Lecture Notes in Computer Science and Engineering, vol. 20, pp. 3–96, Springer, Berlin (2002)
2. Jiaung, W.S., Ho, J.R., Kuo, C.P.: Lattice Boltzmann method for the heat conduction problem with phase change. Numer. Heat Transf. B **39**, 167–187 (2001)
3. Car, R., Parrinello, M.: Unified approach for molecular dynamics and density-functional theory. Phys. Rev. Lett. **55**, 2471–2474 (1985)
4. Esteva, A., Kuprel, B., Novoa, R.A., Ko, J., Swetter, S.M., Blau, H.M., Thrun, S.: Dermatologist-level classification of skin cancer with deep neural networks. Nature **542**, 115–118 (2017)
5. Hicks, J.L., Althoff, T., Sosic, R., Kuhar, P., Bostjancic, B., King, A.C., Leskovec, J., Delp, S.L.: Best practices for analyzing large-scale health data from wearables and smartphone apps. NPJ Digital Med. **2**, 1–12 (2019)
6. d'Humières, D.: Generalized lattice-Boltzmann equations. In: Weaver, D.P., Shizgal, B.D. (eds.) Rarefied Gas Dynamics: Theory and Simulations. Progress in Astronautics & Aeronautics, vol. 59, pp. 450–458. AIAA, Washington, D.C. (1992)
7. Mohamad, A.A.: Lattice Boltzmann Method: Fundamentals and Engineering Applications with Computer Codes. Springer, London (2011)
8. Zhou, J.G.: A lattice Boltzmann method for solute transport. Int. J. Numer. Meth. Fluids **61**, 848–863 (2008)
9. Ross, T.J.: Fuzzy Logic with Engineering Applications. Wiley, New York (2004)
10. Chen, S., Doolen, G.D.: Lattice Boltzmann method for fluid flows. Ann. Rev. Fluid Mech. **30**, 329–364 (1998)
11. Succi, S.: The Lattice Boltzmann Equation for Fluid Dynamics and Beyond. Oxford University Press, New York (2001)

12. Wolf-Gladrow, D.A.: Lattice-Gas Cellular Automata and Lattice Boltzmann Models: An Introduction. Springer, New York (2000)
13. Alexander, W.R., McKinley, L.: Deep Geological Disposal of Radioactive Waste. Elsevier, Amsterdam (2007)
14. Sukop, M., Thorne, D.: Lattice Boltzmann Modeling: An Introduction for Geoscientists and Engineers. Springer, Heidelberg (2006)
15. Wellein, G., Zeiser, T., Hager, G., Donath, S.: On the single processor performance of simple lattice Boltzmann kernels. Comput. Fluids **35**, 910–919 (2006)
16. Raj, K., Prasad, K.K., Bansal, N.K.: Radioactive waste management practices in India. Nucl. Eng. Des. **236**, 914–930 (2006)
17. Steefel, C.I., DePaolo, D.J., Lichtner, P.C.: Reactive transport modeling: an essential tool and a new research approach for the earth sciences. Earth Planet. Sci. Lett. **240**, 539–558 (2005)
18. Zadeh, L.A.: Fuzzy sets. Inf. Control **8**, 338–353 (1965)
19. Klir, G.J., Yuan, B.: Fuzzy Sets and Fuzzy Logic: Theory and Applications. Prentice Hall PTR, New Jersey (1995)
20. Zeigler, D.P.: Boundary conditions for the lattice Boltzmann simulations. J. Stat. Phys. **71**, 1171–1177 (1993)
21. Ayyub, B.M., Klir, G.J.: Uncertainty Modelling and Analysis in Engineering and the Sciences. Chapman and Hall/CRC Press, New York (2006)
22. McKone, T.E., Bogen, K.T.: Predicting the uncertainties in risk assessment. Environ. Sci. Technol. **25**, 1674–1681 (1991)

Chapter 8
Various Generalizations of Fuzzy Sets in the Context of Soft Computing and Decision-Making

Jacob John Sunil

8.1 Preliminaries

Fuzzy sets and logic, which was put forward in 1965 by Zadeh [1] as an alternative to the traditional concepts of classical sets and Aristotelian bi-valued logic, have many applications going up to the leading edge of artificial intelligence and cloud computing. But still, there is room for developments in theoretical as well as practical scenarios for this relatively new concept. This certainly gives a value addition to formal mathematics as a whole.

Apart from the traditional approaches of naive and axiomatic ways, another trouble connected to the primitive view of a set is the vagueness and how to handle it. Classical mathematics needs that all notions, including the idea of a set, must be exact. But this is not the situation in real practice, and vagueness is abundant in practical problems handled by scientists and technologists. That is the reason why philosophers and scientists have become more interested in these notions recently. Fuzzy sets [1] tackle vagueness via membership function. There are other types of structures also available for this purpose, and one important among them is the idea of rough sets [2], which models vagueness using the idea of boundary region of a set. Further, there are many hybrid structures [3] involving two or more such structures.

If one looks at the evolution of mankind, there are various eras with varying importance with a hierarchy as shown: AGRICULTURE → INDUSTRY → INFORMATION → GENETIC-ALGORITHMS → NANO TECHNOLOGY. In this transformation, the problems that we were handling had developed an increased level in magnitude and complexity. Often one needs a good simplification of the level of complexity that supports minimum loss of information. One way of doing this is to permit some degree of uncertainty into the problem by making the statements imprecise. Basically, there are two types of imprecision, vagueness and ambiguity. Loosely

J. J. Sunil (✉)
Department of Mathematics, National Institute of Technology Calicut, Kozhikode, Kerala 673601, India
e-mail: sunil@nitc.ac.in

T. Som et al. (eds.), *Fuzzy, Rough and Intuitionistic Fuzzy Set Approaches for Data Handling*, Forum for Interdisciplinary Mathematics,
https://doi.org/10.1007/978-981-19-8566-9_8

speaking, vagueness is the dilemma of making clear separation and ambiguity is the situation of having two or more alternatives not specified. An initial insight into these ideas can be seen in Black [4].

Historically, the traditional notions of membership and logic go back to Greek philosophy and Aristotelian bi-valued logic. Alternatively, the Law of excluded middle demands "Every proposition must be either TRUE or FALSE". The first objection to this was rised by Plato, and later, Lukasievicz came up with three valued logic and ultimately by Zadeh with fuzzy or infinite valued logic.

After Zadeh, many researchers came up with various generalizations of fuzzy sets which really broadened the scope in representing real-world problems more accurately and precisely. These structures include type-2 fuzzy sets, Interval-valued fuzzy sets, Intuitionistic fuzzy sets, Pythagorean fuzzy sets, Picture fuzzy sets, Spherical fuzzy sets, Fermatean fuzzy sets, Hesitant fuzzy sets, and many more.
We begin with Zadeh [1]'s definition of fuzzy sets.

Definition 1 Let Ψ be a set. A fuzzy set $\mathcal{A}$ in Ψ is characterized by a membership function $\zeta \rightarrow \mu_{\mathcal{A}}(\zeta)$ from Ψ to $I = [0, 1]$. μ is termed as membership value function, and the fuzzy set is denoted by $\mathcal{A} = \{(\zeta, \mu_{\mathcal{A}}(\zeta)) : \zeta \in \Psi\}$.

A major concern in the theory of fuzzy sets is the technique for assigning membership values suitably. Membership functions may be composed by analyzing the problem under deliberation. Further, there are several practicable ways to construct membership functions. Subjective evaluation and extraction, Converted frequencies or probabilities, Physical measurement, and Learning and adaptation are some major methods for the same.

8.2 Type-2 Fuzzy Sets and Systems

Right from the beginning of the theory of fuzzy sets, one major criticism regarding ordinary fuzzy sets, also possible to call type-1 fuzzy sets, was that there is no uncertainty associated with the membership function. Some even said that this contradicts the meaning of fuzzy itself. In 1975, Zadeh himself [5] came up with an answer to this by proposing types of fuzzy sets which are more sophisticated and the ones which generalize the standard fuzzy sets of type-1. The first among these was called type-2 fuzzy set. Instead of a fixed membership function, a fluctuating membership function is used in type-2 fuzzy sets. Later, Zadeh himself [5] extended type-2 fuzzy sets to type- n fuzzy sets. The membership function associated with type-2 fuzzy sets are of three dimensions, where the third dimension is the value of the membership function at any point on its two-dimensional domain, often described as the footprint of uncertainty (FOU).

The membership grade of a type-2 fuzzy set is itself fuzzy. The membership is a function at individual value of the primary variable. The domain of secondary membership function is in [0, 1] (primary membership values), and the range may also be in [0, 1]. Accordingly, for a type-2 fuzzy set, the membership function is dimension of three. This third dimension can provide new degrees of freedom for managing vagueness. So these structures are often useful when it is challenging to evaluate the precise membership for a fuzzy set, as if in situation of modeling a word by a fuzzy set. Recently, type-2 fuzzy sets have become widely established for modeling higher-order uncertainties.

Definition 2 A type-2 fuzzy set $\widetilde{\mathcal{A}}$ in a universal set Ψ is described by a type-2 membership function $\mu_{\dot{\mathcal{A}}}(\zeta, k)$, where $\zeta \in \Psi$, $I = [0, 1]$, and $k \in J_\zeta \subseteq I$. That is, $\widetilde{\mathcal{A}}$ is expressed as

$$\widetilde{\mathcal{A}} = \left\{\left((\zeta, k), \mu_{\widetilde{\mathcal{A}}}(\zeta, k)/\zeta \in \Psi, k \in J_\zeta \subseteq I\right\}$$

where $0 \leq \mu_{\widetilde{\mathcal{A}}}(\zeta, k) \leq 1$.
Also,

$$\widetilde{\mathcal{A}} = \int\limits_{\zeta \in \Psi} \int\limits_{k \in J_\zeta} \frac{\mu_{\widetilde{\mathcal{A}}}(\zeta, k)}{(\zeta, k)} = \int\limits_{\zeta \in \Psi} \frac{\int_{k \in J_\zeta} f_\zeta(k)/k}{\zeta}$$

where $f_\zeta(k) = \mu_{\bar{\mathcal{A}}}(\zeta, k)$.

The secondary membership function, which is also known as secondary grades, is a function from $\Psi \times I$ to I. In the given expression of $\widetilde{\mathcal{A}}$, $\mu_{\widetilde{\mathcal{A}}}(\zeta, k)$ and $f_\zeta(k)$ are all secondary grades. The primary membership of an element in Ψ is the domain of a secondary membership function. In $\widetilde{\mathcal{A}}$, J_ζ is the primary membership of ζ. If all the secondary degrees of a type-2 fuzzy set $\widetilde{\mathcal{A}}$ are touched unity, i.e., $\mu_{\widetilde{\mathcal{A}}}(\zeta, k) = 1$, for all $\zeta \in \Psi$ and for all $k \in J_\zeta \subseteq I$, then $\widetilde{\mathcal{A}}$ is called an interval type-2 fuzzy set.

Basic Operations of Type-2 Fuzzy Sets [6, 7]

Let $\widetilde{\mathcal{A}}$ and $\widetilde{\mathcal{B}}$ be two type-2 fuzzy sets defined on universal set Ψ: i.e.,

$$\widetilde{\mathcal{A}} = \int\limits_{\zeta \in \Psi} \frac{\int_{k \in J_\zeta} f_\zeta(k)/k}{\zeta}$$

$$\widetilde{\mathcal{B}} = \int\limits_{\zeta \in \Psi} \frac{\int_{l \in J_\zeta} g_\zeta(l)/l}{\zeta}$$

Basic operations for the type-2 fuzzy sets are given by,

1. Union : $\widetilde{\mathcal{A}} \cup \widetilde{\mathcal{B}}$

$$\mu_{\widetilde{\mathcal{A}} \cup \widetilde{\mathcal{B}}}(\zeta) = \int\limits_{k \in J_\zeta(k)} \int\limits_{l \in J_\zeta(l)} \frac{\left[f_\zeta(k) \wedge g_\zeta(l)\right]}{(k \vee l)}, \quad \zeta \in \Psi.$$

2. Intersection: $\widetilde{\mathcal{A}} \cap \widetilde{\mathcal{B}}$

$$\mu_{\widetilde{\mathcal{A}}\cap\widetilde{\mathcal{B}}}(\zeta) = \int\limits_{k\in J_\zeta(k) l\in J_\zeta(l)} \frac{[f_\zeta k) \wedge g_\zeta(l)]}{(k \wedge l)}, \quad \zeta \in \Psi.$$

where $\wedge$ and $\vee$ denote the the minimum and the maximum operations, respectively.

3. Complement $(\widetilde{\mathcal{A}})^c$:

$$\mu_{(\widetilde{\mathcal{A}})^c}(\zeta) = \int\limits_{k\in J_\zeta} \frac{f_\zeta(k)}{(1-k)}.$$

Apart from fuzzy sets of type-2, fuzzy sets of order-2 are also available in literature and it is put forward by Zadeh [8]. Here, one is defining fuzzy set over a finite family of fuzzy sets. For example, a fuzzy set representing low outside temperature is possible to represent by a membership function that is specified over the range of real numbers [−59C, 11C] (say). Now, the expression conformable temperature can be considered as an order-2 fuzzy set. A collection of generic information granules that are fuzzy sets can be defined with the following members: low, medium, and high temperature. Each of these is a fuzzy entity created over the collection of real numbers. On this group, one can define an order-2 fuzzy set representing the term comfortable temperature with membership grades [0.3 1.0 0.5]. This means that low temperature has a membership value of 0.3 in the term comfortable temperature and so on. It is observed, however, that the intensity and vigor of the study in this area are limited compared to type-1 and order-1 fuzzy sets or type-2 fuzzy sets.

Interval-Valued Fuzzy Sets [5]

Another useful generalization of fuzzy sets is interval-valued fuzzy sets that instead of single membership grades, there are intervals of feasible membership values. Here, the concept of membership is represented in format of an interval.

8.3 Intuitionistic Fuzzy Sets

The word intuition means instinctive understanding of things or situation which is natural or spontaneous or self-generated. A conscious way of reasoning may not be needed here. As a part of new approaches and theories handling imprecision and uncertainty, Atanassov [9, 10] in 1983 introduced Intuitionistic fuzzy sets (IFS). Simultaneously, it was independently put forward by Takeuti and Titani [11] also. An IFS has two membership functions, one for conveying the degree of membership and the other for conveying degree of non-membership of members of the universe to the IFS. The rationale behind this new idea is the quite intuitionistic fact that an individual nevertheless communicates the degree of non-belongingness exactly as the negation/complement of belongingness. This corresponds to the psychological event that linguistic negation does not every time coincide with logical negation. This is also in-line with the concept of Intuitionism of L Brouwer in the beginning of last century. The association between IFS and other theories modeling vagueness is available in [12].

Definition 3 Let Ψ be a universe. An IFS, P is defined on Ψ is of the form

$$P = \{(\zeta, \mu_P(\zeta), \nu_P(\zeta) \mid \zeta \in \Psi)\}$$

where $\mu_P(\zeta) \in [0, 1]$ is called the degree of membership of ζ in P, $\nu_P(\zeta) \in [0, 1]$ is called the degree of non-membership of ζ in P, and where μ_P and ν_P satisfy:

$$(\forall \zeta \in \Psi) \quad (\mu_P(\zeta) + \nu_P(\zeta) \leq 1).$$

Remark A fuzzy set P on a universe Ψ can be considered as an IFS of the form

$$P = \{\langle \zeta : \mu_P(\zeta), 1 - \mu_P(\zeta) \rangle : \zeta \in \Psi\}$$

The theory of IFSs is well developed and is having various ranges of applications. For a detailed account, we suggest Attasanov [13]. For the remaining of this chapter, we will focus on more recent and advanced versions of generalized fuzzy and Intuitionistic fuzzy structures.

8.4 Pythagorean Fuzzy Sets

Membership grades play a vital contribution in the application of classical fuzzy sets. To improve the ability of fuzzy sets to capture and model user-driven information, many authors put forward nonstandard fuzzy sets such as intuitionistic and interval-valued fuzzy sets. A recent one among them, introduced by Yager [14–16] in 2013, is the one called Pythagorean fuzzy sets (PyFS). The associated membership grades are referred as Pythagorean membership grades. These membership values are given as follows: For any ζ in the universe Ψ, assign pair of values $r(\zeta)$ and $d(\zeta)$, where $r(\zeta) \in [0, 1]$ is called the strength of commitment at ζ and $d(\zeta) \in [0, 1]$ is called the direction of commitment. They actually represent the support for membership of ζ and the assist opposed to membership of ζ in a Pythagorean fuzzy subset of Ψ.

Particularly, the values $\mathcal{P}_Y(\zeta)$ and $\mathcal{P}_N(\zeta)$ are defined using $r(\zeta)$ and $d(\zeta)$ as follows: $\mathcal{P}_Y(\zeta) = r(\zeta)\cos(\theta(\zeta))$ and $\mathcal{P}_N(\zeta) = r(\zeta)\sin(\theta(\zeta))$, where $\theta(\zeta) = (1 - d(\zeta))\frac{\pi}{2}$. $\theta(\zeta)$ is given in radians and $\theta(\zeta)$ belong to $\left[0, \frac{\pi}{2}\right]$. It is observed that if $d(\zeta)$ is nearer to 1, then $\theta(\zeta)$ is closer to 0.

Further, it can be easily proven that $\mathcal{P}_Y(\zeta)$ and $\mathcal{P}_N(\zeta)$ are Pythagorean complements as regards strength of commitment $r(\zeta)$.

For,

$$\mathcal{P}_Y^2(\zeta) + \mathcal{P}_N^2(\zeta) = r^2(\zeta)\cos^2(\theta(\zeta)) + r^2(\zeta)\sin^2(\theta(\zeta))$$

From the well-known fact that $\cos^2(\theta) + \sin^2(\theta) = 1$ we have $\mathcal{P}_Y^2(\zeta) + \mathcal{P}_N^2(\zeta) = r^2(\zeta)$ and hence $\mathcal{P}_Y^2(\zeta) = r^2(\zeta) - \mathcal{P}_N^2(\zeta)$ and $\mathcal{P}_N^2(\zeta) = r^2(\zeta) - \mathcal{P}_Y^2(\zeta)$. Hence, $\mathcal{P}_Y$ and $\mathcal{P}_N$ are Pythagorean complements with respect to $r(\zeta)$.

It is an evident fact that Pythagorean membership grades enable for absence of commitment and vagueness in giving membership grades. Obviously, $r(\zeta) \in [0, 1]$ represents the strength of commitment about membership at point ζ, with larger the value of $r(\zeta)$ stronger the commitment, lesser the uncertainty.

To understand $d(\zeta)$, the direction of the strength, note that $\theta(\zeta) = (1 - d(\zeta))\frac{\pi}{2}$. When $d(\zeta) = 1$, then $\theta(\zeta) = 0$ and $\cos(\theta(\zeta)) = 1$ and $\sin(\theta(\zeta)) = 0$. Thus, $\mathcal{P}_Y(\zeta) = r(\zeta)$ and $\mathcal{P}_N(\zeta) = 0$. Again, if $d(\zeta) = 0$, then $\theta(\zeta) = \pi/2$, and we get $\mathcal{P}_Y(\zeta) = 0$ and $\mathcal{P}_N(\zeta) = 1$. Hence, the changes in $d(\zeta)$ between 0 and 1 represent how fully the strength $r(\zeta)$ is indicating to membership. If $d(\zeta) = 1$, the direction of $r(\zeta)$ is entirely to membership, while $d(\zeta) = 0$ the direction of the strength is totally toward non-membership. Intermediate or in-between value of $d(\zeta)$ is an indication of partial support to both membership and non-membership.

Definition 4 A Pythagorean fuzzy set $\mathcal{P}$ on a fixed universe Ψ is of the form:

$$\mathcal{P} = \{(\zeta, \lambda_{\mathcal{P}}(\zeta), \eta_{\mathcal{P}}(\zeta)) \mid \zeta \in \Psi\}$$

where $\lambda_{\mathcal{P}}(\zeta)$ and $\eta_{\mathcal{P}}(\zeta)$ are functions from Ψ to $[0, 1]$, with $0 \le \lambda_{\mathcal{P}}(\zeta) \le 1$, $0 \le \eta_{\mathcal{P}}(\zeta) \le 1$, and $0 \le \lambda_{\mathcal{P}}^2(\zeta) + \eta_{\mathcal{P}}^2(\zeta) \le 1$, for all $\zeta \in \Psi$, and they denote the degree of membership and degree of non-membership of element $\zeta \in \Psi$ to set $\mathcal{P}$, respectively.

$\pi_{\mathcal{P}}(\zeta) = \sqrt{1 - \lambda_{\mathcal{P}}^2(\zeta) - \eta_{\mathcal{P}}^2(\zeta)}$ is the Pythagorean fuzzy index of element $\zeta \in \Psi$ to set $\mathcal{P}$, representing the degree of indeterminacy of ζ to $\mathcal{P}$ and $0 \le \pi_{\mathcal{P}}(\zeta) \le 1$, for every $\zeta \in \Psi$.

We use PyFS(Ψ) to refer the set of all the Pythagorean fuzzy sets on a universe Ψ.

From the facts that if $p, q \in [0, 1]$ then $p^2 \le p$ and $q^2 \le q$ and if $p + q \le 1$ then $p^2 + q^2 \le 1$, we have,

Theorem 1 *PyFS(Ψ) has a larger membership space than IFS(Ψ).*

Obviously, Pythagorean type membership grades accept a larger choice of membership values than that of intuitionistic ones and hence expand the scope of applications significantly.

Example 1 Consider $\mathcal{P} \in$ PyFS(Ψ) with $\lambda_{\mathcal{P}}(\zeta) = 0.7$ and $\eta_{\mathcal{P}}(\zeta) = 0.5$ for $\Psi = \{\zeta\}$. Clearly, $0.8 + 0.5 \nleq 1$, but $0.8^2 + 0.5^2 \le 1$. Thus, $\pi_{\mathcal{P}}(\zeta) = 0.3317$, also, $(\lambda_{\mathcal{P}}(\zeta))^2 + (\eta_{\mathcal{P}}(\zeta))^2 + (\pi_{\mathcal{P}}(\zeta))^2 = 1$.

Operations on Pythagorean Fuzzy Sets

Definition 5 (i) Complement of $\mathcal{P} \in \text{PyFS}(\Psi)$, $\mathcal{P}^c$ is defined as: $\mathcal{P}^c = \{\langle \zeta, \eta_{\mathcal{P}}(\zeta), \lambda_{\mathcal{P}}(\zeta)\rangle \mid \zeta \in \Psi\}$. Clearly $(\mathcal{P}^c)^c = \mathcal{P}$.

(ii) Union and intersection of $\mathcal{P}, \mathcal{Q} \in \text{PyFS}(\Psi)$, are defined, respectively, as:
$\mathcal{P} \cup \mathcal{Q} = \{\langle \zeta, \max(\lambda_{\mathcal{P}}(\zeta), \lambda_{\mathcal{Q}}(\zeta)), \min(\eta_{\mathcal{P}}(\zeta), \eta_{\mathcal{Q}}(\zeta))\rangle \mid \zeta \in \Psi\}$.
$\mathcal{P} \cap \mathcal{Q} = \{\langle \zeta, \min(\lambda_{\mathcal{P}}(\zeta), \lambda_{\mathcal{Q}}(\zeta)), \max(\eta_{\mathcal{P}}(\zeta), \eta_{\mathcal{Q}}(\zeta))\rangle \mid \zeta \in \Psi\}$.

The following table distinguishes IFS and PyFS.

IF sets	PyF sets
$\mu + \nu \leq 1$	$\lambda + \eta \leq 1$ or $\lambda + \eta \geq 1$
$0 \leq \mu + \nu \leq 1$	$0 \leq \lambda^2 + \eta^2 \leq 1$
$\pi = 1 - (\mu + \nu)$	$\pi = \sqrt{1 - [\lambda^2 + \eta^2]}$
$\mu + \nu + \pi = 1$	$\lambda^2 + \eta^2 + \pi^2 = 1$

8.5 Picture Fuzzy Sets

Even though IFS theory is implemented in many areas, one major drawback is the lacking of neutrality degree concept. If we need to model situations involving human opinions with more than two answer types such as yes, no, refusal, and abstain, the concept of neutrality is prominent. It is more specific in situations involving voting in democracy, where human voters may be divided into many groups. It may be noted that "abstain" means refusing both "agree" and "disagree" but still picks up the vote. Similarly, medical diagnosis is another part where the idea of neutrality plays a vital role. As a remedy to these concerns, Cuong and Kreinovich [17, 18] put forward the concept Picture fuzzy set (PFS) as a straight extension of fuzzy sets and IFS by introducing the notions of positive, negative, and neutral membership degree of an entity.

Definition 6 Let Ψ be a universal st. A PFS $\widetilde{\mathcal{P}}$ on Ψ is of the form

$$\widetilde{\mathcal{P}} = \left\{\left(\zeta, \mu_{\widetilde{\mathcal{P}}}(\zeta), \eta_{\widetilde{\mathcal{P}}}(\zeta), \nu_{\widetilde{\mathcal{P}}}(\zeta)\right) \mid \zeta \in \Psi\right\}$$

where $\mu_{\widetilde{\mathcal{P}}}(\zeta)$, $\eta_{\widetilde{\mathcal{P}}}(\zeta)$, and $\nu_{\widetilde{\mathcal{P}}}(\zeta) \in [0, 1]$ are called the degree of positive, neutral, and negative memberships, respectively, of ζ in $\widetilde{\mathcal{P}}$, with $\mu_{\widetilde{\mathcal{P}}}$, $\eta_{\widetilde{\mathcal{P}}}$, and $\nu_{\widetilde{\mathcal{P}}}$ satisfying:

$$(\forall \zeta \in \Psi) \quad \left(\mu_{\widetilde{\mathcal{P}}}(\zeta) + \eta_{\widetilde{\mathcal{P}}}(\zeta) + \nu_{\widetilde{\mathcal{P}}}(\zeta) \leq 1\right)$$

Also, $\left(1 - \left(\mu_{\widetilde{\mathcal{P}}}(\zeta) + \eta_{\widetilde{\mathcal{P}}}(\zeta) + \nu_{\widetilde{\mathcal{P}}}(\zeta)\right)\right.$ is known as the degree of refusal membership of ζ in $\widetilde{\mathcal{P}}$. Also, $\text{PFS}(\Psi)$ is used to represent the collection of all the PFSs on a universe Ψ.

Definition 7 For $\widetilde{\mathcal{P}}, \widetilde{\mathcal{Q}} \in \text{PFS}(\Psi)$, the definition of subset, union, intersection, and complement are given by,
$\widetilde{\mathcal{P}} \subseteq \widetilde{\mathcal{Q}} \Longleftrightarrow (\forall\zeta \in \Psi, \mu_{\widetilde{\mathcal{P}}}(\zeta) \leq \mu_{\widetilde{\mathcal{Q}}}(\zeta) \text{ and } \eta_{\widetilde{\mathcal{P}}}(\zeta) \leq \eta_{\widetilde{\mathcal{Q}}}(\zeta) \text{ and } \nu_{\widetilde{\mathcal{P}}}(\zeta) \geq \nu_{\widetilde{\mathcal{Q}}}(\zeta))$
$\widetilde{\mathcal{P}} = \widetilde{\mathcal{Q}} \Longleftrightarrow (\widetilde{\mathcal{P}} \subseteq \widetilde{\mathcal{Q}} \text{ and } \widetilde{\mathcal{Q}} \subseteq \widetilde{\mathcal{P}})$
$\widetilde{\mathcal{P}} \cup \widetilde{\mathcal{Q}} = \{(\zeta, \max(\mu_{\widetilde{\mathcal{P}}}(\zeta), \mu_{\widetilde{\mathcal{Q}}}(\zeta)), \min(\eta_{\widetilde{\mathcal{P}}}(\zeta), \eta_{\widetilde{\mathcal{Q}}}(\zeta)), \min(\nu_{\widetilde{\mathcal{P}}}(\zeta), \nu_{\widetilde{\mathcal{Q}}}(\zeta)) \mid \zeta \in \Psi\}$
$\widetilde{\mathcal{P}} \cap \widetilde{\mathcal{Q}} = \{(\zeta, \min(\mu_{\widetilde{\mathcal{P}}}(\zeta), \mu_{\widetilde{\mathcal{Q}}}(\zeta)), \min(\eta_{\widetilde{\mathcal{P}}}(\zeta), \eta_{\widetilde{\mathcal{Q}}}(\zeta)), \max(\nu_{\widetilde{\mathcal{P}}}(\zeta), \nu_{\mathcal{B}}(\zeta)) \mid \zeta \in \Psi\}$
$co(\widetilde{\mathcal{P}}) = \bar{\widetilde{\mathcal{P}}} = \{(\nu_{\widetilde{\mathcal{P}}}(\zeta), \eta_{\widetilde{\mathcal{P}}}(\zeta), \mu_{\widetilde{\mathcal{P}}}(\zeta)) \mid \zeta \in \Psi\}$

8.6 Spherical Fuzzy Sets

Even though Atanassov's construction of IFSs is of exceptional reputation, decision-makers are restricted when specifying values due to the condition on $P(\zeta)$ and $N(\zeta)$ (membership and non-membership values) that $0 \leq P(\zeta) + N(\zeta) \leq 1$. Sometimes, sum of their membership degrees is superior than 1. In such situation, to attain reasonable outcome, IFS fails. So, dealing with such situation, Yager in 2015 established the PyFSs by assigning membership degree say "$P(\zeta)$" along with non-membership degree say "$N(\zeta)$" with $0 \leq P^2(\zeta) + N^2(\zeta) \leq 1$. Further, one can extend the concept of PyFSs to Spherical fuzzy set by assigning neutral membership degree say "$I(\zeta)$" along with positive and negative membership degrees say "$P(\zeta)$" and "$N(\zeta)$" with condition that $0 \leq P^2(\zeta) + I^2(\zeta) + N^2(\zeta) \leq 1$.

Cuong's formation of PFSs is of exceptional fame but decision-makers are somehow limited when specifying values due to the condition on $P(\zeta), I(\zeta)$, and $N(\zeta)$. Sometimes, sum of their membership degrees is superior than 1. In such situation, to attain reasonable outcome, PFS fails. For dealing such situations, Ashraf et al. [19] defined a novel structure called Spherical fuzzy sets (SFSs) which will widen the feasible space of membership degrees $P(\zeta), I(\zeta)$, and $N(\zeta)$ in a better manner than that of PFS. In SFS, membership degrees are satisfying the condition $0 < P^2(\zeta) + I^2(\zeta) + N^2(\zeta) < 1$.

Definition 8 Let Ψ be a universe set. Then, the SFS is of the form

$$J = \{\langle \zeta, P_j(\zeta), I_j(\zeta), N_j(\zeta) \mid \zeta \in \Psi \rangle\}$$

where $P_j : \Psi \to [0, 1]$, $I_j : \Psi \to [0, 1]$, and $N_j : \Psi \to [0, 1]$ are indicated the positive, neutral, and negative membership degrees of each $\zeta \in \Psi$, respectively. In addition, P_j, I_j, and N_j satisfy $0 \leq P_j^2(\zeta) + I_j^2(\zeta) + N_j^2(\zeta) \leq 1 \; \forall\zeta \in \Psi$.

When a decision involves abstinence or refusal rather than yes or no, the usage of SFSs is more suitable in modelling. Further SFS is a straightforward generalization of fuzzy set, IFS and PFS.

Consider the example with $P_j(\zeta) = 0.8$, $I_j(\zeta) = 0.5$ and $N_j(\zeta) = 0.3$ which interrupts the condition that $0 \leq P_j(\zeta) + I_j(\zeta) + N_j(\zeta) \leq 1$ but the square of these values such as, $P_j^2(\zeta) = 0.64, I_j^2(\zeta) = 0.25$, and $N_j^2(\zeta) = 0.09$ is satisfied the condition $0 \leq P_i^2(\zeta) + I_i^2(\zeta) + N_i^2(\zeta) \leq 1$.

If $P_j(\zeta), I_j(\zeta)$, and $N_j(\zeta)$ represent the degrees of positive, neutral, and negative memberships of a SFS, respectively, such that $0 \leq P_j^2(\zeta) + I_j^2(\zeta) + N_j^2(\zeta) \leq 1$, which is nothing but the region covered by a part of unit sphere. By a part of sphere, we indent that we consider the values of $P_j(\zeta), I_j(\zeta)$, and $N_j(\zeta)$ in [0, 1]. The region mentioned as per the inequality

$$P_j(\zeta) \leq \left(1 - I_j^2(\zeta) + N_j^2(\zeta)\right)^{\frac{1}{2}}.$$

This is the justification for the name SFS. Also, if we take $I_j(\zeta) = 0$ in SPSs, then SPSs reduced to PyFSs. Hence, SFSs are direct extensions of PyFSs and also are extensions of PFSs.

Definition 9 Let $\Psi \neq \varphi$ be a universe set. Then, any two SFSs J_1, J_2 can be expressed as;

$$J_1 = \left\{\left\langle\zeta, P_{ji}(\zeta), I_{j_1}(\zeta), N_{ji}(\zeta) \mid \zeta \in \Psi\right\rangle\right\}$$
$$J_2 = \left\{\left\langle\zeta, P_{j_2}(\zeta), I_{j2}(\zeta), N_{j2}(\zeta) \mid \zeta \in \Psi\right\rangle\right\}$$

Union of SFSs J_1 and J_2 in universe set Ψ is defined as

$$J_1 \cup J_2 = \left\{\begin{array}{l} \max\left(P_{j_1}(\zeta), P_{j_2}(\zeta)\right) \\ \min\left(I_{j_1}(\zeta), I_{j_2}(\zeta)\right) \\ \min\left(N_{j_1}(\zeta), N_{j_2}(\zeta)\right) \end{array}\right\}$$

Intersection of SFSs J_1 and J_2 in universe set Ψ is defined as

$$J_1 \cap J_2 = \left\{\begin{array}{l} \min\left(P_{ji}(\zeta), P_{j_2}(\zeta)\right) \\ \min\left(I_{ji}(\zeta), I_{j2}(\zeta)\right), \\ \max\left(N_{j1}(\zeta), N_{j2}(\zeta)\right) \end{array}\right\}$$

The complement of any SFS J_1 in universe set Ψ is defined as

$$J_1^c = \left\{N_{j_1}(\zeta), I_{j_1}(\zeta), P_{j_1}(\zeta)\right\}.$$

8.7 Fermatean Fuzzy Set as a *q*-Rung Orthopair Fuzzy Set

From the discussions above, it is clear that for IFS the sum of membership values and non-membership values is bounded by one and the sum of squares of those is bounded by one for PyFS. As a generalization of these Yager [20] introduced the q -rung orthopair fuzzy sets in which the sum of the qth powers of the membership and non-membership values is bonded by one. Obviously, larger the value of q, more the value of feasible orthopairs which significantly increases the domain of applicability.

In particular, when $q = 3$, q-rung orthopair fuzzy sets are called Fermatean fuzzy sets (FFS).

Definition 10 [21] Let Ψ be a universe of discourse. A Fermatean fuzzy set $\mathcal{F}$ in Ψ is of the form

$$\mathcal{F} = \{\langle \zeta, \alpha_F(\zeta), \beta_F(\zeta) \rangle : \zeta \in \Psi\}$$

where $\alpha_F(\zeta) : \Psi \to [0, 1]$ and $\beta_F(\zeta) : \Psi \to [0, 1]$, with $0 \leq (\alpha_F(\zeta))^3 + (\beta_F(\zeta))^3 \leq 1$ for all $\zeta \in \Psi$. The numbers $\alpha_F(\zeta)$ and $\beta_F(\zeta)$ denote, respectively, the degree of membership and the degree of non-membership of ζ in the set Ψ.

For any FFS $\mathcal{F}$ and $\zeta \in \Psi$, $\pi_F(\zeta) = \sqrt[3]{1 - (\alpha_F(\zeta))^3 - (\beta_F(\zeta))^3}$ is regarded as the degree of indeterminacy of ζ to $\mathcal{F}$.

For simplicity, one may mention the symbol $\mathcal{F} = (\alpha_F, \beta_F)$ for the FFS $\mathcal{F} = \{\langle \zeta, \alpha_F(\zeta), \beta_F(\zeta) \rangle : \zeta \in \Psi\}$.

Definition 11 Let $\mathcal{F} = (\alpha_F, \beta_F)$, $\mathcal{F}_1 = \left(\alpha_{F_1}, \beta_{F_1}\right)$, and $\mathcal{F}_2 = \left(\alpha_{F_2}, \beta_{F_2}\right)$ be three FFSs, then the following operations can be defined:

(i) $\mathcal{F}_1 \cap \mathcal{F}_2 = \left(\min\left\{\alpha_{F_1}, \alpha_{F_2}\right\}, \max\left\{\beta_{F_1}, \beta_{F_2}\right\}\right)$
(ii) $\mathcal{F}_1 \cup \mathcal{F}_2 = \left(\max\left\{\alpha_{F_1}, \alpha_{F_2}\right\}, \min\left\{\beta_{F_1}, \beta_{F_2}\right\}\right)$
(iii) $\mathcal{F}^c = (\beta_F, \alpha_F)$.

8.8 Hesitant Fuzzy Sets

Notion of Hesitant fuzzy sets (HFSs) proposed by Torra [22, 23] is another extension of fuzzy sets. The motivation for this extension is the hesitancy arising in the determination of the membership value of an element. Hesitancy does not arise just because of an error margin or a possibility distribution. It arises because there are some possible values of which there is a hesitation about which one would be the right one. These situations mainly arise in decision-making problems where a group of decision-makers examine the judgment of a scenario. In a HFS, the membership function catches values from the power set of [0, 1]. This allows the use of all the values simultaneously which helps in dealing with the situation effectively. The main difference in HFS theory is conceptual when compared to other extensions of fuzzy sets. HFS theory is having considerable applications in various fields like multi-criteria decision-making, group decision-making, decision support systems, evaluation processes, and clustering algorithms. Zhu et al. [24] extended the HFSs to dual Hesitant fuzzy sets (DHFSs) which deal hesitancy in membership and non-membership functions. Thus, FS, HFS, and IFS are special cases of DHFSs.

Definition 12 A Hesitant fuzzy set (HFS) $\mathcal{H}$ defined on a universe Ψ is characterized by a function defined on Ψ which returns a subset of [0, 1], representing the possible membership values for every member in Ψ. Mathematically:

$$\mathcal{H} = \{\langle \zeta, \mathcal{H}(\zeta) \rangle \mid \zeta \in \Psi\}$$

$\mathcal{H}(\zeta)$ is known as a Hesitant fuzzy element (HFE) and HF(Ψ) denotes the set of all HFSs in Ψ.

Definition 13 Score of a Hesitant fuzzy element is defined as $s(\mathcal{H}(\zeta)) = \frac{1}{l(\mathcal{H}(\zeta))} \sum_{\gamma \in h} \gamma$, where $l(\mathcal{H}(\zeta))$ denotes the number of values occurring in $\mathcal{H}(\zeta)$.
One can define various operations on HFSs based on score.

Definition 14 Consider two HFSs $\mathcal{H}_1$ and $\mathcal{H}_2$ on Ψ.

(i) Subset: $\mathcal{H}_1$ is a Hesitant subset of $\mathcal{H}_2$ ($\mathcal{H}_1 \preceq \mathcal{H}_2$) iff $s\,(\mathcal{H}_1(\zeta)) \leq s\,(\mathcal{H}_2(\zeta))\ \forall \zeta \in \Psi$.
(ii) Equality: $\mathcal{H}_1$ is equal to $\mathcal{H}_2$ ($\mathcal{H}_1 \approx \mathcal{H}_2$) iff $s\,(\mathcal{H}_1(\zeta)) = s\,(\mathcal{H}_2(\zeta))\ \forall \zeta \in \Psi$.
(iii) Complement: Complement of $\mathcal{H}$ is defined as $\mathcal{H}^c : \Psi \to P[0, 1]$ where $\mathcal{H}^c(\zeta) = \bigcup_{\gamma \in \mathcal{H}(\zeta)} \{1 - \gamma\} \forall \zeta \in \Psi$.
(iv) Union: $\mathcal{H}_1 \cup \mathcal{H}_2$ is defined as

$$\begin{aligned}(\mathcal{H}_1 \cup \mathcal{H}_2)(\zeta) &= \left\{\gamma \in \left(\mathcal{H}_1(\zeta) \cup \mathcal{H}_2(\zeta)/\gamma \geq \max\left(\mathcal{H}_1^-, \mathcal{H}_2^-\right)\right\}\right. \\ &= \bigcup_{\gamma_1 \in \mathcal{H}_1, \gamma_2 \in \mathcal{H}_2} \max\{\gamma_1, \gamma_2\}\end{aligned}$$

(v) Intersection: $\mathcal{H}_1 \cap \mathcal{H}_2$ is defined as

$$\begin{aligned}(\mathcal{H}_1 \cap \mathcal{H}_2)(\zeta) &= \left\{\gamma \in \left(\mathcal{H}_1(\zeta) \cup \mathcal{H}_2(\zeta)/\gamma \leq \min\left(\mathcal{H}_1^+, \mathcal{H}_2^+\right)\right\}\right.. \\ &= \bigcup_{\gamma_1 \in \mathcal{H}_1, \gamma_2 \in \mathcal{H}_2} \min\{\gamma_1, \gamma_2\}\end{aligned}$$

(vi) Score-based intersection of $\mathcal{H}_1$ and $\mathcal{H}_2$ ($\mathcal{H}_1 \tilde{\wedge} \mathcal{H}_2$) is defined as

$$(\mathcal{H}_1 \tilde{\wedge} \mathcal{H}_2)(\zeta) = \begin{cases} \mathcal{H}_1(\zeta) & \text{if } \mathcal{H}_1(\zeta) \prec \mathcal{H}_2(\zeta) \\ \mathcal{H}_2(\zeta) & \text{if } \mathcal{H}_2(\zeta) \prec \mathcal{H}_1(\zeta) \\ \mathcal{H}_1(\zeta) \cup \mathcal{H}_2(\zeta) & \text{if } \mathcal{H}_1(\zeta) \approx \mathcal{H}_2(\zeta) \end{cases}$$

(vii) Score-based union of $\mathcal{H}_1$ and $\mathcal{H}_2$ ($\mathcal{H}_1 \tilde{\vee} \mathcal{H}_2$) is defined as

$$(\mathcal{H}_1 \tilde{\vee} \mathcal{H}_2)(\zeta) = \begin{cases} \mathcal{H}_1(\zeta) & \text{if } \mathcal{H}_1(\zeta) \succ \mathcal{H}_2(\zeta) \\ \mathcal{H}_2(\zeta) & \text{if } \mathcal{H}_2(\zeta) \succ \mathcal{H}_1(\zeta) \\ \mathcal{H}_1(\zeta) \cup \mathcal{H}_2(\zeta) & \text{if } \mathcal{H}_1(\zeta) \approx \mathcal{H}_2(\zeta) \end{cases}$$

Definition 15 If $\mathcal{H} \in$ HF(Ψ) and $\alpha \in [0, 1]$, then α-level cut set of HFS $\mathcal{H}$, denoted by $\mathcal{H}_\alpha$ is defined as $\mathcal{H}_\alpha = \{\zeta \in \Psi / s(\mathcal{H}(\zeta)) \geq \alpha\}$. $\mathcal{H}_{\alpha+} = \{\zeta \in \Psi / s(\mathcal{H}(\zeta)) > \alpha\}$ is termed strong α-level cut set of $\mathcal{H}$.

Definition 16 (i) A Hesitant fuzzy relation $\mathcal{R}$ from $\mathcal{U}$ to $\mathcal{V}$ is defined as a Hesitant fuzzy subset of $\mathcal{U} \times \mathcal{V}$

i.e.,

$$\mathcal{R} : \mathcal{U} \times \mathcal{V} \rightarrow P[0, 1]$$

(ii) The complement $\mathcal{R}^C$ of Hesitant fuzzy relation $\mathcal{R}$ from $\mathcal{U}$ to $\mathcal{V}$ is defined as $\mathcal{R}^C(\zeta, \rho) = [\mathcal{R}(\zeta, \rho)]^C = \bigcup_{\gamma \in \mathcal{R}(\zeta, \rho)} \{1 - \gamma\}$

(iii) Inverse of $\mathcal{R}$ is a Hesitant fuzzy relation from $\mathcal{V}$ to $\mathcal{U}$ defined as $\mathcal{R}^{-1}(\rho, \zeta) = \mathcal{R}(\zeta, \rho)$.

Definition 17 For a Hesitant fuzzy relation $\mathcal{R}$ defined on a universe $\mathcal{U}$, we may define

$\mathcal{R}_\alpha = \{(s, t) \in \mathcal{U} \times \mathcal{U} : s\,(h_{\mathcal{R}}(s, t)) \geq \alpha\}$
$\mathcal{R}_\alpha(s) = \{t \in \mathcal{U} : s\,(h_{\mathcal{R}}(s, t)) \geq \alpha\}, \forall \alpha \in [0, 1]$
$\mathcal{R}_{\alpha+} = \{(s, t) \in \mathcal{U} \times \mathcal{U} : s\,(h_{\mathcal{R}}(s, t)) > \alpha\}$
$\mathcal{R}_{\alpha+}(s) = \{t \in \mathcal{U} : s\,(h_{\mathcal{R}}(s, t)) > \alpha\}, \forall \alpha \in [0, 1]$.

8.9 Applications

Given an information system, one of the most significant tasks in decision-making problems as well as artificial intelligence applications is knowledge extraction from these systems. This section is an attempt to present and study information system (IS) in the Hesitant fuzzy environment. Here, we discuss an IS with a set of possible membership values. An illustrative study is given in which Hesitant fuzzy membership values, which are families of sets, are obtained from values of attributes. Further, the concepts of reduct and core together with indiscernibility matrix are provided in Hesitant fuzzy setup. The discussion given below on Hesitant fuzzy ISs is from Deepak and John [25].

Hesitant Fuzzy Information Systems

A quadruple HFIS $= \langle \Psi, A, \Gamma, \wp \rangle$ is regarded a Hesitant Fuzzy Information System (HFIS) where the set $\Psi \neq \emptyset$ is the universal set containing objects, $A \neq \emptyset$ is set of attributes, Γ is the group of attribute values $\Gamma = \bigcup_{a \in A} \Gamma_a$, Γ_a is the collection of all feasible values of $a \in A$, and $\wp : \Psi \times A \rightarrow \mathcal{P}(\mathcal{P}(\Gamma))$ is a function in such a way $\forall \zeta \in \Psi, a \in A, \wp(\zeta, a) \in \mathcal{P}(\mathcal{P}(\Gamma))$.

A Hesitant Fuzzy Decision System is a quadruple HFDS $= \langle \Psi, \mathcal{C} \cup \{d\}, \Gamma, \wp \rangle$, where $\Psi \neq \phi$ is the universal set of objects, $\mathcal{C} \neq \phi$ is a finite set of conditional attributes, d is a decision attribute with $\mathcal{C} \cap \{d\} = \phi$; $\Gamma = \Gamma_c \cup \Gamma_d$ where Γ_c is a set of conditional attribute values, Γ_d is the set of decision attribute values; and $\wp : \Psi \times \{\mathcal{C} \bigcup \{d\}\} \rightarrow \mathcal{P}(\mathcal{P}(\Gamma))$ such that $\wp : \Psi \times \mathcal{C} \rightarrow \mathcal{P}\left(\mathcal{P}\left(\Gamma_c\right)\right)$; $\wp : \Psi \times \{d\} \rightarrow \Gamma_d$ is a mapping which indicate for every conditional attribute-object pair a group of attribute

Table 8.1 A hesitant fuzzy information system

Ψ	a_1	a_2	a_3
ζ_1	$\{r, s\}$	$\{1, 2\}, \{3, 4\}$	$\{r, s\}, \{t, u\}$
ζ_2	$\{r, s, \zeta\}, \{v, w\}, \{t, u\}$	$\{5, 6\}, \{1, 2\}$	$\{t, u, z\}, \{v, w\}$
ζ_3	$\{t, u\}, \{r, v\}$	$\{1, 2\}$	$\{r, s, \zeta\}, \{t, u, z, j\}$
ζ_4	$\{r, s, \rho\}, \{t, u, j\}, \{v, \zeta\}$	$\{1, 2\}, \{3, 4\}$	$\{r, s, \rho\}, \{t, u, k\}$
ζ_5	$\{v, w, z\}, \{\zeta, \rho\}$	$\{5, 6\}$	$\{v, w, l\}$

value sets and for the decision attribute we have one and only one value from the attribute value domain.

Note For all $a \in A$ and $\zeta \in \Psi, \wp(\zeta, a) = a(\zeta) \in \mathcal{P}(\mathcal{P}(\Gamma))$, and $a(\zeta) = \left\{\Gamma^a_{\zeta_1}, \Gamma^a_{\zeta_2}, \ldots, \Gamma^a_{\zeta_n}\right\}$ for some $n \in \mathbb{N}$ where each $\Gamma^a_{\zeta_i} \in \mathcal{P}(\Gamma)$ and $\Gamma^a_{\zeta_i} \cap \Gamma^a_{\zeta_j} = \phi$ for any $i \neq j; i, j \leq n$.
i.e., for a given $\zeta \in \Psi$, elements of $a(\zeta)$ are mutually disjoint.

Definition 18 For every $a \in A$, a Hesitant fuzzy relation $\widetilde{R}_a$ on HFIS $S = \langle \Psi, A, \Gamma, \wp \rangle$, can be defined as

$$\widetilde{R}_a(\zeta, \rho) = \bigcup_{\substack{\Gamma^a_{\zeta_i} \in a(\zeta), \Gamma^a_{\rho j} \in a(\rho) \\ \Gamma^a_{\zeta_i} \cap \Gamma^a_{\rho_j} \neq \phi}} \frac{\left|\Gamma^a_{\zeta_i} \cap \Gamma^a_{\rho_j}\right|}{\left|\Gamma^a_{\zeta_i} \cup \Gamma^a_{\rho_j}\right|}$$

and

$$\widetilde{R}_a(\zeta, \rho) = \{0\} \text{ if } \Gamma^a_{\zeta_i} \cap \Gamma^a_{\rho j} = \phi \; \forall i, j$$

For an attribute subset $\widetilde{\mathcal{Q}}$ of A, a Hesitant fuzzy relation $\widetilde{R}_{\mathcal{B}}$ is defined as

$$\widetilde{R}_{\mathcal{B}}(\zeta, \rho) = \widehat{\bigwedge_{b \in \mathcal{B}} \widetilde{R}_b(\zeta, \rho)}$$

Example 2 An example of a HFIS is shown in Table 8.1.

Now,

$$\widetilde{R}_{\{a_1\}} = \begin{bmatrix} \widetilde{R}_{a_1} & \zeta_1 & \zeta_2 & \zeta_3 & \zeta_4 & \zeta_5 \\ \zeta_1 & \{1\} & & & & \\ \zeta_2 & \left\{\frac{2}{3}\right\} & \{1\} & & & \\ \zeta_3 & \left\{\frac{1}{3}\right\} & \left\{\frac{1}{4}, \frac{1}{3}, 1\right\} & \{1\} & & \\ \zeta_4 & \left\{\frac{2}{3}\right\} & \left\{\frac{1}{4}, \frac{1}{3}, \frac{1}{2}, \frac{2}{3}\right\} & \left\{\frac{1}{4}, \frac{1}{3}, \frac{2}{3}\right\} & \{1\} & \\ \zeta_5 & \{0\} & \left\{\frac{1}{4}, \frac{2}{3}\right\} & \left\{\frac{1}{4}\right\} & \left\{\frac{1}{4}, \frac{1}{3}\right\} & \{1\} \end{bmatrix}$$

$$\widetilde{R}_{\{a_2\}} = \begin{bmatrix} \widetilde{R}_{a_1} & \zeta_1 & \zeta_2 & \zeta_3 & \zeta_4 & \zeta_5 \\ \zeta_1 & \{1\} & & & & \\ \zeta_2 & \{1\} & \{1\} & & & \\ \zeta_3 & \{1\} & \{1\} & \{1\} & & \\ \zeta_4 & \{1\} & \{1\} & \{1\} & \{1\} & \\ \zeta_5 & \{0\} & \{1\} & \{0\} & \{0\} & \{1\} \end{bmatrix}$$

$$\widetilde{R}_{\{a_3\}} = \begin{bmatrix} \widetilde{R}_{a_1} & \zeta_1 & \zeta_2 & \zeta_3 & \zeta_4 & \zeta_5 \\ \zeta_1 & \{1\} & & & & \\ \zeta_2 & \left\{\frac{2}{3}\right\} & \{1\} & & & \\ \zeta_3 & \left\{\frac{1}{2}, \frac{2}{3}\right\} & \left\{\frac{3}{4}\right\} & \{1\} & & \\ \zeta_4 & \left\{\frac{2}{3}\right\} & \left\{\frac{1}{2}\right\} & \left\{\frac{1}{2}, \frac{2}{5}\right\} & \{1\} & \\ \zeta_5 & \{0\} & \left\{\frac{2}{3}\right\} & \{0\} & \{0\} & \{1\} \end{bmatrix}$$

There can be values of the attribute in the IS which may not have any effect on the classification of the system into equivalence classes. The reduct of an IS is defined as the minimal subset of attributes which produces the same classification as that of the initial set of attributes. Also, reduct is not unique and the intersection of all reducts will form the core.

Definition 19 For a HFIS $S = \langle \Psi, \mathcal{C}, \Gamma, \wp \rangle$, $\mathcal{C}_1 \subseteq \mathcal{C}$ is a reduct of $\mathcal{C}$ iff

1. $\forall \zeta, \rho \in \Psi \;\; \widetilde{R}_{\mathcal{C}}(\zeta, \rho) \approx \widetilde{R}_{\mathcal{C}_1}(\zeta, \rho)$.
2. For any $\mathcal{C}_1' \subset \mathcal{C}_1, \exists \zeta, \rho \in \Psi$ such that $s\left(\widetilde{R}_{\mathcal{C}}(\zeta, \rho)\right) \neq s\left(\widetilde{R}_{\mathcal{C}_1'}(\zeta, \rho)\right)$.

Note If $s\left(\widetilde{R}_{\mathcal{C}}(\zeta, \rho)\right) \neq s\left(\widetilde{R}_{\mathcal{C}_1^{\gamma}}(\zeta, \rho)\right)$, then it will be denoted by $\widetilde{R}_{\mathcal{C}}(\zeta, \rho) \not\approx \widetilde{R}_{\mathcal{C}_1}(\zeta, \rho)$.

Definition 20 The core of $\mathcal{C}$ denoted by Core $(\mathcal{C})$ is the intersection of all reducts, i.e., Core $(\mathcal{C}) = \bigcap \text{Red}(\mathcal{C})$.

Definition 21 Let $S = \langle \Psi, \mathcal{C}, \Gamma, \wp \rangle$ be a HFIS. For attributes $\mathcal{C}$, universe set Ψ, and a set of relations $\left\{\widetilde{R}_{\mathfrak{d}} \mid \mathfrak{d} \in \mathcal{C}\right\}$, the discernibility matrix $\mathfrak{D}$ is defined as

$$\mathfrak{D}_{n\times n} = \left(\mathfrak{D}_{ij}\right)_{n\times n} = \begin{bmatrix} \mathfrak{D}_{11} & \mathfrak{D}_{12} & \cdots & \mathfrak{D}_{1n} \\ \mathfrak{D}_{21} & \mathfrak{D}_{22} & \cdots & \mathfrak{D}_{2n} \\ \vdots & \vdots & \ddots & \vdots \\ \mathfrak{D}_{n1} & \mathfrak{D}_{n2} & \cdots & \mathfrak{D}_{nn} \end{bmatrix}$$

where $n = |\Psi|$ and $\mathfrak{D}_{ij}$ is a set of attributes where $\mathfrak{d} \in \mathfrak{D}_{ij}$ iff $\widetilde{R}_{\mathfrak{d}}\left(\zeta_i, \zeta_j\right) \approx \widetilde{R}_{\mathcal{C}}\left(\zeta_i, \zeta_j\right)$ $\left(\zeta_i, \zeta_j \in \Psi \text{ and } \mathfrak{d} \in A\right)$. It is also demonstrated by $\mathfrak{d} \in \mathfrak{D}_{ij}$ iff $\widetilde{R}_{\mathfrak{d}}\left(\zeta_i, \zeta_j\right) \approx \widetilde{\wedge} \widetilde{R}_{\mathfrak{d} \in \mathcal{C}} \widetilde{R}_{\mathfrak{d}}$ $\left(\zeta_i, \zeta_j\right)$. Thus, $\mathfrak{D}_{ij}$ is the collection of entire attributes with lowest possible degrees of indiscernibility among elements ζ_i and ζ_j.

Theorem 2 *[25] For the HFIS $S = \langle \Psi, \mathcal{C}, \Gamma, \wp \rangle, \mathcal{C}_1 \subseteq \mathcal{C}$ is a reduct of $\mathcal{C}$ iff $\mathcal{C}_1$ is the minimal set agreeable $\forall i, j \;\; \mathcal{C}_1 \cap \mathfrak{D}_{ij} \neq \phi$.*

Theorem 3 *[25] For a HFIS* $= \langle \Psi, \mathcal{C}, \Gamma, \wp, \rangle$, *set of all single element entries of the discernibility matrix is Core, i.e.,* $\eth \in \mathcal{C}$ *is in Core* $(\mathcal{C})$ *iff there exists an* $\mathfrak{D}_{ij}$ *such that* $\mathfrak{D}_{ij} = \{\eth\}$.

Definition 22 Let $S = \langle \Psi, \Gamma, \Gamma, \wp \rangle$ be a HFIS and $B \subseteq A$. A discernibility function d for S is a Boolean function of k Boolean variables $\beta_1^*, \beta_2^*, \ldots, \beta_k^*$ corresponding to the attributes $\beta_1, \beta_2, \cdots, \beta_k$
$(\beta_i \in B; i \in \{1, 2, \ldots k\})$ and is given by

$$d\left(B^*\right) = d\left(\beta_1^*, \beta_2^*, \ldots, \beta_k^*\right) = \bigwedge \left\{ \bigvee \mathfrak{D}_{ij}^* : \mathfrak{D}_{ij} \in \mathfrak{D}_{n \times n} \right\}$$

where $\bigvee \mathfrak{D}_{ij}^*$ is the disjunction of all variables β^* such that $\beta \in \mathfrak{D}_{ij}$ and $\wedge$ is the conjunction.

Theorem 4 *All constituents in the minimal disjunctive normal form of the function* $d\ (B^*)$ *are all reducts of* B.

Example 3 Given

$$\mathfrak{D} = \begin{bmatrix} & \zeta_1 & \zeta_2 & \zeta_3 & \zeta_4 & \zeta_5 \\ \zeta_1 & \{\beta_1, \beta_2, \beta_3\} & & & & \\ \zeta_2 & \{\beta_1, \beta_3\} & \{\beta_1, \beta_2, \beta_3\} & & & \\ \zeta_3 & \{\beta_1\} & \{\beta_1\} & \{\beta_1, \beta_2, \beta_3\} & & \\ \zeta_4 & \{\beta_1, \beta_3\} & \{\beta_1\} & \{\beta_1\} & \{\beta_1, \beta_2, \beta_3\} & \\ \zeta_5 & \{\beta_1, \beta_2, \beta_3\} & \{\beta_1\} & \{\beta_2, \beta_3\} & \{\beta_2, \beta_3\} & \{\beta_1, \beta_2, \beta_3\} \end{bmatrix}$$

The discernibility function is

$$\begin{aligned} d\left(\beta_1^*, \beta_2^*, \beta_3^*\right) &= \left(\beta_1^* \vee \beta_2^* \vee \beta_3^*\right) \wedge \left(\beta_1^* \vee \beta_3^*\right) \wedge \left(\beta_1^*\right) \wedge \left(\beta_2^* \vee \beta_3^*\right) \\ &= \left(\beta_1^* \wedge \beta_2^*\right) \vee \left(\beta_1^* \wedge \beta_3^*\right) \end{aligned}$$

Thus, the reducts are $\{\beta_1, \beta_2\}$ and $\{\beta_1, \beta_3\}$. Here, the core is $\{\beta_1\}$.

Decision-Making Problems

Fuzzy sets, its various extensions, and generalizations are very useful and have been applied successfully in decision-making, evaluation, and clustering problems especially in the context of soft computing. To provide assessments over various alternatives in multi-criteria, multi-expert decision-making problems, experts use these extensions enormously. Various decision support systems can also be modified effectively using these. Evaluation problems often deal with information which is uncertain or vague. These problems which are provoked mainly by hesitation can be effectively dealt with extensions of fuzzy sets, especially HFSs. To get a better understanding, a typical decision-making algorithm in the picture fuzzy environment is provided below from [18].

For simplicity, we consider the following single criterion decision-making problem. Without any difficulty, the same can be extended to multi-criteria problems with single or multiple experts also.

Consider a finite set of alternatives given as $\mathbf{S} = \{\alpha_1, \alpha_2, \ldots, \alpha_n\}$. Suppose that the alternatives have evaluations as per the relevant criterion provided in the form of a PFS as $E = \{e(\alpha_1), \ldots, e(\alpha_n)\}$, where for all i,
$e(\alpha_i) = (\mu(\alpha_i), \eta(\alpha_i), \nu(\alpha_i)), 0 \leq \mu(\alpha_i), \eta(\alpha_i), \quad \nu(\alpha_i) \leq 1, \mu(\alpha_i) + \eta(\alpha_i) + \nu(\alpha_i) \leq 1$.

A typical decision problem is to sort out the alternatives and get an optimal solution. For this purpose, the following algorithm based on score function can be used.

Algorithm
Step 1 Define three score functions on $\mathbf{S}$ as

$$s_1(\alpha_i) = \mu(\alpha_i), \forall_i,$$
$$s_2(\alpha_i) = \eta(\alpha_i), \forall_i,$$
$$s_3(\alpha_i) = \mu(\alpha_i) + \eta(\alpha_i) - \nu(\alpha_i), \forall_i$$

Step 2 Using score functions define three orders on $\mathbf{S}$ as

(i) $\alpha_i \geq_1 \alpha_k$ iff $s_1(\alpha_i) \geq s_1(s_k)$,
(ii) $\alpha_i \geq_2 \alpha_k$ iff $s_2(\alpha_i) \geq s_2(\alpha_k)$,
(iii) $\alpha_i \geq_3 \alpha_k$ iff $s_3(\alpha_i) \geq s_3(\alpha_k)$

Step 3 Using the orders defined in step 2, define an aggregation order on $\mathbf{S}$ for ranking and then choose the best solution.

8.10 Conclusions

There does exist a vast range of structures that model or describe various kinds of non-probabilistic problems involving incomplete as well as inaccurate information or data. Inspired by the pioneer work of Zadeh [1], there is tremendous amount of work on generalizations and extensions of fuzzy sets, and some of them are discussed in the chapter briefly. Nevertheless, the list is complete and ever expanding. Further, there are possible hybridizations of these structures with other theories of uncertainty dealing such as Soft sets [26], Rough sets [27], and Multisets [28]. Theoretical studies such as topological and algebraic structures on these extensions are also promising and worth studying.

References

1. Zadeh, L.A.: Fuzzy sets. Inf. Control **8**(3), 338–353 (1965)
2. Polkowski, L.: Rough Sets: Mathematical Foundations. Springer, Berlin (2002)

3. John, S.J.: Handbook of Research on Generalized and Hybrid Set Structures and Applications for Soft Computing. IGI Global Publications, USA (2016)
4. Black, M.: Vagueness. An exercise in logical analysis. Philos. Sci. **4**(4), 427–455 (1937)
5. Zadeh, L.A.: The concept of a linguistic variable and its application to approximate reasoning-1. Inf. Sci. **8**, 199–249 (1975)
6. Karnik, N.N., Mendel, J.M.: Operations on type 2 fuzzy sets. Fuzzy Sets Syst. **122**(2), 327–348 (2001)
7. Mendel, J.M., John, R.B.: Type-2 fuzzy sets made simple. IEEE Trans. Fuzzy Syst. **10**(2), 117–127 (2002)
8. Zadeh, L.A.: Towards a theory of fuzzy information granulation and its centrality in human reasoning and fuzzy logic. Fuzzy Sets Syst. **90**, 111–117 (1997)
9. Atanassov, K.T.: Intuitionistic fuzzy sets. Fuzzy Sets Syst. **20**, 87–96 (1986)
10. Atanassov, K.T.: Intuitionistic fuzzy sets. VII ITKRs Session, Sofia (deposed in Central Science-Technical Library of Bulgarian Academy of Science, 1697/84) (in Bulgarian) (1983)
11. Takeuti, G., Titani, S.: Intuitionistic fuzzy logic and intuitionistic fuzzy set theory. J. Symbolic Logic **49**(3), 851–866 (1984)
12. Deschrijver, G., Kerre, E.E.: On the position of intuitionistic fuzzy set theory in the framework of theories modelling imprecision. Inf. Sci. **177**(8), 1860–1866 (2007)
13. Atanassov, K.T.: Intuitionistic Fuzzy Sets Theory and Applications. Springer, Berlin (1999)
14. Yager, R.R.: Pythagorean membership grades in multicriteria decision making. Technical report MII-3301. Machine Intelligence Institute, Iona College, New Rochelle (2013)
15. Yager, R.R.: Pythagorean fuzzy subsets. In: Proceedings of the Joint IFSA World Congress NAFIPS Annual Meeting, pp. 57–61 (2013)
16. Yager, R.R.: Pythagorean membership grades in multicriteria decision making. IEEE Trans. Fuzzy Syst. **22**(4), 958–965 (2014)
17. Cuong, B.C., Kreinovich,V.: Picture fuzzy sets—a new concept for computational intelligence problems. In: Proceedings of the Third World Congress on Information and Communication Technologies, p. 809 (2013)
18. Cuong, B.C.: Picture fuzzy sets. J. Comput. Sci. Cybern. **30**, 409–420 (2014)
19. Ashraf, S., Abdullah, S., Mahmood, T., Ghani, F., Mahmood, T.: Spherical fuzzy sets and their applications in multi-attribute decision making problems. J. Intell. Fuzzy Syst. **36**, 2829–2844 (2019)
20. Yager, R.R.: Generalized orthopair fuzzy sets. IEEE Trans. Fuzzy Syst. **25**, 1222–1230 (2017)
21. Senapati, T., Yager, R.R.: Fermatean fuzzy sets. J. Ambient Intell. Human Comput. **11**, 663–674 (2020)
22. Torra, V.: Hesitant fuzzy sets. Int. J. Intell. Syst. **25**, 529–539 (2010)
23. Xu, Z.: Hesitant Fuzzy Sets Theory. Studies in Fuzziness and Soft Computing. Springer, Cham (2014)
24. Zhu, B., Xu, Z., Xia, M.: Dual hesitant fuzzy sets. J. Appl. Math. (Article ID 879629), 13 p (2012)
25. Deepak, D., John, S.J.: Information systems on hesitant fuzzy sets. Int. J. Rough Sets Data Anal. **3**(1) (2016)
26. John, S.J.: Soft Sets: Theory and Applications. Springer, Cham (2020)
27. Pawlak, Z.: Rough sets. Int. J. Comput. Inf. Sci. **11**, 341–356 (1982)
28. Yager, R.R.: On the theory of bags. Int. J. Gen. Syst. **13**(1), 23–37 (1986)

Chapter 9
A Linear Diophantine Fuzzy Soft Set-Based Decision-Making Approach Using Revised Max-Min Average Composition Method

G Punnam Chander and Sujit Das

9.1 Introduction

In real life, it is necessary to choose the better alternative in emergency decision-making under uncertain and imprecise environments for the best results or to avoid worsening the situation. Solving decision-making problems in uncertain or imprecise environments with the multi-attribute decision-making methods and fuzzy set theory demonstrates a significant role. The incertitude of emergency situations and the presence of different alternatives makes it challenging for decision-makers to provide concise decision-making evaluations. To solve the decision-making problems, it is important to create decision-making solutions in imprecise conditions that reduce environmental damage while also enhancing efficiency. Fuzzy set-based decision-making approaches in uncertain circumstances have shown their applicability in a wide variety of fields like environmental planning, medical sciences, military services, socio-economic and environmental development, etc. A robust decision-making approach is found to be suitable to handle the problems and conclude with the limited and imprecise information in intuitive environments. Employing fuzzy set theory to manage uncertainty unveiled a new research platform in decision-making studies [1]. In recent decades, handling imprecise and realistic multi-attribute decision-making (MADM) problems with fuzzy sets and their several extensions has become an interesting and useful research study. The intrinsic capabilities of MADM techniques to evaluate the better alternative among a considerable number of alternatives with a variety of attributes are well established.

G. Punnam Chander · S. Das (✉)
Department of Computer Science and Engineering, National Institute of Technology, Warangal, Warangal, India
e-mail: sujit.das@nitw.ac.in

G. Punnam Chander
e-mail: punnamchander@student.nitw.ac.in

T. Som et al. (eds.), *Fuzzy, Rough and Intuitionistic Fuzzy Set Approaches for Data Handling*, Forum for Interdisciplinary Mathematics,
https://doi.org/10.1007/978-981-19-8566-9_9

The use of various fuzzy sets in MADM has made some notable contributions, as shown below. In 1965, Zadeh [2] introduced fuzzy set (FS) theory, which has been used in a variety of decision-making methods. As a valuable addition to the FS, Atanassov [3] fostered an intuitionistic fuzzy set (IFS) to account for the importance of non-membership degrees to handle imprecision better than the FS with its non-membership, membership, and indeterminacy grades. Yager and Abbasov [4] further enhanced IFS with the Pythagorean fuzzy set (PFS), which broadens the range of membership and non-membership grades by the sum of the squares of membership and non-membership grades. In comparison with IFS, PFS manages imprecision better. With the incorporation of the qth reference parameter to the membership and non-membership grades, the structural space of q-rung orthopair fuzzy sets [5] (q-ROFS) is widened more comparatively, allowing the expert to grade without reluctance. Riaz and Hashmi [6] introduced a novel fuzzy set called the linear diophantine fuzzy set (LDFS), which outperforms IFS, PFS, and q-ROFS. By integrating reference parameters in the LDFS representation, the structural space of non-membership and membership grades is effectively expanded. In some circumstances, fuzzy sets, q-ROFS, PFS, and IFS are constrained to describe the non-membership and membership grades by experts and decision-makers, which has an impact on selecting the best decision or alternative. If the problem representation is bound to be limited in uncertain circumstances, the problem might be handled in an approximate manner and cannot be resolved properly with limited information. The significance of reference parameters is that it allows experts and decision-makers to freely choose membership and non-membership grades. These reference factors can also be utilized to categorize the problem in a physical sense. For example, the problem information is classified by reference parameters, which determine how much of the problem must still be addressed, and the non-membership and membership grades determine the factor included for a given problem. In comparison with the structural space in FS, IFS, PFS, and q-ROFS, this makes LDFS more efficient in capturing problem information and enhances its use of structural space. In terms of structural space of bounds in uncertain situations, Table 9.1 [6] compares the LDFS to several current fuzzy sets and highlights its limitations and characteristics. With its parameterization feature, we can see that LDFS has more space in problem representation without any limits. In few cases, IFS, PFS, and q-ROFS fail to manage the fuzzy values with their respective conditions to lie between 0 and 1, whereas LDFS with the reference parameters α and β with the condition $0 \leq \alpha(\mathfrak{U}_F(G)) + \beta(\mathfrak{N}_F(G)) \leq 1$ ranges between 0 and 1 in most of the cases. LDFSS contributes to the proposed decision-making approach along with max-min average composition method for effective decision-making.

Russian Scholar Molodtsov [7] introduced soft set theory as a completely comprehensive theory for modelling imprecision. By implementing this theory in different directions, Maji et al. [8, 9] dealt with a variety of uncertain and imprecise problems, which was extended by Pei and Miao [10], and Chen et al. [11]. Yang and Ji [12] used the matrix form of a fuzzy soft set to tackle a variety of decision-making problems. Using an interval-valued Pythagorean fuzzy set, Chander and Das [13] presented a similarity measure in the application of medical diagnostics. The fuzzy

Table 9.1 Analysis of LDFS in comparison with existing fuzzy techniques

Sets	Comments	Parameterization
FS	Unable to manage with non-membership grades	X
IFS	Unable to manage with the condition $\mathfrak{U}_F(G) + \mathfrak{N}_F(G) > 1$	X
PFS	Unable to manage with the condition $\mathfrak{U}_F(G) + \mathfrak{N}_F(G) > 1$	X
q-ROFS	Unable to manage with the smaller values of "q" with the condition, $(\mathfrak{U}_F(G))^q + (\mathfrak{N}_F(G))^q > 1$ and for $\mathfrak{U}_F(G) = 1$, $\mathfrak{N}_F(G) = 1$	X
LDFS	It manages to deal with the condition, $(\mathfrak{U}_F(G))^q + (\mathfrak{N}_F(G))^q \leq 1$ as well as all other conditions that do not hold for FS, IFS, PFS, and q-ROFS. LDFS works under the influence of reference parameters (α, β). $\mathfrak{U}_F$ and $\mathfrak{N}_F$ can be chosen at will from [0, 1]	✓

soft matrix theory and its applications were extended by Neog and Sut [14]. Intuitionistic fuzzy soft set matrix model was propounded by Chetia and Das [15, 16]. Various definitions for intuitionistic fuzzy soft matrices were proposed by Rajeshwari and Dhanalakshmi [17, 18]. Shanmugasundaram et al. [19] proposed a new decision-making methodology based on intuitionistic fuzzy soft matrix (IFSM) and a revised max-min average composition method, as well as demonstrated its usefulness in the selection of students depending on their abilities for the recruitment. This method uses the revised max-min average composition method based on intuitionistic fuzzy soft matrices. It implies some restrictions to the decision-maker by limiting the membership grades, which can affect the selection of the best alternative in the decision-making process. Applying Pythagorean fuzzy soft sets, multi-criteria group decision-making problems (MCGDM) based on the methodologies of TOPSIS and VIKOR were addressed by Naeem et al. [20]. M-polar fuzzy soft rough sets were described by Akram et al. [21], and its applicability in MADM issues was demonstrated. Das et al. [22] suggested an algorithmic method to predict unknown information in a fuzzy soft set that is incomplete. Riaz et al. [23] developed LDFS to include soft rough LDFS sets (LDFSRS) and demonstrated its use in sustainable material handling equipment. Chander and Das [24] presented a differential evolutionary optimization-based decision-making method based on an interval-valued Pythagorean fuzzy set and compared DE to the particle swarm optimization technique in decision-making. Ejegwa [25] presented a Pythagorean fuzzy relation in terms of max-min composition as a decision-making technique for determining the suitability of employment in career placement for candidates based on academic achievement. Krishnakumar et al. [26] proposed a new decision-making method based on q-ROFS with evidence-based Bayes approximation and validated it in a green supplier selection application. In soft set theory, the problem information is represented in approximate manner without any restrictions on the problem parameters. The priority attribute strategy can be used to prioritize the attributes for the alternatives which in turn affects the decision-making approach and rendering the soft set theory better feasible and rational in decision-making methods.

The motivation for this work comes from the fact that the structural space of bounds of LDFS is much larger than that of other fuzzy sets like FS, IFS, PFS, and q-ROFS. Experts in grading membership and non-membership values have no hesitation about providing the problem information. As a result, LDFS successfully accommodates uncertainty by incorporating the reference parameters to non-membership and membership grades, whereas other fuzzy sets do not have this kind of reference parameters concepts. LDFSS is a hybridization of LDFS and soft set which is significant for making decisions in uncertain situations with its ability to systemize membership and non-membership grades with respect to priority criteria. The integration of reference parameters to the LDFSS widens the problem space available to the decision-maker, allowing him/her to assess membership and non-membership grades without constraint. As a result, inclusion of reference parameters and priority criteria in decision-making problems through LDFSS improves its efficiency and decisiveness in uncertain environments. Other fuzzy sets, on the other hand, cannot deal with the priority to the attributes in decision-making in particular. The choice of priority attribute strategy for the attributes of the alternatives renders the soft set theory better feasible and realistic in decision-making methods. In the recent study, weighted aggregation operators, geometric weighted aggregation operators, etc., have been employing for obtaining the aggregated fuzzy values. However, the attribute with higher value may influence the decision-making. With the use of LDFSS, the max-min average composition method can be used as a feasible technique in MADM environments to obtain the average score matrix of the alternatives for the problem information, which can result in decisive and effective decision-making. The revised max-min average composition method, which was earlier proposed employing intuitionistic fuzzy soft matrices, has its own limits and restrictions in the problem representation's structural space of bounds. To improve the aggregation operation and structural scope for the effectiveness of decision-making, the revised max-min average composition technique was adopted and used with LDFSS in the proposed decision-making approach. This motivated us to propose LDFSS theory in uncertain decision-making, which can effectively handle uncertainty in dubious environments while ensuring reliable and effective decision-making.

The objective of this paper is to propose a decision-making approach based on LDFSS that uses the revised max-min average composition method to overcome the limitations and uncertainties associated with structural space of bounds for describing the problem using reference parameters.

- In this paper, we propose a multi-attribute decision-making approach for choosing the best alternative in uncertain situations based on LDFSS and revised max-min average composition method.
- Initially, decision-makers/experts present the problem information in the form of linear diophantine fuzzy soft matrices (LDFSMs). LDFSM is the matrix representation of LDFSS.
- Next the revised max-min average composition method is used to aggregate the LDFSMs into a combined LDFSM.

- Finally, we compute the score values of the alternatives in order to rank them. The object with the maximum score value is chosen as the best alternative.

Moreover, we have shown the applicability of the proposed approach using a real-time case study and evaluated the consistency of the proposed approach using three different LDFS-based score functions where all of the three score functions produced almost similar ranking order. To the best of our knowledge, many of the decision-making approaches have not justified the decision-making as consistent. However, LDFSS using max-min average composition method can ensure consistent decision-making by choosing the same alternative for three different score functions.

The rest of the paper is organized as follows. Section 9.2 describes few basic concepts related to fuzzy set theory and methodologies. A linear diophantine fuzzy soft set (LDFSS)-based MADM approach using max-min average composition method is presented in Sect. 9.3. Section 9.4 shows the real-time case study with a comparative analysis. Section 9.5 concludes the work.

9.2 Preliminaries

This section covers the preliminary information related to the proposed approach.

Definition 9.2.1 [2]: If F be a fuzzy set on a non-empty discourse space S = {$\mathcal{G}_1$, $\mathcal{G}_2$, …, $\mathcal{G}_n$}, then F is defined as

$$F = \{\mathcal{G}, \mathfrak{U}_F(\mathcal{G}) \parallel \mathcal{G} \in S\} \tag{9.1}$$

Here $\mathfrak{U}_F(\mathcal{G})$ represents the grade of membership of the entity $\mathcal{G}$ in the fuzzy set F, and it belongs to [0, 1].

Definition 9.2.2 [3]: If F an intuitionistic fuzzy set on a non-empty discourse space S = {$\mathcal{G}_1$, $\mathcal{G}_2$, …, $\mathcal{G}_n$}, then F is defined as

$$F = \{\mathcal{G}, \mathfrak{U}_F(\mathcal{G}), \mathfrak{N}_F(\mathcal{G}) \parallel \mathcal{G} \in S\} \tag{9.2}$$

$\mathfrak{U}_F(\mathcal{G})$ and $\mathfrak{N}_F(\mathcal{G})$ denote the grade of membership and non-membership of entity $\mathcal{G}$, respectively, where $\mathfrak{U}_F(\mathcal{G})$ and $\mathfrak{N}_F(\mathcal{G})$ range between [0 and 1], with a condition $0 \leq \mathfrak{U}_F(\mathcal{G}) + \mathfrak{N}_F(\mathcal{G}) \leq 1$ for all $\mathcal{G} \in S$. The grade of indeterminacy or hesitation π for the fuzzy set F is $\pi_F = 1 - \mathfrak{U}_F(\mathcal{G}) - \mathfrak{N}_F(\mathcal{G})$. Figure 9.1 depicts IFS graphically.

Definition 9.2.3 [4]: If F a Pythagorean fuzzy set on a non-empty discourse space S = {$\mathcal{G}_1$, $\mathcal{G}_2$, …, $\mathcal{G}_n$}, then F is distinguished as

$$F = \{\mathcal{G}, \mathfrak{U}_F(\mathcal{G}), \mathfrak{N}_F(\mathcal{G}) \parallel \mathcal{G} \in S\} \tag{9.3}$$

$\mathfrak{U}_F(\mathcal{G})$ and $\mathfrak{N}_F(\mathcal{G})$ denote the grade of membership and non-membership of entity $\mathcal{G}$, respectively, where $\mathfrak{U}_F(\mathcal{G})$ and $\mathfrak{N}_F(\mathcal{G})$ range between [0 and 1], with a condition 0

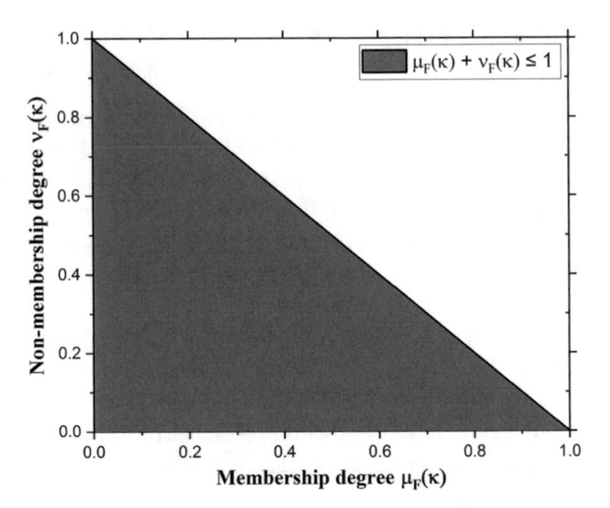

Fig. 9.1 Intuitionistic fuzzy set

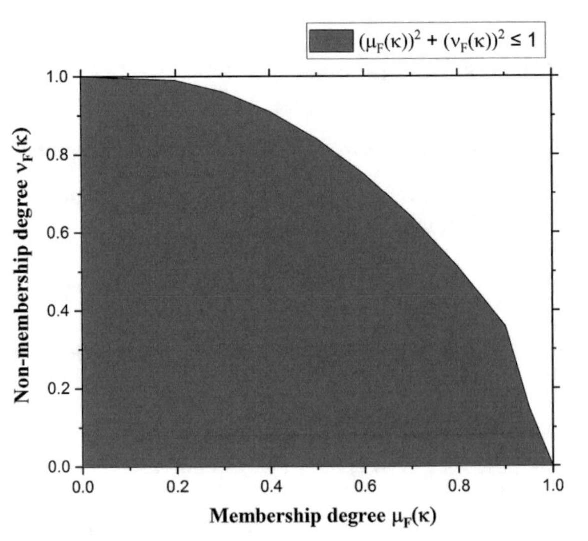

Fig. 9.2 Pythagorean fuzzy set

$\leq (\mathfrak{U}_F(\mathcal{G}))^2 + (\mathfrak{N}_F(\mathcal{G}))^2 \leq 1$ for all $\mathcal{G} \in S$. The grade of indeterminacy or hesitation π for the fuzzy set F is $\pi_F = \sqrt{1 - (\mathfrak{U}_F(\mathcal{G}))^2 - (\mathfrak{N}_F(\mathcal{G}))^2}$. Figure 9.2 depicts PFS graphically.

Definition 9.2.4 [5]: If F a q-rung orthopair fuzzy set on a non-empty discourse space $S = \{\mathcal{G}_1, \mathcal{G}_2, \ldots, \mathcal{G}_n\}$, then F is distinguished as

$$F = \{\mathcal{G}, \mathfrak{U}_F(\mathcal{G}), \mathfrak{N}_F(\mathcal{G}) \parallel \mathcal{G} \in S\} \tag{9.4}$$

$\mathfrak{U}_F(\mathcal{G})$ and $\mathfrak{N}_F(\mathcal{G})$ denote the grade of membership and non-membership of entity $\mathcal{G}$, respectively, where $\mathfrak{U}_F(\mathcal{G})$ and $\mathfrak{N}_F(\mathcal{G})$ range between [0 and 1], with a condition

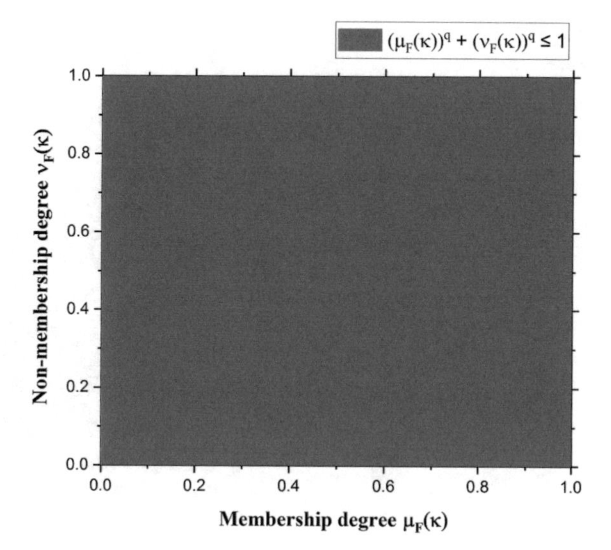

Fig. 9.3 q-rung orthopair fuzzy set

$0 \leq (\mathfrak{U}_F(\mathcal{G}))^q + (\mathfrak{N}_F(\mathcal{G}))^q \leq 1$; $q \geq 1$, for all $\mathcal{G} \in S$. The grade of indeterminacy or hesitation π for the fuzzy set F is $\pi_F = \sqrt[q]{1 - (\mathfrak{U}_F(\mathcal{G}))^q - (\mathfrak{N}_F(\mathcal{G}))^q}$. Figure 9.3 depicts q-ROFS graphically.

Definition 9.2.5 [6]: If F a linear diophantine fuzzy set on a non-empty discourse space $S = \{\mathcal{G}_1, \mathcal{G}_2, \ldots, \mathcal{G}_n\}$, then F is distinguished as

$$F = \{\mathcal{G}, (\mathfrak{U}_F(\mathcal{G}), \mathfrak{N}_F(\mathcal{G})), (\alpha, \beta) \parallel \mathcal{G} \in S\} \tag{9.5}$$

$\mathfrak{U}_F(\mathcal{G})$ and $\mathfrak{N}_F(\mathcal{G})$, respectively, denote the grade of membership and non-membership of entity $\mathcal{G}$ in F, and α and β are the reference parameters with a condition $\mathfrak{U}_F(\mathcal{G}) \in [0, 1]$, $\mathfrak{N}_F(\mathcal{G}) \in [0, 1]$, $\alpha, \beta \in [0, 1]$, $0 \leq \alpha(\mathfrak{U}_F(\mathcal{G})) + \beta(\mathfrak{N}_F(\mathcal{G})) \leq 1$, and $0 \leq \alpha + \beta \leq 1$ for all $\mathcal{G} \in S$. The grade of indeterminacy or hesitation π for the fuzzy set F is $\pi_F = 1 - \alpha(\mathfrak{U}_F(\mathcal{G})) + \beta(\mathfrak{N}_F(\mathcal{G}))$. Figure 9.4 depicts LDFS graphically.

Definition 9.2.6 [7]: Let U be a non-empty universe of discourse and E be the set of attributes, we consider $A \subseteq E$, then the pair (F, A) is called a soft set over U when F is a mapping of E into the set of all subsets of the set U, i.e. $F: E \to P(U)$, where $P(U)$ is the power set of U.

9.3 Proposed Approach

We propose an LDFSS-based decision-making approach based on a revised max-min average composition method in this section. The proposed method ensures effective decision-making by allowing the decision-maker to prioritize the attributes of the

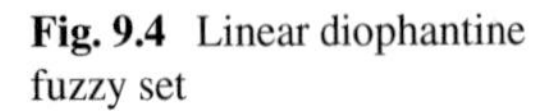

Fig. 9.4 Linear diophantine fuzzy set

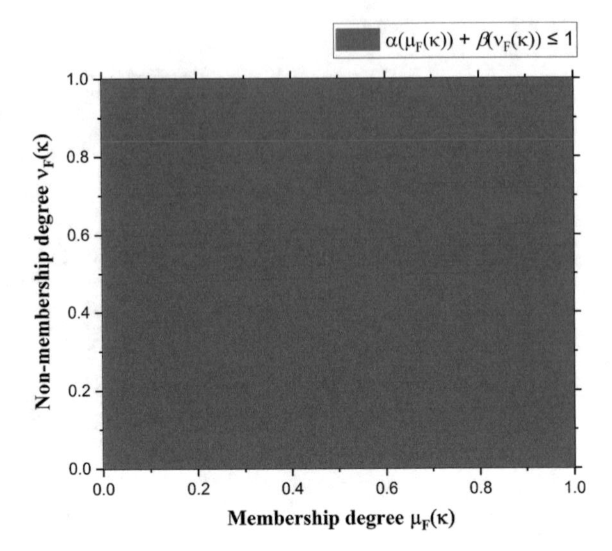

problem information based on the characteristics of the LDFSS and revised max-min composition.

Definition 9.3.1: Let U be a non-empty universe of discourse and E be the set of attributes. Let LDFS(U) denote the set of all LDFS of U and $A \subseteq E$, then the pair (F, A) is called a linear diophantine fuzzy soft set (LDFSS) over U if and only if F is a mapping of E into the set of all subsets of the set U, i.e. F: $E \rightarrow$ LDFS(U), where LDFS(U) be the power set of U. Alternatively, it can also be stated as follows:

$$(F, A) = (A, F(A)) : A \in E, F(A) \in \text{LDFS}(U)$$

Definition 9.3.2: Let (F, A) be an LDFSS in non-empty universe of discourse U, where $U = \mathcal{G}_1, \mathcal{G}_2, \ldots, \mathcal{G}_m$, and $E = C_1, C_2, \ldots, C_n$ be the set of attributes, then the LDFSS (F, A), $A \subseteq E$ is represented using linear diophantine fuzzy soft matrix (LDFSM) in the form $D_{m\times n} = (\mathcal{G}_i) \times (C_j)_{m\times n}$ as

$$(D_{ij})_{m\times n} = \begin{cases} (\mathcal{G}, (\mathfrak{U}_j(\mathcal{G}_i), \mathfrak{N}_j(\mathcal{G}_i)), (\alpha_j(\mathcal{G}_i), \beta_j(\mathcal{G}_i))) & A \in E \\ (0, 1) & A \notin E \end{cases} \tag{9.6}$$

Here $\mathfrak{U}(\mathcal{G}, C)$, $\mathfrak{N}(\mathcal{G}, C) \in [0, 1]$ are the membership and non-membership grades of the decision matrix $\mathcal{G} \times C$, and $\alpha(\mathcal{G}, C)$, $\beta(\mathcal{G}, C) \in [0, 1]$ are the reference parameters to the corresponding membership and non-membership grades, respectively, with the condition $0 \leq \alpha(\mathcal{G}, C)\mathfrak{U}(\mathcal{G}, C) + \beta(\mathcal{G}, C)\mathfrak{N}(\mathcal{G}, C) \leq 1$ and $0 \leq \alpha(\mathcal{G}, C) + \beta(\mathcal{G}, C) \leq 1$, where $i = 1, 2, \ldots, m$ and $j = 1, 2, \ldots, n$.

Example 1: Let $U = \{\mathcal{G}_1, \mathcal{G}_2, \mathcal{G}_3, \mathcal{G}_4\}$ and $E = \{C_1, C_2, C_3\}$ respectively be the set of alternatives and attributes. Say, LDFSS (F, E) is considered as $(F, A) =$

$\{\{C_1, (\mathcal{G}_1, \mathcal{G}_2, \mathcal{G}_3)\}, \{C_2, (\mathcal{G}_2, \mathcal{G}_4)\}, \{C_3, (\mathcal{G}_1, \mathcal{G}_2, \mathcal{G}_3, \mathcal{G}_4)\}\}$, $A \in E$ which is shown below.

$$(F, A) = F(C_1) = \{(\mathcal{G}_1, \langle 0.86, 0.34\rangle, \langle 0.75, 0.24\rangle), (\mathcal{G}_2, \langle 0.75, 0.34\rangle, \langle 0.60, 0.24\rangle), (\mathcal{G}_3, \langle 0.56, 0.44\rangle, \langle 0.48, 0.26\rangle, (\mathcal{G}_4, \langle 0, 0\rangle, \langle 0, 0\rangle)\},$$

$$F(C2) = \{(\mathcal{G}_1, \langle 0, 0\rangle, \langle 0, 0\rangle), (\mathcal{G}_2, \langle 0.46, 0.74\rangle, \langle 0.28, 0.60\rangle), (\mathcal{G}_3, \langle 0, 0\rangle, \langle 0, 0\rangle), (\mathcal{G}_4, \langle 0.99, 0.21\rangle, \langle 0.88, 0.08\rangle)\},$$

$$F(C3) = \{(\mathcal{G}_1, \langle 0.78, 0.35\rangle, \langle 0.65, 0.25\rangle), (\mathcal{G}_2, \langle 0.45, 0.41\rangle, \langle 0.32, 0.27\rangle), (\mathcal{G}_3, \langle 0.78, 0.59\rangle, \langle 0.61, 0.49\rangle), (\mathcal{G}_4, \langle 0.86, 0.35\rangle, \langle 0.75, 0.24\rangle)\}.$$

The corresponding LDFSM is represented as follows:

$$\begin{pmatrix} (\langle 0.86, 0.34\rangle, \langle 0.75, 0.24\rangle) & (\langle 0, 0\rangle, \langle 0, 0\rangle) & (\langle 0.78, 0.35\rangle, \langle 0.65, 0.25\rangle) \\ (\langle 0.75, 0.34\rangle, \langle 0.60, 0.24\rangle) & (\langle 0.46, 0.74\rangle, \langle 0.28, 0.60\rangle) & (\langle 0.45, 0.41\rangle, \langle 0.32, 0.27\rangle) \\ (\langle 0.56, 0.44\rangle, \langle 0.48, 0.26\rangle) & (\langle 0, 0\rangle, \langle 0, 0\rangle) & (\langle 0.78, 0.59\rangle, \langle 0.61, 0.49\rangle) \\ (\langle 0, 0\rangle, \langle 0, 0\rangle) & (\langle 0.99, 0.21\rangle, \langle 0.88, 0.08\rangle) & (\langle 0.86, 0.35\rangle, \langle 0.75, 0.24\rangle) \end{pmatrix}$$

Definition 9.3.3: Let $(B_1)_{m\times n} = [p_{ij}]$ and $(B_2)_{m\times n} = [q_{ij}]$ be two LDFSSs, then addition and subtraction of LDFSMs of B_1 and B_2 are defined as follows:

$$B_1 + B_2 = \{\max[\mathfrak{U}_{B1}(p_{ij}), \mathfrak{U}_{B2}(q_{ij})], \min[\mathfrak{N}_{B1}(p_{ij}), \mathfrak{N}_{B2}(q_{ij})], \max[\alpha_{B1}(p_{ij}), \alpha_{B2}(q_{ij})], \min[\beta_{B1}(p_{ij}), \beta_{B2}(q_{ij})]\}$$

$$B_1 - B_2 = \{\min[\mathfrak{U}_{B1}(p_{ij}), \mathfrak{U}_{B2}(q_{ij})], \max[\mathfrak{N}_{B1}(p_{ij}), \mathfrak{N}_{B2}(q_{ij})], \min[\alpha_{B1}(p_{ij}), \alpha_{B2}(q_{ij})], \max[\beta_{B1}(p_{ij}), \beta_{B2}(q_{ij})]\}$$

Definition 9.3.4: Let $(B_1)_{m\times n} = [p_{ij}] = (\langle \mathfrak{U}_{B_1}(p_{ij}), \mathfrak{N}_{B_1}(p_{ij})\rangle, \langle \alpha_{B1}(p_{ij}), \beta_{B1}(p_{ij})\rangle)$ be a LDFSS, then the complement of LDFSM $(B_1^c)_{m\times n}$ is defined as,

$$(B_1^c)_{m\times n} = [p_{ij}]_{m\times n} = (\langle \mathfrak{N}_{B_1}(p_{ij}), \mathfrak{U}_{B_1}(p_{ij})\rangle, \langle \beta_{B_1}(p_{ij}), \alpha_{B_1}(p_{ij})\rangle)$$

Revised Fuzzy Max-Min Average Composition Method

Definition 9.3.5: [19]: If B_1 and B_2 are linear diophantine fuzzy soft sets on a non-empty discourse space $U = \mathcal{G}_1, \mathcal{G}_2, \ldots, \mathcal{G}_m$, then revised fuzzy max-min average composition for linear diophantine fuzzy soft set is distinguished as

$$B_1 \phi B_2 = \left\{ \text{Max} \left\{ \frac{\mathfrak{U}_{B1}(\mathcal{G}) + \mathfrak{U}_{B2}(\mathcal{G})}{2} \right\}, \text{Min} \left\{ \frac{\mathfrak{N}_{B1}(\mathcal{G}) + \mathfrak{N}_{B2}(\mathcal{G})}{2} \right\}, \right.$$
$$\left. \text{Max} \left\{ \frac{\alpha_{B1}(\mathcal{G}) + \alpha_{B2}(\mathcal{G})}{2} \right\}, \text{Min} \left\{ \frac{\beta_{B1}(\mathcal{G}) + \beta_{B2}(\mathcal{G})}{2} \right\} \right\} \tag{9.7}$$

Example 2: Assume $B_1 = \begin{pmatrix} (\langle 0.86, 0.34\rangle, \langle 0.75, 0.24\rangle) \\ (\langle 0.75, 0.34\rangle, \langle 0.60, 0.24\rangle) \end{pmatrix}$, $B_2 = \begin{pmatrix} (\langle 0.78, 0.35\rangle, \langle 0.65, 0.25\rangle) \\ (\langle 0.45, 0.41\rangle, \langle 0.32, 0.27\rangle) \end{pmatrix}$ be two LDFSMs, then the following operations can be calculated as follows:

$$B_1 + B_2 = \begin{pmatrix} (\langle 0.86, 0.34\rangle, \langle 0.75, 0.24\rangle) \\ (\langle 0.75, 0.34\rangle, \langle 0.60, 0.24\rangle) \end{pmatrix}$$

$$B_1 - B_2 = \begin{pmatrix} (\langle 0.78, 0.35\rangle, \langle 0.65, 0.25\rangle) \\ (\langle 0.45, 0.41\rangle, \langle 0.32, 0.27\rangle) \end{pmatrix}$$

$$(B_1)^c = \begin{pmatrix} (\langle 0.34, 0.86\rangle, \langle 0.24, 0.75\rangle) \\ (\langle 0.34, 0.75\rangle, \langle 0.24, 0.60\rangle) \end{pmatrix}$$

$$B_1 \phi B_2 = \begin{pmatrix} (\langle 0.82, 0.345\rangle, \langle 0.7, 0.245\rangle) \\ (\langle 0.6, 0.375\rangle, \langle 0.46, 0.255\rangle) \end{pmatrix}$$

Procedure

Consider $U = (\mathcal{G}_1, \mathcal{G}_2, \ldots, \mathcal{G}_m), C = (C_1, C_2, \ldots, C_n), \text{DM}^l = (\text{DM}^1, \text{DM}^2, \ldots, \text{DM}^k)$ be the respective set of alternatives, attributes, and decision-makers, and an LDFSM for decision-maker $\text{DM}^l, l = 1, 2, \ldots, k$ is represented by $D^l_{m\times n} = (\mathfrak{U}_F(\mathcal{G}), \mathfrak{N}_F(\mathcal{G}), \alpha, \beta)$ as shown below.

$$\begin{pmatrix} (\mathfrak{U}_{11}(\mathcal{G}), \mathfrak{N}_{11}(\mathcal{G}), (\alpha_{11}, \beta_{11})) & (\mathfrak{U}_{12}(\mathcal{G}), \mathfrak{N}_{12}(\mathcal{G}), (\alpha_{12}, \beta_{12})) & \cdots & (\mathfrak{U}_{1n}(\mathcal{G}), \mathfrak{N}_{1n}(\mathcal{G}), (\alpha_{1n}, \beta_{1n})) \\ (\mathfrak{U}_{21}(\mathcal{G}), \mathfrak{N}_{21}(\mathcal{G}), (\alpha_{21}, \beta_{21})) & (\mathfrak{U}_{22}(\mathcal{G}), \mathfrak{N}_{22}(\mathcal{G}), (\alpha_{22}, \beta_{22})) & \cdots & (\mathfrak{U}_{2n}(\mathcal{G}), \mathfrak{N}_{2n}(\mathcal{G}), (\alpha_{2n}, \beta_{2n})) \\ \vdots & \vdots & \cdots & \vdots \\ (\mathfrak{U}_{m1}(\mathcal{G}), \mathfrak{N}_{m1}(\mathcal{G}), (\alpha_{m1}, \beta_{m1})) & (\mathfrak{U}_{m2}(\mathcal{G}), \mathfrak{N}_{m2}(\mathcal{G}), (\alpha_{m2}, \beta_{m2})) & \cdots & (\mathfrak{U}_{mn}(\mathcal{G}), \mathfrak{N}_{mn}(\mathcal{G}), (\alpha_{mn}, \beta_{mn})) \end{pmatrix}$$

Here $(\mathfrak{U}_F(\mathcal{G}), \mathfrak{N}_F(\mathcal{G}), (\alpha_F, \beta_F))$ is called LDFVs and $0 \le \alpha(\mathfrak{U}_F(\mathcal{G})) + \beta(\mathfrak{N}_F(\mathcal{G})) \le 1$ and $0 \le \alpha(\mathcal{G}) + \beta(\mathcal{G}) \le 1$, for all $\mathcal{G} \in U$.

The proposed approach is presented below in step-wise manner.

Step 1: The LDFSMs $(D^l_{ij})_{m\times n}$ are constructed based on the choice attributes of the experts or the decision-makers $\text{DM}^l, l = 1, 2, \ldots, k$.

$$(D^l_{ij})_{m\times n} = \begin{cases} (\langle \mathfrak{U}^l_j(\mathcal{G}_i), \mathfrak{N}^l_j(\mathcal{G}_i)\rangle, \langle \alpha^l_j(\mathcal{G}_i), \beta^l_j(\mathcal{G}_i)\rangle) & A \in E \\ (0, 1) & A \notin E \end{cases} \tag{9.8}$$

Here, E is the set of attributes, and A is the choice attributes of the LDFSSM.
Step 2: Then compute the aggregated LDFSM represented by R based on revised max-min average composition method using below Eq. (9.9).

$$B_1 \phi B_2 = \left\{ \text{Max} \left\{ \frac{\mathfrak{U}_{D^l}(\mathcal{G}) + \mathfrak{U}_{D^l}(\mathcal{G})}{2} \right\}, \text{Min} \left\{ \frac{\mathfrak{N}_{D^l}(\mathcal{G}) + \mathfrak{N}_{D^l}(\mathcal{G})}{2} \right\}, \right. \\ \left. \text{Max} \left\{ \frac{\alpha_{D^l}(\mathcal{G}) + \alpha_{D^l}(\mathcal{G})}{2} \right\}, \text{Min} \left\{ \frac{\beta_{D^l}(\mathcal{G}) + \beta_{D^l}(\mathcal{G})}{2} \right\} \right\} \quad (9.9)$$

Step 3: Compute the score values of the alternatives $\mathcal{G}_i$, $i = 1, 2, \ldots, m$, based on the aggregated LDFSM R, by using one of the score functions such as the expected score function (ESF), the quadratic score function (QSF), and the score function (SF), given, respectively, in Eqs. (9.10), (9.11), and (9.12), where ESF $(\mathcal{G}_i) \in [0, 1]$ and QSF and SF $(\mathcal{G}_i) \in [-1, 1]$.

$$\text{ESF}(\mathcal{G}_i) = \frac{1}{2}\left[\frac{(\mathfrak{U}(\mathcal{G}_i) - \mathfrak{N}(\mathcal{G}_i) + 1)}{2} + \frac{(\alpha(\mathcal{G}_i) - \beta(\mathcal{G}_i) + 1)}{2}\right] \quad (9.10)$$

$$\text{QSF}(\mathcal{G}_i) = \frac{1}{2}\left[((\mathfrak{U}(\mathcal{G}_i))^2 - (\mathfrak{N}(\mathcal{G}_i))^2) + ((\alpha(\mathcal{G}_i))^2 - (\beta(\mathcal{G}_i))^2)\right] \quad (9.11)$$

$$\text{SF}(\mathcal{G}_i) = \frac{1}{2}\left[(\mathfrak{U}(\mathcal{G}_i) - \mathfrak{N}(\mathcal{G}_i)) + (\alpha(\mathcal{G}_i) - \beta(\mathcal{G}_i))\right] \quad (9.12)$$

Step 4: The better alternative is chosen based on the larger score value.

Flow chart: Selection for the best alternative by using LDFSS and max-min average composition method (Fig. 9.5).

9.4 Case Study and Comparative Analysis

This section describes the proposed approach using a real-time case study on selection of vehicle for a better environment, performance, and maintenance followed by a comparative analysis. Vehicles play an important role in a country's growth by meeting the needs of people and goods. It serves a purpose in terms of economic development. As a result, it is a necessary component for long-term energy and economic development. Vehicles, on the other hand, have a considerable impact on the environment by emitting gases and liquids as a result of fuel consumption and hence contribute to air pollution. Therefore, it is important to select a vehicle that has a minimal environmental effect, high performance, and low maintenance requirements, as well as safety measures.

This case study demonstrates the selection of a vehicle from various kinds of vehicles such as battery electric vehicles (BEVs), plug-in hybrid electric vehicles

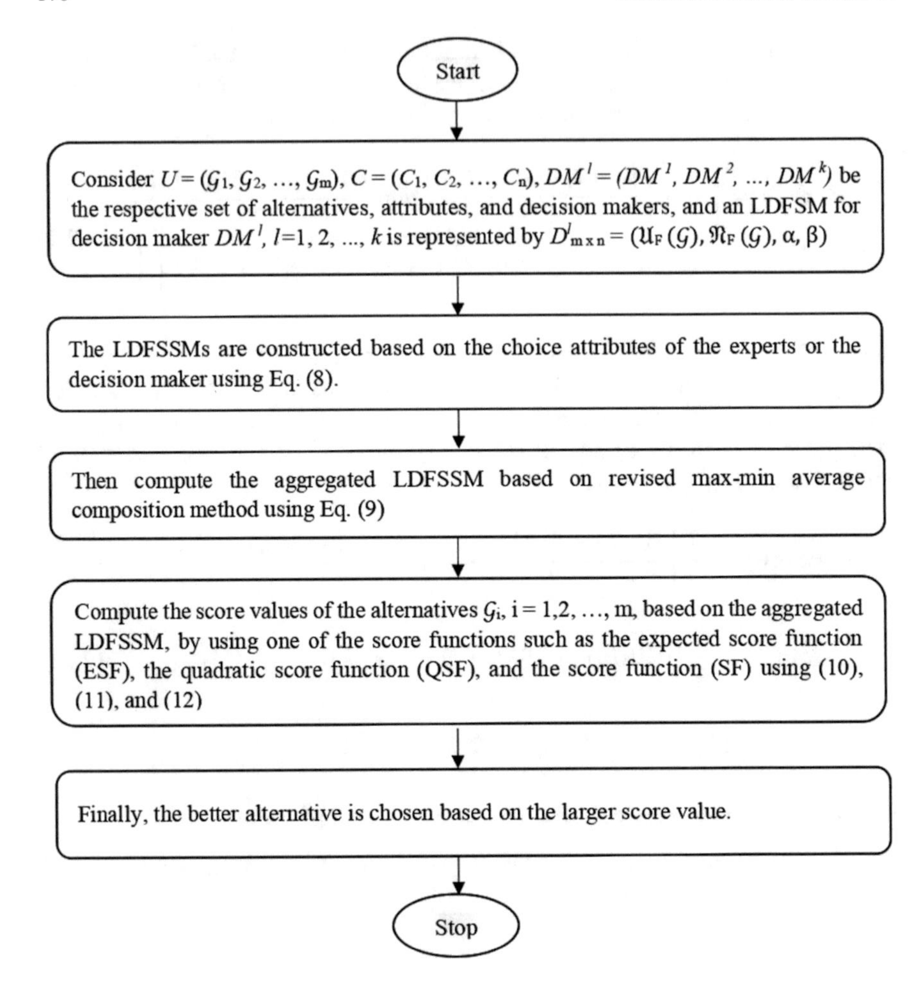

Fig. 9.5 Flowchart of the proposed approach

(PHEVs), hybrid electric vehicles (HEVs), petrol, diesel, bio-diesel, and compressed natural gas (CNG) vehicles, and their description is given in Table 9.2. The selection of the vehicle is based on the multiple attributes such as cost, performance and ride, mileage, features and technology, and safety and comfort as given in Table 9.3 along with the description.

This case study considers the opinions of three decision-makers regarding a set of seven vehicles (alternatives) $\mathcal{G}_i = \{\mathcal{G}_1, \mathcal{G}_2, \mathcal{G}_3, \mathcal{G}_4, \mathcal{G}_5, \mathcal{G}_6, \mathcal{G}_7\}$ = {CNG vehicles, PHEVs, HEVs, bio-diesel, diesel, petrol, BEVs} and five factors or criteria $C_n = \{C_1, C_2, C_3, C_4, C_5\}$ = {cost, performance, mileage, features and technology, safety and comfort}, where the decision-makers present their opinions in terms of LDFSMs. A brief description about the vehicles and factors is, respectively, given in Tables 9.2 and 9.3. Among these vehicles (alternatives), the better vehicle is selected based on

Table 9.2 Brief description about the vehicles

Symbol	Alternatives	Description	Types
$\mathcal{G}_1$	CNG	It has a gasoline internal combustion engine and is equipped with a fuel tank. It emits fewer carbon emissions into the atmosphere	Aircraft, helicopters, bus, trucks, vans, cars, bikes, bicycles, etc.
$\mathcal{G}_2$	PHEVs	A PHEV is powered with a charging cable plug-in from an electric power source. It is internally attached with internal combustion engine power generator	Bus, utility trucks, trains, military vans, cars, bikes, etc.
$\mathcal{G}_3$	HEVs	A HEV is powered by both rechargeable electric battery packs and internal combustion engine. It also has a fuel tank and internal combustion engine	Bus, light trucks, vans, high-performance cars, bikes, aircraft, etc.
$\mathcal{G}_4$	Bio-diesel	It is a renewable and domestically derived from plants, animal fats, and fatty acid esters. It has high oxygen content more than petrol and diesel, which reduces the air pollution	Bus, vans, cars, bikes, bicycles, etc.
$\mathcal{G}_5$	Petrol	It has a spark-ignition internal combustion engine that runs on gasoline and other fuels. The fuel tank holds gasoline until the engine needs it. It produces carbon dioxide and a little amount of toxic emissions	Bus, trucks, boats, vans, cars, bikes, etc.
$\mathcal{G}_6$	Diesel	It uses an internal combustion engine and works by compressing air and combustion gases. It emits less carbon dioxide emissions	Bus, trucks, boats, vans, cars, bikes, bicycles, etc.
$\mathcal{G}_7$	BEVs	A BEV is powered by rechargeable electric battery packs, which stores chemical energy with no fuel tank and internal combustion engine	Bus, trucks, boats, vans, cars, bikes, bicycles, etc.

five factors or criteria $C_n = \{C_1, C_2, C_3, C_4, C_5\}$ for precise and efficient decision-making by the group of three decision-makers, where $1 \leq i \leq 7$, $1 \leq n \leq 5$.

Step 1: Initially, the group of three decision-makers provide their opinions about the vehicles $\mathcal{G} = \{\mathcal{G}_1, \mathcal{G}_2, \mathcal{G}_3, \mathcal{G}_4, \mathcal{G}_5, \mathcal{G}_6, \mathcal{G}_7\}$ and their criteria/attribute $C = \{C_1, C_2, C_3, C_4, C_5\}$ based on their preferred attribute represented in the form of LDFSMs $(D^l)_{7\times 5}$

Table 9.3 Brief description about the criteria

Symbol	Criteria	Description
C_1	Cost	This aspect includes the vehicle's cost, which you want to be as minimal as possible, and it is determined by factors such as service, maintenance, features, and technology
C_2	Performance	It defines the vehicle's mobility against all forces and constraints as a function of its engine horsepower
C_3	Mileage	This feature calculates the vehicle's mileage and is an important factor to consider when choosing a vehicle
C_4	Features and technology	This attribute describes a vehicle's technical evolution. It also redefines the vehicle's driving experience
C_5	Safety and comfort	This attribute describes a vehicle's safety and comfort for the passengers while driving

$= (\mathfrak{U}(\mathcal{G}_i), \mathfrak{N}(\mathcal{G}_i), \alpha(\mathcal{G}_i), \beta(\mathcal{G}_i))$, where $l = 1, 2, 3$. Decision-maker DM^1 prefers the criteria set/factors $p = \{C_1, C_2, C_4, C_5\}$, $p \subseteq C$ to express his/her opinion about the vehicles, and the corresponding LDFSM is given below.

Similarly, decision-makers DM^2 and DM^3 prefer the criteria set/factors $p = \{C_1, C_2, C_3, C_4\}$ and $p = \{C_1, C_2, C_3, C_5\}$, $p \subseteq C$, respectively, to express their opinion about the vehicles, and the corresponding LDFSMs are given below.

Step 2: Revised max-min average composition method is applied on the three LDFSMs mentioned above in Step 1 using (9.9) in order to get the aggregated LDFSM R as shown below. $R = \{\mathcal{G}_1, (\langle 0.63, 0\rangle, \langle 0.41, 0.406\rangle), \mathcal{G}_2, (\langle 0.74, 0\rangle, \langle 0.41, 0\rangle), \mathcal{G}_3, (\langle 0.71, 0\rangle, \langle 0.51, 0\rangle), \mathcal{G}_4, (\langle 0.69, 0.136\rangle, \langle 0.61, 0.137\rangle), \mathcal{G}_5, (\langle 0.553, 0\rangle, \langle 0.28, 0\rangle), \mathcal{G}_6, (\langle 0.78, 0.273\rangle, \langle 0.41, 0.14\rangle), \mathcal{G}_7, (\langle 0.81, 0\rangle, \langle 0.49, 0\rangle)\}$.

Step 3: Finally, we calculate the score values of the vehicles $\mathcal{G}_i = \{\mathcal{G}_1, \mathcal{G}_2, \mathcal{G}_3, \mathcal{G}_4, \mathcal{G}_5, \mathcal{G}_6, \mathcal{G}_7\}$ = {CNG vehicles, PHEVs, HEVs, bio-diesel, diesel, petrol, BEVs} based on the aggregated LDFSM R obtained in Step 2, by using one of the score functions given in Eqs. (9.10), (9.11), and (9.12). The computed results for the score values are given in Table 9.5.

Step 4: Since the alternative $\mathcal{G}_7$, i.e. BEV vehicle has the highest score value with 0.825, 0.65, and 0.448 for ESF, SF, and QSF score functions, respectively, among all other vehicles, it is chosen as the best vehicle.

For the given case study, the application of the LDFSS and max-min average composition method provides the $\mathcal{G}_7$ vehicle as the better vehicle for the three different score functions as given in Table 9.4. We have also compared the proposed approach with the IFSM-based revised max-min approach [19], where the reference parameters are not considered. The comparative results are given in Table 9.5, and it shows that both the approaches yield $\mathcal{G}_7$ as the best option. It is also observed that all of the three score functions produce almost similar order of ranking, which determines the consistency of the proposed approach.

Table 9.4 Computed results

Score functions	$\mathcal{G}_1$	$\mathcal{G}_2$	$\mathcal{G}_3$	$\mathcal{G}_4$	$\mathcal{G}_5$	$\mathcal{G}_6$	$\mathcal{G}_7$
ESF	0.759	0.787	0.805	0.756	0.708	0.694	**0.825**
QSF	0.518	0.575	0.61	0.513	0.417	0.388	**0.65**
SF	0.281	0.357	0.382	0.405	0.192	0.341	**0.448**

Table 9.5 Comparative analysis

Methods	Order of ranking
IFSM method [19]	$\mathcal{G}_7 > \mathcal{G}_1 > \mathcal{G}_2 > \mathcal{G}_3 > \mathcal{G}_5 > \mathcal{G}_4 > \mathcal{G}_6$
Proposed method using ESF	$\mathcal{G}_7 > \mathcal{G}_3 > \mathcal{G}_2 > \mathcal{G}_1 > \mathcal{G}_4 > \mathcal{G}_5 > \mathcal{G}_6$
Proposed method using SF	$\mathcal{G}_7 > \mathcal{G}_3 > \mathcal{G}_2 > \mathcal{G}_1 > \mathcal{G}_4 > \mathcal{G}_5 > \mathcal{G}_6$
Proposed method using QSF	$\mathcal{G}_7 > \mathcal{G}_4 > \mathcal{G}_3 > \mathcal{G}_2 > \mathcal{G}_6 > \mathcal{G}_1 > \mathcal{G}_5$

LDFSM given by DM^1 of the form $(\mathfrak{U}(\mathcal{G}_i), \mathfrak{N}(\mathcal{G}_i), \alpha(\mathcal{G}_i), \beta(\mathcal{G}_i))$;

$$\mathrm{DM}^1 = \begin{bmatrix} (\langle 0.73, 0.41\rangle, \langle 0.31, 0.13\rangle) & (\langle 0.63, 0.53\rangle, \langle 0.13, 0.23\rangle) & (\langle 0, 0\rangle, \langle 0, 0\rangle) & (\langle 0.63, 0.53\rangle, \langle 0.31, 0.36\rangle) & (\langle 0.63, 0.53\rangle, \langle 0.31, 0.36\rangle) \\ (\langle 0.63, 0.43\rangle, \langle 0.41, 0.42\rangle) & (\langle 0.74, 0.32\rangle, \langle 0.63, 0.21\rangle) & (\langle 0, 0\rangle, \langle 0, 0\rangle) & (\langle 0, 0\rangle, \langle 0, 0\rangle) & (\langle 0.63, 0.59\rangle, \langle 0.61, 0.21\rangle) \\ (\langle 0.71, 0.34\rangle, \langle 0.51, 0.31\rangle) & (\langle 0, 0\rangle, \langle 0, 0\rangle) & (\langle 0, 0\rangle, \langle 0, 0\rangle) & (\langle 0.69, 0.38\rangle, \langle 0.41, 0.31\rangle) & (\langle 0, 0\rangle, \langle 0, 0\rangle) \\ (\langle 0.69, 0.59\rangle, \langle 0.61, 0.21\rangle) & (\langle 0.81, 0.51\rangle, \langle 0.31, 0.42\rangle) & (\langle 0, 0\rangle, \langle 0, 0\rangle) & (\langle 0.73, 0.49\rangle, \langle 0.28, 0.41\rangle) & (\langle 0.69, 0.73\rangle, \langle 0.31, 0.31\rangle) \\ (\langle 0, 0\rangle, \langle 0, 0\rangle) & (\langle 0.83, 0.41\rangle, \langle 0.42, 0.31\rangle) & (\langle 0, 0\rangle, \langle 0, 0\rangle) & (\langle 0.83, 0.49\rangle, \langle 0.18, 0.41\rangle) & (\langle 0.73, 0.41\rangle, \langle 0.31, 0.13\rangle) \\ (\langle 0.63, 0.59\rangle, \langle 0.41, 0.31\rangle) & (\langle 0.78, 0.43\rangle, \langle 0.38, 0.41\rangle) & (\langle 0, 0\rangle, \langle 0, 0\rangle) & (\langle 0.58, 0.49\rangle, \langle 0.31, 0.42\rangle) & (\langle 0.72, 0.41\rangle, \langle 0.51, 0.21\rangle) \\ (\langle 0.81, 0.58\rangle, \langle 0.49, 0.31\rangle) & (\langle 0.73, 0.68\rangle, \langle 0.43, 0.49\rangle) & (\langle 0, 0\rangle, \langle 0, 0\rangle) & (\langle 0, 0\rangle, \langle 0, 0\rangle) & (\langle 0, 0\rangle, \langle 0, 0\rangle) \end{bmatrix}$$

LDFSM given by DM^2 of the form $(\mathfrak{U}(\mathcal{G}_i), \mathfrak{N}(\mathcal{G}_i), \alpha(\mathcal{G}_i), \beta(\mathcal{G}_i))$;

$$\mathrm{DM}^2 = \begin{bmatrix} (\langle 0.73, 0.41\rangle, \langle 0.31, 0.13\rangle) & (\langle 0.63, 0.53\rangle, \langle 0.13, 0.23\rangle) & (\langle 0, 0\rangle, \langle 0, 0\rangle) & (\langle 0.63, 0.53\rangle, \langle 0.31, 0.36\rangle) & (\langle 0, 0\rangle, \langle 0, 0\rangle) \\ (\langle 0, 0\rangle, \langle 0, 0\rangle) & (\langle 0.74, 0.32\rangle, \langle 0.63, 0.21\rangle) & (\langle 0, 0\rangle, \langle 0, 0\rangle) & (\langle 0.71, 0.41\rangle, \langle 0.43, 0.28\rangle) & (\langle 0, 0\rangle, \langle 0, 0\rangle) \\ (\langle 0.71, 0.34\rangle, \langle 0.51, 0.31\rangle) & (\langle 0, 0\rangle, \langle 0, 0\rangle) & (\langle 0, 0\rangle, \langle 0, 0\rangle) & (\langle 0, 0\rangle, \langle 0, 0\rangle) & (\langle 0, 0\rangle, \langle 0, 0\rangle) \\ (\langle 0.69, 0.59\rangle, \langle 0.61, 0.21\rangle) & (\langle 0, 0\rangle, \langle 0, 0\rangle) & (\langle 0.83, 0.41\rangle, \langle 0.32, 0.41\rangle) & (\langle 0, 0\rangle, \langle 0, 0\rangle) & (\langle 0, 0\rangle, \langle 0, 0\rangle) \\ (\langle 0.72, 0.41\rangle, \langle 0.51, 0.21\rangle) & (\langle 0.83, 0.41\rangle, \langle 0.42, 0.31\rangle) & (\langle 0.73, 0.41\rangle, \langle 0.31, 0.42\rangle) & (\langle 0.83, 0.49\rangle, \langle 0.28, 0.41\rangle) & (\langle 0, 0\rangle, \langle 0, 0\rangle) \\ (\langle 0.63, 0.59\rangle, \langle 0.41, 0.31\rangle) & (\langle 0.78, 0.43\rangle, \langle 0.38, 0.41\rangle) & (\langle 0.63, 0.48\rangle, \langle 0.28, 0.17\rangle) & (\langle 0.58, 0.49\rangle, \langle 0.31, 0.42\rangle) & (\langle 0, 0\rangle, \langle 0, 0\rangle) \\ (\langle 0.81, 0.58\rangle, \langle 0.49, 0.31\rangle) & (\langle 0.73, 0.68\rangle, \langle 0.43, 0.49\rangle) & (\langle 0, 0\rangle, \langle 0, 0\rangle) & (\langle 0, 0\rangle, \langle 0, 0\rangle) & (\langle 0, 0\rangle, \langle 0, 0\rangle) \end{bmatrix}$$

LDFSM given by DM^3 of the form $(\mathfrak{U}(\mathcal{G}_i), \mathfrak{N}(\mathcal{G}_i), \alpha(\mathcal{G}_i), \beta(\mathcal{G}_i))$;

$$\mathrm{DM}^3 = \begin{bmatrix} (\langle 0, 0\rangle, \langle 0, 0\rangle) & (\langle 0.63, 0.53\rangle, \langle 0.13, 0.23\rangle) & (\langle 0, 0\rangle, \langle 0, 0\rangle) & (\langle 0, 0\rangle, \langle 0, 0\rangle) & (\langle 0.69, 0.59\rangle, \langle 0.61, 0.21\rangle) \\ (\langle 0.63, 0.43\rangle, \langle 0.41, 0.42\rangle) & (\langle 0.74, 0.32\rangle, \langle 0.63, 0.21\rangle) & (\langle 0, 0\rangle, \langle 0, 0\rangle) & (\langle 0, 0\rangle, \langle 0, 0\rangle) & (\langle 0.63, 0.59\rangle, \langle 0.61, 0.21\rangle) \\ (\langle 0.71, 0.34\rangle, \langle 0.51, 0.31\rangle) & (\langle 0, 0\rangle, \langle 0, 0\rangle) & (\langle 0.71, 0.41\rangle, \langle 0.31, 0.41\rangle) & (\langle 0, 0\rangle, \langle 0, 0\rangle) & (\langle 0, 0\rangle, \langle 0, 0\rangle) \\ (\langle 0.69, 0.59\rangle, \langle 0.61, 0.21\rangle) & (\langle 0.81, 0.51\rangle, \langle 0.31, 0.42\rangle) & (\langle 0, 0\rangle, \langle 0, 0\rangle) & (\langle 0, 0\rangle, \langle 0, 0\rangle) & (\langle 0.69, 0.73\rangle, \langle 0.31, 0.31\rangle) \\ (\langle 0.72, 0.41\rangle, \langle 0.51, 0.21\rangle) & (\langle 0.83, 0.41\rangle, \langle 0.42, 0.31\rangle) & (\langle 0.73, 0.41\rangle, \langle 0.31, 0.42\rangle) & (\langle 0, 0\rangle, \langle 0, 0\rangle) & (\langle 0.73, 0.41\rangle, \langle 0.31, 0.13\rangle) \\ (\langle 0.63, 0.59\rangle, \langle 0.41, 0.31\rangle) & (\langle 0, 0\rangle, \langle 0, 0\rangle) & (\langle 0.63, 0.48\rangle, \langle 0.28, 0.17\rangle) & (\langle 0, 0\rangle, \langle 0, 0\rangle) & (\langle 0.72, 0.41\rangle, \langle 0.51, 0.21\rangle) \\ (\langle 0.81, 0.58\rangle, \langle 0.49, 0.31\rangle) & (\langle 0.73, 0.68\rangle, \langle 0.43, 0.49\rangle) & (\langle 0, 0\rangle, \langle 0, 0\rangle) & (\langle 0, 0\rangle, \langle 0, 0\rangle) & (\langle 0, 0\rangle, \langle 0, 0\rangle) \end{bmatrix}$$

9.5 Conclusion

Applicability of the LDFSS-based proposed approach is enhanced by removing the limits and ambiguities for defining membership and non-membership grades given by experts or decision-makers and embraces more imprecision in decision-making. The proposed approach is based on the concepts of LDFSSs and max-min average composition method which ensure consistent decision-making. LDFSS is significant enough for making decisions in uncertain situations with its ability to systemize membership and non-membership grades along with the priority criteria in the decision-making process. The revised max-min average composition method aggregates the LDFSS matrices. Then, the score values of the alternatives are computed from the aggregated LDFSM, and alternative with larger score is selected. The applicability of the proposed approach has been illustrated with a case study related to vehicle selection. The consistency of our approach is justified by performing a comparative analysis using three different score functions. The comparative analysis mentioned in Table 9.5 shows the validity of the proposed approach. In future, this study can be enhanced in many ways by employing aggregation operators, evidential reasoning methodology, nonlinear programming, distance and similarity measures, and other relevant techniques.

References

1. De, A., Kundu, P., Das, S., Kar, S.: A ranking method based on interval type-2 fuzzy sets for multiple attribute group decision making. Soft Comput. **24**, 131–154 (2020)
2. Zadeh, L.A.: Fuzzy sets. Inf. Control **8**, 338–353 (1965)
3. Atanassov, K.T.: More on intuitionistic fuzzy sets. Fuzzy Sets Syst. **33**, 37–45 (1989)
4. Yager, R.R.: Pythagorean membership grades in multicriteria decision making. IEEE Trans. Fuzzy Syst. **22**, 958–965 (2013)
5. Yager, R.R.: Generalized orthopair fuzzy sets. IEEE Trans. Fuzzy Syst. **25**, 1222–1230 (2016)
6. Riaz, M., Hashmi, M.R.: Linear Diophantine fuzzy set and its applications towards multi-attribute decision making problems. J. Intell. Fuzzy Syst. **37**, 5417–5439 (2019)
7. Molodtsov, D.: Soft set theory-first results. Comput. Math. Appl. **37**(4–5), 19–31 (1999)
8. Maji, P.K., Biswas, R., Roy, A.R.: Fuzzy soft sets. J. Fuzzy Math. **9**(3), 589–602 (2001)
9. Maji, P.K., Roy, A.R., Biswas, R.: On intuitionistic fuzzy soft sets. J. Fuzzy Math. **12**(3), 669–683 (2004)
10. Pei, D., Miao, D.: From soft sets to information systems. In: Proceedings of the IEEE International Conference on Granular Computing, vol. 2, pp. 617–621 (2005)
11. Chen, D., Tsang, E.C.C., Yeung, D.S., Wang, X.: The parameterization reduction of soft sets and its applications. Comput. Math. Appl. **49**(5–6), 757–763 (2005)
12. Yang, Y., Ji, C.: Fuzzy soft matrices and their applications. Artif. Intell. Comput. Intell. **7002**, 618–627 (2011)
13. Chander, G.P., Das, S.: Decision making using interval-valued Pythagorean fuzzy set-based similarity measure. In: Intelligent Computing and Communication Systems, pp. 269–277. Springer, Singapore (2021)
14. Borah, M.J., Neog, T.J., Sut, D.K.: Fuzzy soft matrix theory and its decision making. Int. J. Mod. Eng. Res. **2**, 121–127 (2012)
15. Chetia, B., Das, P.K.: On fuzzy soft matrix theory. J. Assam Acad. Math. **2**, 71–83 (2010)

16. Chetia, B., Das, P.K.: Some results of intuitionistic fuzzy soft matrix theory. Adv. Appl. Sci. Res. Pelagia Res. Libr. **3**(1), 412–423 (2012)
17. Rajarajeswari, P., Dhanalakshmi, P.: Intuitionistic fuzzy soft matrix theory and its application in medical diagnosis. Ann. Fuzzy Math. Inform. **2**, 1–11 (2012)
18. Rajarajeswari, P., Dhanalakshmi, P.: Similarity measures of intuitionistic fuzzy soft sets and their application in medical diagnosis. Int. J. Math. Arch. **5**(5), 143–149 (2014)
19. Shanmugasundaram, P., Seshaiah, C.V., Rathi, K.: Revised max-min average composition method for decision making using intuitionistic fuzzy soft matrix theory. Adv. Fuzzy Syst. (2014)
20. Naeem, K., Riaz, M., Peng, X., Afzal, D.: Pythagorean fuzzy soft MCGDM methods based on TOPSIS, VIKOR and aggregation operators. J. Intell. Fuzzy Syst. **37**, 6937–6757 (2019)
21. Akram, M., Ali, G., Alshehri, N.O.: A new multi-attribute decision making method based on m-polar fuzzy soft rough sets. Symmetry **9**, 271 (2017)
22. Das, S., Ghosh, S., Kar, S., Pal, T.: An algorithmic approach for predicting unknown information in incomplete fuzzy soft set. Arab. J. Sci. Eng. **42**, 3563–3571 (2017). https://doi.org/10.1007/s13369-017-2591-2
23. Riaz, M., Hashmi, M.R., Kalsoom, H., Pamucar, D., Chu, Y.M.: Linear Diophantine fuzzy soft rough sets for the selection of sustainable material handling equipment. Symmetry **12**(8), 1215 (2020)
24. Chander, G.P., Das, S.: Multi-attribute decision making using interval-valued Pythagorean fuzzy set and differential evolutionary algorithm. In: 2021 IEEE International Conference on Fuzzy Systems (FUZZ-IEEE), Luxembourg, pp. 1–6 (2021)
25. Ejegwa, P.A.: Pythagorean fuzzy set and its application in career placements based on academic performance using max-min-max composition. Complex Intell. Syst. **5**(2), 165–175 (2019)
26. Krishankumar, R., Gowtham, Y., Ahmed, I., Ravichandran, K.S., Kar, S.: Solving green supplier selection problem using q-rung orthopair fuzzy-based decision framework with unknown weight information. Appl. Soft Comput. **94**, 106431 (2020)

Chapter 10
Recent Developments in Fuzzy Dynamic Data Envelopment Analysis and Its Applications

Rajinder Kaur and Jolly Puri

10.1 Introduction

Data envelopment analysis (DEA) is a performance evaluation tool. It is a nonparametric technique based on a linear programming approach to estimate the relative efficiencies of similar decision-making units (DMUs) in terms of multiple inputs–outputs. The DMUs can be educational institutions, banks, bank branches, hospitals, etc. Charnes et al. [1] initially proposed DEA in terms of constant returns to scale and later on extended by Banker et al. [2] to introduce variable returns to scale in DEA. The wide literature on DEA models can be seen in Cooper et al. [3], Tone [4], Tone and Tsutsui [5], Li et al. [6], Kao [7], Emrouznejad and Yang [8], and Contreras [9]. Despite all these extensions and the immense literature on DEA models, it has two key limitations: (i) It measures the performance statically in a particular period and ignores interrelationship present between consecutive periods, and (ii) it entails crisply defined input and output data. However, observed data values are imprecise or vague in real-life applications, e.g., data for customer satisfaction cannot be defined crisply. Fuzzy dynamic DEA (FDDEA) is found to be an emerging area that enables to evaluate a DMU's efficiency by considering interrelationship in the form of carryovers between consecutive periods. Gholizadeh et al. [10] incorporated fuzzy data in dynamic DEA for the first time to measure the efficiency of the investment corporations in the stock exchange. The present study presents a review of fuzzy dynamic DEA (FDDEA) during the last decade by classifying the studies into four categories

R. Kaur · J. Puri (✉)
School of Mathematics, Thapar Institute of Engineering and Technology,
Patiala, Punjab 147001, India
e-mail: jolly.puri@thapar.edu

R. Kaur
e-mail: rkaur_phd19@thapar.edu

T. Som et al. (eds.), *Fuzzy, Rough and Intuitionistic Fuzzy Set Approaches for Data Handling*, Forum for Interdisciplinary Mathematics,
https://doi.org/10.1007/978-981-19-8566-9_10

(i) theoretical development of FDDEA models with different fuzzy sets, (ii) FDDEA with network structure, (iii) applications of FDDEA approach, and (iv) integration of FDDEA with other operations research and/or artificial intelligence techniques that facilitates practical situations. As per the available literature, the present work seems to be the first review on FDDEA.

Section 10.2 presents an overview of dynamic DEA models followed by a review of FDDEA in Sect. 10.3. Section 10.4 classifies the FDDEA studies into four categories for a systematic review of the FDDEA. Section 10.5 concludes the present study.

10.2 Overview of Dynamic DEA

Sengupta [11–14] introduced the term dynamic efficiency to overcome the limitation of DEA for not incorporating time effect into the analysis and evaluated the performance of DMUs over different time periods connected through intermediates or links. Nemoto and Goto [15, 16] presented a dynamic approach in which inputs are categorized as variable inputs and quasi-fixed inputs, and later, this approach has been extended by many authors in literature. Many of the studies in literature allocated different weights to the intermediates according to their role of input or output in production system. Based on the idea of Kao [17] of assigning the same weights to the same factor, Kao [18] introduced a relational model to evaluate efficiency in dynamic environment when all the periods are linked through intermediates which are assigned the same weights no matter which period they belong to and are acting as an input or output in that period. Figure 10.1 presents a simple dynamic structure for kth DMU over q periods connected through intermediates or links. Consider n number of DMUs for evaluation over q periods, and each DMU consumes l number of inputs to produce s number of outputs. Let $X_{ij} = \sum_{t=1}^{q} X_{ij}^{(t)}$ and $Y_{gj} = \sum_{t=1}^{q} Y_{gj}^{(t)}$ be the ith ($i = 1, \ldots, l$) system input and gth ($g = 1, 2, \ldots, s$) system output for DMU$_j$ ($j = 1, \ldots, n$), respectively, where $X_{ij}^{(t)}$ and $Y_{gj}^{(t)}$, respectively, denote the lth input and gth output of DMU$_j$ in period t, and $Z_{dj}^{(t)}$ ($d = 1, 2, \ldots, h$) acts as an intermediate between the two successive periods t and $t + 1$ ($t = 1, \ldots, q - 1$). Let the initial and final links for DMU$_j$ be denoted by $Z_{dj}^{(0)}$ and $Z_{dj}^{(q)}$, respectively. Model-1 [18] presents an output-oriented model to evaluate dynamic efficiency of DMU$_k$ for the structure depicted in Fig. 10.1.

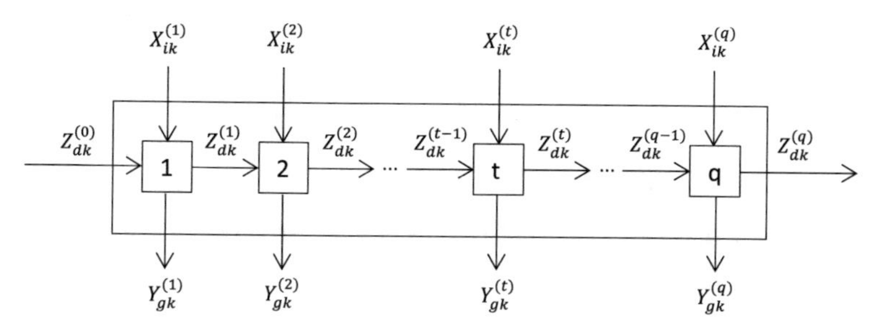

Fig. 10.1 Dynamic structure of DMU_k over q periods

Model-1

$$1/E_k^S = \min \sum_{i=1}^{l} v_i X_{ik} + \sum_{d=1}^{h} w_d Z_{dk}^{(0)}$$

$$\text{s.t.} \quad \sum_{g=1}^{s} u_g Y_{gk} + \sum_{d=1}^{h} w_d Z_{dk}^{(q)} = 1$$

$$\left(\sum_{i=1}^{l} v_i X_{ij} + \sum_{d=1}^{h} w_d Z_{dj}^{(0)}\right) - \left(\sum_{g=1}^{s} u_g Y_{gj} + \sum_{d=1}^{h} w_d Z_{dj}^{(q)}\right) \geq 0, \quad j = 1, \ldots, n,$$

$$\left(\sum_{i=1}^{l} v_i X_{ij}^{(t)} + \sum_{d=1}^{h} w_d Z_{dj}^{(t-1)}\right) - \left(\sum_{g=1}^{s} u_g Y_{gj}^{t} + \sum_{d=1}^{h} w_d Z_{dj}^{(t)}\right) \geq 0,$$

$$j = 1, \ldots, n; \quad t = 1, \ldots, q,$$

$$v_i \geq \epsilon; \quad u_g \geq \epsilon; \quad w_d \geq \epsilon,$$

where $\epsilon > 0$ is a non-Archimedean infinitesimal.

By using the optimal weights ($v_i^* \ \forall i$, $u_g^* \ \forall g$, $w_d^* \ \forall d$) derived from the Model-1, system efficiency (E_k^S) and period efficiencies ($E_k^{(t)}$) for DMU_k are defined as

$$E_k^S = \frac{\sum_{g=1}^{s} u_g^* Y_{gk} + \sum_{d=1}^{h} w_d^* Z_{dk}^{(q)}}{\sum_{i=1}^{l} v_i^* X_{ik} + \sum_{d=1}^{h} w_d^* Z_{dk}^{(0)}} \tag{10.1}$$

$$E_k^{(t)} = \frac{\sum_{g=1}^{s} u_g^* Y_{gj}^{t} + \sum_{d=1}^{h} w_d^* Z_{dj}^{(t)}}{\sum_{i=1}^{l} v_i^* X_{ij}^{(t)} + \sum_{d=1}^{h} w_d^* Z_{dj}^{(t-1)}} \quad \forall t = 1, \ldots, q. \tag{10.2}$$

Since the second set of constraints in Model-1 is redundant as it can be obtained by taking summation over the constraints corresponding to all periods, so by using this relation and Eqs. (10.1) and (10.2), a relationship has been established between system efficiencies and the period efficiencies which is defined as follows:

$$1 - E_k^S = \sum_{t=1}^{q} \left(1 - E_k^{(t)}\right) \alpha^{(t)}, \tag{10.3}$$

where $\alpha^{(t)} = \left(\sum_{i=1}^{l} v_i^* X_{ik}^{(t)} + \sum_{d=1}^{h} w_d^* Z_{dk}^{(t-1)}\right) / \left(\sum_{i=1}^{l} v_i^* X_{ik} + \sum_{d=1}^{h} w_d^* Z_{dk}^{(0)}\right)$, i.e., the complement of the system efficiency $(1 - E_k^S)$ can be written as linear combination of the period efficiencies $(1 - E_k^{(t)})$.

10.3 Fuzzy Dynamic DEA

While dealing with real-life problems, it is not always possible to collect precise or crisp data, as in the case of customer satisfaction. The uncertainty or imprecision may exist in the form of interval numbers, linguistic data, ordinal data, or fuzzy numbers. This section is devoted to an overview of fuzzy set theory and its use in dynamic DEA.

10.3.1 Fuzzy Set Theory

Definition 1 [19] A fuzzy set $\tilde{A}$ in a universe of discourse X is defined by $\tilde{A} = \{(x, \mu_{\tilde{A}}(x)) : x \in X\}$, where $\mu_{\tilde{A}} : X \to [0, 1]$ is the membership function of $\tilde{A}$ and $\mu_{\tilde{A}}(x)$ represents the degree of belongingness of x in $\tilde{A}$.

Definition 2 [20] The support of a fuzzy set $\tilde{A}$, denoted by $S(\tilde{A})$, is a crisp set defined by

$$S(\tilde{A}) = \{x | \mu_{\tilde{A}}(x) \geq 0\}.$$

Definition 3 [20] A fuzzy set $\tilde{A}$ in universe of discourse X is said to be convex if and only if

$$\mu_{\tilde{A}}(\lambda x + (1 - \lambda)y) \geq \min(\mu_{\tilde{A}}(x), \mu_{\tilde{A}}(y)), \text{ for all } x, y \in X \text{ and } 0 \leq \lambda \leq 1.$$

Definition 4 [20] Let $\tilde{A}$ be a fuzzy set in universe of discourse X. Then, it is said to be normal if $\mu_{\tilde{A}}(x) = 1$ for some $x \in X$.

Definition 5 [20] Let $\tilde{A}$ be a fuzzy set in universe of discourse X. Then, it is said to be a fuzzy number if it is both convex and normal.

Definition 6 [21] Let $\tilde{A}$ be a fuzzy set in universe of discourse X. Then, an α-cut of $\tilde{A}$, denoted by $\tilde{A}_\alpha$, is defined as $\tilde{A}_\alpha = \{x \in X | \mu_{\tilde{A}}(x) \geq \alpha\}$. It is a crisp set of all those elements of X having membership degree greater than or equal to α in $\tilde{A}$.

Definition 7 [22] An *l-r* fuzzy number, denoted by $\tilde{A} = (\underline{n}, \overline{n}, \rho, \phi)_{lr}$, is a fuzzy number with membership function $\mu_{\tilde{A}}$ given by

$$\mu_{\tilde{A}}(x) = \begin{cases} l\left(\dfrac{\underline{n} - x}{\rho}\right), & \underline{n} - \rho \leq x \leq \underline{n}, \\ 1, & \underline{n} \leq x \leq \overline{n}, \\ r\left(\dfrac{x - \overline{n}}{\phi}\right), & \overline{n} \leq x \leq \overline{n} + \phi, \\ 0, & \text{otherwise,} \end{cases}$$

where $l : [0, 1] \to [0, 1]$ and $r : [0, 1] \to [0, 1]$ are non-increasing continuous shape functions with $l(0) = r(0) = 1$ and $l(1) = r(1) = 0$, $[\underline{n}, \overline{n}]$ is the peak of $\tilde{A}$, and ρ, ϕ are positive scalars.

Definition 8 [19] A triangular fuzzy number $\tilde{A} = (a_1, a_2, a_3)$ is a fuzzy number with membership function $\mu_{\tilde{A}}$ defined as

$$\mu_{\tilde{A}}(x) = \begin{cases} \dfrac{x - a_1}{a_2 - a_1}, & a_1 < x \leq a_2, \\ \dfrac{x - a_3}{a_2 - a_3}, & a_2 \leq x < a_3, \\ 0, & \text{otherwise.} \end{cases}$$

Definition 9 [23] A fuzzy set $\tilde{\tilde{A}}$ is said to be a type-2 fuzzy set if membership function of its elements is of type-1 fuzzy set.

Definition 10 [24] Let $\tilde{\tilde{A}}$ be a type-2 fuzzy set in a universe of discourse X with membership function denoted by $\mu_{\tilde{\tilde{A}}}$, then $\tilde{\tilde{A}}$ is said to be an interval type-2 fuzzy set if $\mu_{\tilde{\tilde{A}}}(x, v) = 1$, for all $x \in X, v \in l_x \subseteq [0, 1]$.

Definition 11 [24] A trapezoidal interval type-2 fuzzy set $\tilde{\tilde{A}}$ is defined as

$$\begin{aligned} \tilde{\tilde{A}} &= (\tilde{A}^U, \tilde{A}^L) \\ &= ((d_1^U, d_2^U, d_3^U, d_4^U, h_1(\tilde{A}^U), h_2(\tilde{A}^U)), (d_1^L, d_2^L, d_3^L, d_4^L, h_1(\tilde{A}^L), h_2(\tilde{A}^L))), \end{aligned}$$

where $\tilde{A}^U$ and $\tilde{A}^L$ denote the trapezoidal upper and lower membership functions, respectively, and $h_l(\tilde{A}^U) \in [0, 1]$, $h_l(\tilde{A}^L) \in [0, 1]$ are the membership values of d_{l+1}^U and d_{l+1}^L $(l = 1, 2)$, respectively.

10.3.2 Fuzzy Set Theory and Dynamic DEA

Dynamic DEA is used to evaluate efficiency while keeping in mind the interdependence of periods represented by carryovers from one period to the subsequent period. However, the data for inputs and outputs as well as carryovers are not always in precise or crisp form like customer satisfaction, and environmental pollution [25], airport reputation, and social responsibility [26]. Zadeh [27] introduced fuzzy numbers to represent various imprecise data forms. Since then, several authors have incorporated the concept of fuzzy in DEA and other performance measuring techniques [19, 28]. Let the data for all inputs, outputs, and links be fuzzy numbers and $\tilde{X}_{ik}$, $\tilde{Y}_{gk}$, and $\tilde{Z}_{dk}$ denote the ith fuzzy input, gth fuzzy output, and dth fuzzy link for DMU$_k$, respectively, then dynamic fuzzy efficiency is evaluated by using the following model.

Model-2

$$1/\tilde{E}_k^S = \min \sum_{i=1}^{l} v_i \tilde{X}_{ik} + \sum_{d=1}^{h} w_d \tilde{Z}_{dk}^{(0)}$$

$$\text{s.t.} \quad \sum_{g=1}^{s} u_g \tilde{Y}_{gk} + \sum_{d=1}^{h} w_d \tilde{Z}_{dk}^{(q)} = 1,$$

$$\left(\sum_{i=1}^{l} v_i \tilde{X}_{ij} + \sum_{d=1}^{h} w_d \tilde{Z}_{dj}^{(0)} \right) - \left(\sum_{g=1}^{s} u_g \tilde{Y}_{gj} + \sum_{d=1}^{h} w_d \tilde{Z}_{dj}^{(q)} \right) \geq 0, \quad j = 1, \ldots, n,$$

$$\left(\sum_{i=1}^{l} v_i \tilde{X}_{ij}^{(t)} + \sum_{d=1}^{h} w_d \tilde{Z}_{dj}^{(t-1)} \right) - \left(\sum_{g=1}^{s} u_g \tilde{Y}_{gj}^{t} + \sum_{d=1}^{h} w_d \tilde{Z}_{dj}^{(t)} \right) \geq 0,$$

$$j = 1, \ldots, n; \quad t = 1, \ldots, q,$$

$$v_i \geq \epsilon; \quad u_g \geq \epsilon; \quad w_d \geq \epsilon,$$

where ϵ is a non-Archimedean infinitesimal and v_i, u_g, and w_d are the respective multipliers for ith fuzzy input, gth fuzzy output, and dth fuzzy link.

10.4 Classification of FDDEA Studies

The literature on Fuzzy dynamic DEA can be classified into four categories: (i) Theoretical development of FDDEA models with different fuzzy sets, (ii) FDDEA with network structure, (iii) applications of FDDEA approach, and (iv) integration of FDDEA with other techniques are discussed in detail in subsequent sections.

10.4.1 Theoretical Development of FDDEA Models with Different Fuzzy Sets

Nemoto and Goto [15, 16] proposed an approach in which inputs are classified into two categories, (i) variable inputs and (ii) quasi-fixed inputs, which cannot be immediately adjusted without acquiring an adjustment cost [15, 29]. Based on this idea of Nemoto and Goto [15, 16], Chiang and Tzeng [30] developed a multi-objective DEA model in a dynamic framework which is further extended by Jafarian-Moghaddam and Ghoseiri [21] in fuzzy environment. To solve the given model, they reduced it to a single-objective model by using the membership function suggested in Zimmermann [20].

Kordrostami et al. [31] and Keikha-Javan et al. [32] presented dynamic network DEA models to evaluate interval overall and interval period efficiencies for the whole system and each subunit. The subunits are connected in a parallel structure to each other where the inputs and outputs are not known precisely and are known in the form of interval numbers. Kordrostami et al. [31] also derived a relationship between the interval system efficiency and interval subunit efficiencies in a dynamic environment. Keikha-Javan [32] classified carryovers as desirable carryovers and undesirable carryovers to reflect the interdependence of periods more realistically. Soleimani-damaneh [22] provided a theoretical discussion on fuzzy dynamic DEA approaches for incorporating imprecise data.

Ghobadi et al. [33] extended the models presented by Emrouznejad and Yang [8] and Jahanshahloo et al. [34] to deal with fuzzy inputs–outputs in a dynamic environment and presented an inverse dynamic DEA model to evaluate efficiency when data are in the form of LR fuzzy numbers.

As while evaluating efficiency using the DEA model, it is possible for more than one DMUs to be regarded as efficient, so to further rank these efficient DMUs, Andersen and Petersen [35] presented a concept of super-efficiency in DEA. Li et al. [36] further extended it to incorporate dynamic factors and interval data. Yaghoubi et al. [37] presented a dynamic random fuzzy data envelopment analysis (DRF-DEA) model using a common set of weights methodology with mean chance constraints to evaluate efficiency when the inputs–outputs data are in the form of random triangular fuzzy numbers with normal distribution and to deal with the same type of data, Yaghoubi and Amiri [38] presented a multi-objective stochastic fuzzy DEA model with a common set of weights under mean chance constraints to evaluate efficiency in a dynamic environment. Further, the DDEA model of Emrouznejad and Yang [8] has been extended by Yen and Chiou [39] to handle fuzzy data and is solved by embedding the fuzzy DEA approach of Lan et al. [40].

Zhou et al. [25] developed a goal sequence with the help of a benchmarking model based on dynamic DEA in an uncertain environment (triangular fuzzy number) and used α-cut approach to measure efficiency and presented a layering scheme for the suppliers. Ebrahimi et al. [41] developed a slacks-based approach in dynamic network DEA with free disposal hull in which four types of carryovers (good, bad, discretionary, and non-discretionary carryovers) are considered for the interdependence of two consecutive periods when the data for all the variables are interval numbers. The main feature of their study is that all the inefficient DMUs are projected to the existing DMUs on the frontier. Bansal and Mehra [42] introduced a directional distance function-based model, namely the interval dynamic network DEA model, to estimate efficiency when the data for inputs and outputs are available in the form of integers, intervals, or negative data. Both optimistic and pessimistic approaches were followed to evaluate interval efficiencies when the periods are connected by the desirable and undesirable carryovers.

10.4.2 FDDEA with Network Structure

Although dynamic DEA incorporates the time factor, there is still a limitation that it ignores the internal structure of a DMU. To deal with the issue, many researchers studied dynamic DEA with different types of network structures, which can be seen in Hashimoto et al. [43], Avkiran and McCrystal [44], Tone and Tsutsui [45], Khalili-Damghani et al. [46], and Omrani and Soltanzadeh [47].

Kordrostami et al. [31] and Keikha-Javan et al. [32] presented DNDEA models to study the internal structure of DMUs in a dynamic environment and evaluated interval overall and interval period efficiencies for the whole system and each subunit where the subunits are connected in parallel to each other, and the data are in an imprecise form, particularly interval form. Kordrostami et al. [31] provided a relationship between the interval system efficiency and the interval efficiency of subunits in a manner that the interval dynamic efficiency of all the systems can be derived by taking the sum or average of the interval dynamic efficiency of its subunits.

Considering into account the complexity of structures present in real-life problems, Zadeh [23] introduced type-2 fuzzy sets. Olfat et al. [26] extended dynamic network slacks-based measure (DNSBM) to deal with trapezoidal interval type-2 fuzzy data with undesirable inputs–outputs. Let there be n DMUs with three nodes connected through links, and periods are connected through carryovers as depicted in Fig. 10.2. The trapezoidal interval type-2 fuzzy data are transformed into the interval data by deriving its lower and upper bounds from Eqs. (10.4) and (10.5), respectively. Let $\tilde{\tilde{A}}$ be an interval type-2 fuzzy number [26] written as

$$\begin{aligned}\tilde{\tilde{A}} &= (\tilde{A}^U, \tilde{A}^L)\\ &= ((d_1^U, d_2^U, d_3^U, d_4^U, h_1(\tilde{A}^U), h_2(\tilde{A}^U)), (d_1^L, d_2^L, d_3^L, d_4^L, h_1(\tilde{A}^L), h_2(\tilde{A}^L))),\end{aligned}$$

then lower (M^L) and upper (M^U) bounds of transformed interval number are defined as

$$M^L = \frac{1}{6}(d_1^U + 2d_2^U)h_1^U + \frac{1}{6}(d_1^L + 2d_2^L)h_1^L \tag{10.4}$$

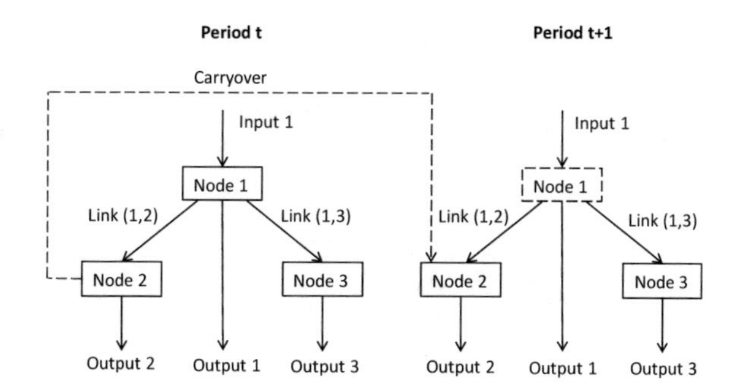

Fig. 10.2 Dynamic network structure with three nodes over two periods

$$M^U = \frac{1}{6}(d_4^U + 2d_3^U)h_2^U + \frac{1}{6}(d_4^L + 2d_3^L)h_2^L \tag{10.5}$$

After transforming all the data into interval numbers, Olfat et al. [26] presented an approach to evaluate the upper bound and lower bound of system efficiency for DMU_o in which they considered constraints related to every input (x), link (z_{link}), carryovers (z_{carry}), and output (y) for all nodes. While writing the constraints, all the undesirable inputs are considered as desirable outputs, whereas all the undesirable outputs are treated as desirable inputs, respectively. Also, a link from node a to node b is both an input (desirable) to node b and output (desirable) of node a. So two sets of constraints related to desirable inputs–outputs for each node (k) and time period (t) are presented in Eqs. (10.6) and (10.7), respectively.

$$\sum_{j=1}^{n} x_j^{Lt}\lambda_{kj}^t + s_{ki}^{t-\text{in}} = x_o^{Lt}, \quad \forall t;\ k = 1, 2, 3;\ i = 1, \ldots, n_{\text{in}}, \tag{10.6}$$

$$\sum_{j=1}^{n} y_j^{Ut}\lambda_{kj}^t + s_{kr}^{t-\text{out}} = y_o^{Ut}, \quad \forall t;\ k = 1, 2, 3;\ r = 1, \ldots, n_{\text{out}}. \tag{10.7}$$

Equations (10.8) and (10.9) depict the continuity of links between nodes and continuity of carryovers between two consecutive periods, respectively. Equation (10.10) represents the variable returns to scale, and Eq. (10.11) represents the non-negativity of weights and slacks.

$$\sum_{j=1}^{n} z_{\text{link } j}^{U(\text{out})}\lambda_{bj}^t = \sum_{j=1}^{n} z_{\text{link } j}^{L(\text{in})}\lambda_{aj}^t, \quad \forall t,\ \forall j \tag{10.8}$$

$$\sum_{j=1}^{n} z_{\text{carry } j}^{U(\text{out})}\lambda_{bj}^t = \sum_{j=1}^{n} z_{\text{carry } j}^{L(\text{in})}\lambda_{aj}^{t+1}, \quad \forall t,\ \forall j \tag{10.9}$$

$$\sum_{j=1}^{n} \lambda_{kj}^t = 1, \quad \forall k;\ \forall t, \tag{10.10}$$

$$\forall s \geq 0;\ \forall \lambda \geq 0. \tag{10.11}$$

The upper bound of system efficiency can be obtained from the objective function defined in Eq. (10.12) subject to the constraints given by Eqs. (10.6)–(10.11).

$$E_o^U = \min\left[\frac{\sum_{t=1}^{T} W^t\left[\sum_{k=1}^{3} w^k\left[1-\frac{1}{n_{\text{in}}+l_{\text{in}}+c_{\text{in}}}\left(\sum_{i=1}^{n_{\text{in}}}\frac{s_{ki}^{t-\text{in}}}{x_o^{Lt}}+\sum_{l=1}^{l_{\text{in}}}\frac{s_{kl}^{t-\text{in}}}{z_{\text{link }o}^{Lt}}+\sum_{c=1}^{c_{\text{in}}}\frac{s_{kc}^{(t,t+1)-\text{in}}}{z_{\text{carry }o}^{L(t,t+1)}}\right)\right]\right]}{\sum_{t=1}^{T} W^t\left[\sum_{k=1}^{3} w^k\left[1+\frac{1}{n_{\text{out}}+l_{\text{out}}+c_{\text{out}}}\left(\sum_{r=1}^{n_{\text{out}}}\frac{s_{kr}^{t-\text{out}}}{y_o^{Ut}}+\sum_{l=1}^{l_{\text{out}}}\frac{s_{kl}^{t-\text{out}}}{z_{\text{link }o}^{Ut}}+\sum_{c=1}^{c_{\text{out}}}\frac{s_{kc}^{(t,t+1)-\text{out}}}{z_{\text{carry }o}^{U(t,t+1)}}\right)\right]\right]}\right] \tag{10.12}$$

Upper bound efficiencies of node k for DMU$_o$ in period t ($E_{ko}^{U(t)}$) and for whole time interval (E_{ko}^{U}) are evaluated using Eqs. (10.13) and (10.14).

$$E_{ko}^{U(t)} = \min\frac{1-\frac{1}{n_{\text{in}}+l_{\text{in}}+c_{\text{in}}}\left(\sum_{i=1}^{n_{\text{in}}}\frac{s_{ki}^{t-\text{in}}}{x_o^{Lt}}+\sum_{l=1}^{l_{\text{in}}}\frac{s_{kl}^{t-\text{in}}}{z_{\text{link }o}^{Lt}}+\sum_{c=1}^{c_{\text{in}}}\frac{s_{kc}^{(t,t+1)-\text{in}}}{z_{\text{carry }o}^{L(t,t+1)}}\right)}{1+\frac{1}{n_{\text{out}}+l_{\text{out}}+c_{\text{out}}}\left(\sum_{r=1}^{n_{\text{out}}}\frac{s_{kr}^{t-\text{out}}}{y_o^{Ut}}+\sum_{l=1}^{l_{\text{out}}}\frac{s_{kl}^{t-\text{out}}}{z_{\text{link }o}^{Ut}}+\sum_{c=1}^{c_{\text{out}}}\frac{s_{kc}^{(t,t+1)-\text{out}}}{z_{\text{carry }o}^{U(t,t+1)}}\right)} \tag{10.13}$$

$$E_{ko}^{U} = \min\frac{\sum_{t=1}^{T} W^t\left[1-\frac{1}{n_{\text{in}}+l_{\text{in}}+c_{\text{in}}}\left(\sum_{i=1}^{n_{\text{in}}}\frac{s_{ki}^{t-\text{in}}}{x_o^{Lt}}+\sum_{l=1}^{l_{\text{in}}}\frac{s_{kl}^{t-\text{in}}}{z_{\text{link }o}^{Lt}}+\sum_{c=1}^{c_{\text{in}}}\frac{s_{kc}^{(t,t+1)-\text{in}}}{z_{\text{carry }o}^{L(t,t+1)}}\right)\right]}{\sum_{t=1}^{T} W^t\left[1+\frac{1}{n_{\text{out}}+l_{\text{out}}+c_{\text{out}}}\left(\sum_{r=1}^{n_{\text{out}}}\frac{s_{kr}^{t-\text{out}}}{y_o^{Ut}}+\sum_{l=1}^{l_{\text{out}}}\frac{s_{kl}^{t-\text{out}}}{z_{\text{link }o}^{Ut}}+\sum_{c=1}^{c_{\text{out}}}\frac{s_{kc}^{(t,t+1)-\text{out}}}{z_{\text{carry }o}^{U(t,t+1)}}\right)\right]} \tag{10.14}$$

In a similar way, lower bound of system efficiency can be evaluated from objective function defined in Eq. (10.15) subject to the constraints given by Eqs. (10.16) and (10.17) with Eqs. (10.8)–(10.11) and lower bounds of interval efficiencies of node k in period t ($E_{ko}^{L(t)}$) and for whole time interval (E_{ko}^{L}) are evaluated using Eqs. (10.18) and (10.19).

$$E_o^L = \min\left[\frac{\sum_{t=1}^{T} W^t\left[\sum_{k=1}^{3} w^k\left[1-\frac{1}{n_{\text{in}}+l_{\text{in}}+c_{\text{in}}}\left(\sum_{i=1}^{n_{\text{in}}}\frac{s_{ki}^{t-\text{in}}}{x_o^{Ut}}+\sum_{l=1}^{l_{\text{in}}}\frac{s_{kl}^{t-\text{in}}}{z_{\text{link }o}^{Ut}}+\sum_{c=1}^{c_{\text{in}}}\frac{s_{kc}^{(t,t+1)-\text{in}}}{z_{\text{carry }o}^{U(t,t+1)}}\right)\right]\right]}{\sum_{t=1}^{T} W^t\left[\sum_{k=1}^{3} w^k\left[1+\frac{1}{n_{\text{out}}+l_{\text{out}}+c_{\text{out}}}\left(\sum_{r=1}^{n_{\text{out}}}\frac{s_{kr}^{t-\text{out}}}{y_o^{Lt}}+\sum_{l=1}^{l_{\text{out}}}\frac{s_{kl}^{t-\text{out}}}{z_{\text{link }o}^{Lt}}+\sum_{c=1}^{c_{\text{out}}}\frac{s_{kc}^{(t,t+1)-\text{out}}}{z_{\text{carry }o}^{L(t,t+1)}}\right)\right]\right]}\right] \tag{10.15}$$

$$\sum_{j=1}^{n} x_j^{Lt}\lambda_{kj}^t + s_{ki}^{t-\text{in}} = x_o^{Ut},\quad \forall t;\ k = 1, 2, 3;\ i = 1, \ldots, n_{\text{in}}, \tag{10.16}$$

$$\sum_{j=1}^{n} y_j^{Ut}\lambda_{kj}^t + s_{kr}^{t-\text{out}} = y_o^{Lt},\quad \forall t;\ k = 1, 2, 3;\ r = 1, \ldots, n_{\text{out}}. \tag{10.17}$$

$$E_{ko}^{L(t)} = \min \frac{1 - \frac{1}{n_{\text{in}} + l_{\text{in}} + c_{\text{in}}} \left(\sum_{i=1}^{n_{\text{in}}} \frac{s_{ki}^{t-\text{in}}}{x_o^{Ut}} + \sum_{l=1}^{l_{\text{in}}} \frac{s_{kl}^{t-\text{in}}}{z_{\text{link } o}^{Ut}} + \sum_{c=1}^{c_{\text{in}}} \frac{s_{kc}^{(t,t+1)-\text{in}}}{z_{\text{carry } o}^{U(t,t+1)}} \right)}{1 + \frac{1}{n_{\text{out}} + l_{\text{out}} + c_{\text{out}}} \left(\sum_{r=1}^{n_{\text{out}}} \frac{s_{kr}^{t-\text{out}}}{y_o^{Lt}} + \sum_{l=1}^{l_{\text{out}}} \frac{s_{kl}^{t-\text{out}}}{z_{\text{link } o}^{Lt}} + \sum_{c=1}^{c_{\text{out}}} \frac{s_{kc}^{(t,t+1)-\text{out}}}{z_{\text{carry } o}^{L(t,t+1)}} \right)} \tag{10.18}$$

$$E_{ko}^{L} = \min \frac{\sum_{t=1}^{T} W^t \left[1 - \frac{1}{n_{\text{in}} + l_{\text{in}} + c_{\text{in}}} \left(\sum_{i=1}^{n_{\text{in}}} \frac{s_{ki}^{t-\text{in}}}{x_o^{Ut}} + \sum_{l=1}^{l_{\text{in}}} \frac{s_{kl}^{t-\text{in}}}{z_{\text{link } o}^{Ut}} + \sum_{c=1}^{c_{\text{in}}} \frac{s_{kc}^{(t,t+1)-\text{in}}}{z_{\text{carry } o}^{U(t,t+1)}} \right) \right]}{\sum_{t=1}^{T} W^t \left[1 + \frac{1}{n_{\text{out}} + l_{\text{out}} + c_{\text{out}}} \left(\sum_{r=1}^{n_{\text{out}}} \frac{s_{kr}^{t-\text{out}}}{y_o^{Lt}} + \sum_{l=1}^{l_{\text{out}}} \frac{s_{kl}^{t-\text{out}}}{z_{\text{link } o}^{Lt}} + \sum_{c=1}^{c_{\text{out}}} \frac{s_{kc}^{(t,t+1)-\text{out}}}{z_{\text{carry } o}^{L(t,t+1)}} \right) \right]} \tag{10.19}$$

Olfat and Pishdar [48] presented an extended version of DNSBM to evaluate both optimistic and pessimistic efficiencies with interval type-2 fuzzy data in the presence of undesirable inputs–outputs.

Although dynamic DEA measures the efficiency of a DMU by taking into account the interdependence of periods, ignoring the internal structure of DMUs may produce misleading results. Tone and Tsutsui [45] proposed a slacks-based dynamic network DEA model to compute system and period efficiencies when there exist four types of links (as input link, as output link, free link, and fixed link) between subdivisions of a DMU, and similarly, the periods are connected through four types of carryovers, namely desirable, undesirable, free, and fixed carryovers. A dynamic network structure with subdivisions linked to each other through intermediate links and periods connected through carryovers is shown in Fig. 10.3. Soltanzadeh and Omrani [49] introduced a dynamic network DEA (DNDEA) model to evaluate efficiency using the α-cut approach when the data for inputs, outputs, and links are of type-1 fuzzy data. They extended the dynamic DEA model proposed by Omrani and Soltanzadeh [47] in the presence of fuzzy data.

Nomenclature

n :	Number of DMUs ($j = 1, \ldots, n$)
K:	Number of divisions in a DMU ($k = 1, \ldots, K$)
p:	Number of time periods ($t = 1, \ldots, p$)
m_k :	Number of inputs of kth division and $i^k \in \{1, 2, \ldots, m_k\}$
s_k :	Number of outputs of kth division and $r^k \in \{1, 2, \ldots, s_k\}$
l_k :	Number of links from kth division to the next division and $l^k \in \{1, 2, \ldots, l_k\}$
d_k :	Number of carryovers at kth division from period t to $t+1$ and $d^k \in \{1, 2, \ldots, d_k\}$

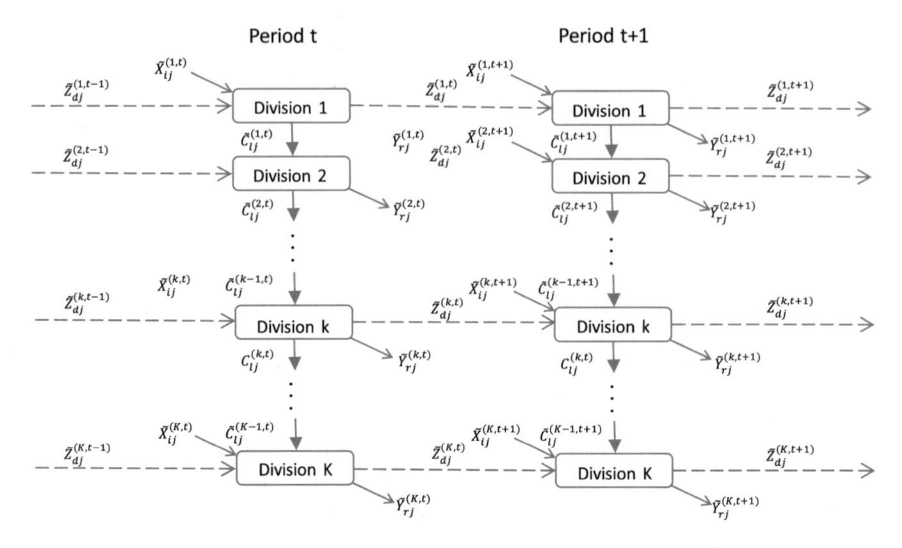

Fig. 10.3 Dynamic structure with K divisions over two periods t and $t+1$

$\tilde{X}_{ij}^{(k,t)}$: ith fuzzy input of DMU$_j$ for kth division in period t where $i = 1, 2, \ldots, m_k, \ldots, m$

$\tilde{Y}_{gj}^{(k,t)}$: gth fuzzy output of DMU$_j$ for kth division in period t where $r = 1, 2, \ldots, s_k, \ldots, s$

$\tilde{C}_{lj}^{(k,t)}$: ith fuzzy intermediate link of DMU$_j$ for kth to $(k+1)$th division in period t where $l = 1, 2, \ldots, l_k, \ldots, L$

$\tilde{Z}_{dj}^{(k,t)}$: dth fuzzy carry-over of DMU$_j$ for kth division from period t to $t+1$ where $d = 1, 2, \ldots, d_k, \ldots, D$

Let $\tilde{X}_{io} = \sum_{t=1}^{p} \sum_{k=1}^{K} \tilde{X}_{io}^{(k,t)}$ and $\tilde{Y}_{ro} = \sum_{t=1}^{p} \sum_{k=1}^{K} \tilde{Y}_{ro}^{(k,t)}$ be the ith fuzzy system input and rth fuzzy system output for DMU$_o$ and the initial fuzzy carryovers as inputs to division k from period t_0 and the final fuzzy carryovers as outputs from division k at period p be denoted by $\tilde{Z}_{do}^{(k,t_0)}$ and $\tilde{Z}_{do}^{(k,p)}$, respectively. Then efficiency of DMU$_o$ over p periods is evaluated using following model:

Model-3

$$E_o^S = \max \sum_{r=1}^{s} u_r \tilde{Y}_{ro} + \sum_{k=1}^{K} \sum_{d=1}^{D} f_d \tilde{Z}_{do}^{(k,p)}$$

$$\text{s.t.} \sum_{i=1}^{m} v_i \tilde{X}_{io} + \sum_{k=1}^{K} \sum_{d=1}^{D} f_d \tilde{Z}_{do}^{(k,t_0)} = 1$$

$$\sum_{r=1}^{s} u_r \tilde{Y}_{rj} + \sum_{k=1}^{K} \sum_{d=1}^{D} f_d \tilde{Z}_{dj}^{(k,p)} - \sum_{i=1}^{m} v_i \tilde{X}_{ij} - \sum_{k=1}^{K} \sum_{d=1}^{D} f_d \tilde{Z}_{dj}^{(k,t_0)} \leq 0 \ \forall j,$$

$$\sum_{r \in r^1} u_r \tilde{Y}_{rj}^{(1,t)} + \sum_{l \in l^1} w_l \tilde{C}_{lj}^{(1,t)} + \sum_{d \in d^1} f_d \tilde{Z}_{dj}^{(1,t)} - \sum_{i \in i^1} v_i \tilde{X}_{ij}^{(1,t)} - \sum_{d \in d^{(1)}} f_d \tilde{Z}_{dj}^{(1,t-1)} \leq 0 \forall j;\ \forall t;\ k = 1,$$

$$\sum_{r \in r^k} u_r \tilde{Y}_{rj}^{(k,t)} + \sum_{l \in l^k} w_l \tilde{C}_{lj}^{(k,t)} + \sum_{d \in d^k} f_d \tilde{Z}_{dj}^{(k,t)} - \sum_{i \in i^k} v_i \tilde{X}_{ij}^{(k,t)} - \sum_{l \in l^k} w_l \tilde{C}_{lj}^{(k-1,t)} - \sum_{d \in d^{(k)}} f_d \tilde{Z}_{dj}^{(k,t-1)} \leq 0\ \forall j;\ \forall t;\ k = 2, \ldots, K-1,$$

$$\sum_{r \in r^K} u_r \tilde{Y}_{rj}^{(K,t)} + \sum_{d \in d^K} f_d \tilde{Z}_{dj}^{(K,t)} - \sum_{i \in i^K} v_i \tilde{X}_{ij}^{(K,t)} - \sum_{l \in l^k} w_l \tilde{C}_{lj}^{(K-1,t)} - \sum_{d \in d^{(K)}} f_d \tilde{Z}_{dj}^{K(t-1)} \leq 0\ \forall j;\ \forall t;\ k = K,$$

$$v_i \geq \epsilon\ \forall i;\ u_r \geq \epsilon\ \forall r;\ f_d \geq \epsilon\ \forall d;\ w_l \geq \epsilon\ \forall l.$$

It is obvious that the efficiency obtained from fuzzy numbers will also be a fuzzy number. Let $\mu_{\tilde{X}_{ij}}$, $\mu_{\tilde{Y}_{rj}}$, $\mu_{\tilde{Z}_{dj}}$, and $\mu_{\tilde{C}_{lj}}$ be the membership functions of $\tilde{X}_{ij}$, $\tilde{Y}_{rj}$, $\tilde{Z}_{dj}$, and $\tilde{C}_{lj}$, respectively. Then, membership function $\mu_{\tilde{E}_k}$ for system efficiency of division k of DMU$_o$ denoted by $\tilde{E}_k$ is given by

$$\mu_{\tilde{E}_o}(e) = \sup_{x,y,z,c} \min_{i.r.d.l} \{\mu_{\tilde{X}_{ij}}^{(k,t)}(x_{ij}^{(k,t)}), \mu_{\tilde{Y}_{rj}}^{(k,t)}(y_{rj}^{(k,t)}), \mu_{\tilde{Z}_{dj}}^{(k,t)}(z_{dj}^{(k,t)}), \mu_{\tilde{C}_{lj}}^{(k,t)}(c_{lj}^{(k,t)}) | e = E_k(x_{ij}^{(k,t)}, y_{rj}^{(k,t)}, z_{dj}^{(k,t)}, c_{lj}^{(k,t)})\}$$

Soltanzadeh and Omrani [49] used α-cut approach to solve Model-3. The α-cuts for $\tilde{X}_{ij}$, $\tilde{Y}_{rj}$, $\tilde{Z}_{dj}$, and $\tilde{C}_{lj}$ are defined as follows:

$$(X_{ij}^{(k,t)})_\alpha = \left[\min_{X_{ij}^{(k,t)}} \left\{ X_{ij}^{(k,t)} \in S(\tilde{X}_{ij}^{(k,t)}) | \mu_{\tilde{X}_{ij}}^{(k,t)}(X_{ij}^{(k,t)}) \geq \alpha \right\}, \max_{X_{ij}^{(k,t)}} \left\{ X_{ij}^{(k,t)} \in S(\tilde{X}_{ij}^{(k,t)}) | \mu_{\tilde{X}_{ij}}^{(k,t)}(X_{ij}^{(k,t)}) \geq \alpha \right\} \right] = \left[(X_{ij}^{(k,t)})_\alpha^L, (X_{ij}^{(k,t)})_\alpha^U \right]$$

$$(Y_{rj}^{(k,t)})_\alpha = \left[\min_{Y_{rj}^{(k,t)}} \left\{ Y_{rj}^{(k,t)} \in S(\tilde{Y}_{rj}^{(k,t)}) | \mu_{\tilde{Y}_{rj}}^{(k,t)}(Y_{rj}^{(k,t)}) \geq \alpha \right\}, \max_{Y_{rj}^{(k,t)}} \left\{ Y_{rj}^{(k,t)} \in S(\tilde{Y}_{rj}^{(k,t)}) | \mu_{\tilde{Y}_{rj}}^{(k,t)}(Y_{rj}^{(k,t)}) \geq \alpha \right\} \right] = \left[(Y_{rj}^{(k,t)})_\alpha^L, (Y_{rj}^{(k,t)})_\alpha^U \right]$$

$$(Z_{dj}^{(k,t)})_\alpha = \left[\min_{Z_{dj}^{(k,t)}} \left\{ Z_{dj}^{(k,t)} \in S(\tilde{Z}_{dj}^{(k,t)}) | \mu_{\tilde{Z}_{dj}}^{(k,t)}(Z_{dj}^{(k,t)}) \geq \alpha \right\}, \max_{Z_{dj}^{(k,t)}} \left\{ Z_{dj}^{(k,t)} \in S(\tilde{Z}_{dj}^{(k,t)}) | \mu_{\tilde{Z}_{dj}}^{(k,t)}(Z_{dj}^{(k,t)}) \geq \alpha \right\} \right]$$
$$= \left[(Z_{dj}^{(k,t)})_\alpha^L, (Z_{dj}^{(k,t)})_\alpha^U \right]$$

$$(C_{lj}^{(k,t)})_\alpha = \left[\min_{C_{lj}^{(k,t)}} \left\{ C_{lj}^{(k,t)} \in S(\tilde{C}_{lj}^{(k,t)}) | \mu_{\tilde{C}_{lj}}^{(k,t)}(C_{lj}^{(k,t)}) \geq \alpha \right\}, \max_{C_{lj}^{(k,t)}} \left\{ C_{lj}^{(k,t)} \in S(\tilde{C}_{lj}^{(k,t)}) | \mu_{\tilde{C}_{lj}}^{(k,t)}(C_{lj}^{(k,t)}) \geq \alpha \right\} \right]$$
$$= \left[(C_{lj}^{(k,t)})_\alpha^L, (C_{lj}^{(k,t)})_\alpha^U \right]$$

After using an α-cut approach and some transformations, Model-4(a) and Model-4(b) were presented to measure the lower and upper bounds of system efficiency $\tilde{E}_o$ for each α.

Model-4(a)

$$(E_o^S)_\alpha^U = \max \ E_o^S = \frac{\sum_{r=1}^{s} u_r (Y_{ro})_\alpha^U + \sum_{k=1}^{K} \sum_{d=1}^{D} \hat{z}_{do}^{(k,p)}}{\sum_{i=1}^{m} v_i (X_{io})_\alpha^L + \sum_{k=1}^{K} \sum_{d=1}^{D} \hat{z}_{do}^{(k,t_0)}}$$

$$\text{s.t.} \ \frac{\sum_{r=1}^{s} u_r (Y_{ro})_\alpha^U + \sum_{k=1}^{K} \sum_{d=1}^{D} \hat{z}_{do}^{(k,p)}}{\sum_{i=1}^{m} v_i (X_{io})_\alpha^L + \sum_{k=1}^{K} \sum_{d=1}^{D} \hat{z}_{do}^{(k,t_0)}} \leq 1,$$

$$\frac{\sum_{r=1}^{s} u_r (Y_{rj})_\alpha^L + \sum_{k=1}^{K} \sum_{d=1}^{D} \hat{z}_{dj}^{(k,p)}}{\sum_{i=1}^{m} v_i (X_{ij})_\alpha^U + \sum_{k=1}^{K} \sum_{d=1}^{D} \hat{z}_{do}^{(k,t_0)}} \leq 1; \ \ \forall j \neq o,$$

$$\frac{\sum_{r \in r^1} u_r (Y_{ro}^{(1,t)})_\alpha^U + \sum_{l \in l^1} \hat{c}_{lo}^{(1,t)} + \sum_{d \in d^1} \hat{z}_{do}^{(1,t)}}{\sum_{i \in i^1} v_i (X_{io}^{(1,t)})_\alpha^L + \sum_{d \in d^{(1)}} \hat{Z}_{do}^{(1,t-1)}} \leq 1; \ \ k = 1,$$

$$\frac{\sum_{r \in r^1} u_r (Y_{rj}^{(1,t)})_\alpha^L + \sum_{l \in l^1} \hat{c}_{lj}^{(1,t)} + \sum_{d \in d^1} \hat{z}_{dj}^{(1,t)}}{\sum_{i \in i^1} v_i (X_{ij}^{(1,t)})_\alpha^U + \sum_{d \in d^{(1)}} \hat{Z}_{dj}^{(1,t-1)}} \leq 1; \ \forall j \neq o; \ k = 1,$$

$$\frac{\sum_{r \in r^k} u_r (Y_{ro}^{(k,t)})_\alpha^U + \sum_{l \in l^k} \hat{c}_{lo}^{(k,t)} + \sum_{d \in d^k} \hat{z}_{do}^{(k,t)}}{\sum_{i \in i^k} v_i (X_{io}^{(k,t)})_\alpha^L + \sum_{l \in l^k} \hat{c}_{lo}^{(k-1,t)} + \sum_{d \in d^{(k)}} \hat{Z}_{do}^{(k,t-1)}} \leq 1; \ \forall k \neq 1, K,$$

$$\frac{\sum_{r \in r^k} u_r (Y_{rj}^{(k,t)})_\alpha^L + \sum_{l \in l^k} \hat{c}_{lj}^{(k,t)} + \sum_{d \in d^k} \hat{z}_{dj}^{(k,t)}}{\sum_{i \in i^k} v_i (X_{ij}^{(k,t)})_\alpha^U + \sum_{l \in l^k} \hat{c}_{lj}^{(k-1,t)} + \sum_{d \in d^{(k)}} \hat{Z}_{dj}^{(k,t-1)}} \leq 1; \ \forall j \neq o; \ \forall k \neq 1, K,$$

$$\frac{\sum_{r \in r^K} u_r (Y_{ro}^{(K,t)})_\alpha^U + \sum_{d \in d^k} \hat{z}_{do}^{(K,t)}}{\sum_{i \in i^K} v_i (X_{io}^{(K,t)})_\alpha^L + \sum_{l \in l^K} \hat{c}_{lo}^{(K-1,t)} + \sum_{d \in d^{(K)}} \hat{Z}_{do}^{K(t-1)}} \leq 1; \ k = K,$$

$$\frac{\sum_{r\in r^K} u_r (Y_{ro}^{(K,t)})_\alpha^L + \sum_{d\in d^k} \hat{z}_{do}^{(K,t)}}{\sum_{i\in i^K} v_i (X_{io}^{(K,t)})_\alpha^U + \sum_{l\in l^K} \hat{c}_{lo}^{(K-1,t)} + \sum_{d\in d^{(K)}} \hat{Z}_{do}^{K(t-1)}} \leq 1\ \forall j \neq o;\ k = K,$$

$$w_l (C_{lj}^{(k,t)})_\alpha^L \leq \hat{c}_{lj}^{(k,t)} \leq w_l (C_{lj}^{(k,t)})_\alpha^U;\ \forall k;\ \forall t$$

$$f_d (Z_{dj}^{(k,t)})_\alpha^L \leq \hat{z}_{dj}^{(k,t)} \leq f_d (Z_{dj}^{(k,t)})_\alpha^U;\ \forall k;\ \forall t$$

$$v_i \geq \epsilon\ \forall i;\ u_r \geq \epsilon\ \forall r;\ f_d \geq \epsilon\ \forall d;\ w_l \geq \epsilon\ \forall l.$$

Upper bounds of system and process efficiencies in each α-cut by using optimal weights derived from Model-4(a) are defined as follows:

$$(E_o^S)_\alpha^U = \frac{\sum_{r=1}^{s} u_r^* (Y_{ro})_\alpha^U + \sum_{k=1}^{K} \sum_{d=1}^{D} \hat{z}_{do}^{(k,p)}}{\sum_{i=1}^{m} v_i^* (X_{io})_\alpha^L + \sum_{k=1}^{K} \sum_{d=1}^{D} \hat{z}_{do}^{(k,t_0)}}$$

$$(E_o^{S(t)})_\alpha^U = \frac{\sum_{k=1}^{K} \sum_{r=1}^{s} u_r^* (Y_{ro}^{(k,t)})_\alpha^U + \sum_{k=1}^{K} \sum_{d=1}^{D} \hat{z}_{do}^{(k,t)}}{\sum_{k=1}^{K} \sum_{i=1}^{m} v_i^* (X_{io}^{(k,t)})_\alpha^L + \sum_{k=1}^{K} \sum_{d=1}^{D} \hat{z}_{do}^{(k,t-1)}}$$

$$(E_o^{(1,t)})_\alpha^U = \frac{\sum_{r\in r^1} u_r^* (Y_{ro}^{(1,t)})_\alpha^U + \sum_{l\in l^1} \hat{c}_{lo}^{(1,t)} + \sum_{d\in d^1} \hat{z}_{do}^{(1,t)}}{\sum_{i\in i^1} v_i^* (X_{io}^{(1,t)})_\alpha^L + \sum_{d\in d^1} \hat{Z}_{do}^{(1,t-1)}}$$

$$(E_o^{(k,t)})_\alpha^U = \frac{\sum_{r\in r^k} u_r^* (Y_{ro}^{(k,t)})_\alpha^U + \sum_{l\in l^k} \hat{c}_{lo}^{(k,t)} + \sum_{d\in d^k} \hat{z}_{do}^{(k,t)}}{\sum_{i\in i^k} v_i^* (X_{io}^{(k,t)})_\alpha^L + \sum_{d\in d^k} \hat{Z}_{do}^{(k,t-1)}};\ k = 2, \ldots, K-1$$

$$(E_o^{(K,t)})_\alpha^U = \frac{\sum_{r\in r^K} u_r^* (Y_{ro}^{(K,t)})_\alpha^L + \sum_{d\in d^k} \hat{z}_{do}^{(K,t)}}{\sum_{i\in i^K} v_i^* (X_{io}^{(K,t)})_\alpha^U + \sum_{l\in l^K} \hat{c}_{lo}^{(K-1,t)} + \sum_{d\in d^K} \hat{Z}_{do}^{K(t-1)}}$$

Model-4(b)

$$(E_o^S)_\alpha^L = \min\ \theta - \epsilon \left(\sum_{i=1}^{m} s_i^- + \sum_{r=1}^{s} s_r^+ + \sum_{l=1}^{L} \sum_{k=1}^{K} s_{lk} + \sum_{d=1}^{D} \sum_{k=1}^{K} s_{dk} \right)$$

$$\text{s.t.} \sum_{t=1}^{p} \sum_{k=1}^{K} \left(\lambda_o^{(k,t)} \cdot (X_{io}^{(k,t)})_\alpha^U + \sum_{j=1, j\neq k}^{n} \lambda_j^{(k,t)} \cdot (X_{ij}^{(k,t)})_\alpha^L \right) + s_i^- = \theta (X_{io})_\alpha^U;\ \forall i,$$

$$\sum_{t=1}^{p} \sum_{k=1}^{K} \left(\lambda_o^{(k,t)} \cdot (Y_{ro}^{(k,t)})_\alpha^L + \sum_{j=1, j\neq k}^{n} \lambda_j^{(k,t)} \cdot (Y_{rj}^{(k,t)})_\alpha^U \right) - s_r^+ = (Y_{ro})_\alpha^L;\ \forall r,$$

$$\sum_{t=1}^{p} \sum_{j=1}^{n} \lambda_j^{(k,t)} \cdot c_{lj}^{(k,t)} - \sum_{t=1}^{p} \sum_{j=1}^{n} \lambda_j^{k+(1,t)} \cdot c_{lj}^{(k,t)} - s_{lk} = 0;\ \forall k;\ l \in l^k,$$

$$\sum_{t=1}^{p}\sum_{j=1}^{n}\lambda_j^{(k,t)}\cdot\left(z_{dj}^{(k,t)}-z_{dj}^{(k,t-1)}\right)+z_{do}^{(1,p)}-s_{dk}=\theta z_{do}^{(1,1)};\ \forall k;\ d\in d^k,$$
$$(C_{lj})_\alpha^L\le c_{lj}\le(C_{lj})_\alpha^U;\ \ l=1,\ldots,L,$$
$$(Z_{dj})_\alpha^L\le z_{dj}\le(Z_{dj})_\alpha^U;\ \ d=1,\ldots,D,$$
$$\lambda_j^{(k,t)},s_i^-,s_r^+,s_{lk},s_{dk}\ge 0,\theta\text{ is free};\ \forall j;\ \forall k;\ \forall t.$$

By using optimal weights derived from Model-4(b), lower bounds of system and process efficiencies in each α-cut are defined as follows:

$$(E_o^S)_\alpha^L=\frac{\sum_{r=1}^{s}u_r^*(Y_{ro})_\alpha^L+\sum_{k=1}^{K}\sum_{d=1}^{D}f_d^*\hat{z}_{do}^{(k,p)^*}}{\sum_{i=1}^{m}v_i^*(X_{io})_\alpha^U+\sum_{k=1}^{K}\sum_{d=1}^{D}f_d^*\hat{z}_{do}^{(k,t_0)^*}}$$

$$(E_o^{S(t)})_\alpha^L=\frac{\sum_{k=1}^{K}\sum_{r=1}^{s}u_r^*(Y_{ro}^{(k,t)})_\alpha^L+\sum_{k=1}^{K}\sum_{d=1}^{D}f_d^*\hat{z}_{do}^{(k,t)^*}}{\sum_{k=1}^{K}\sum_{i=1}^{m}v_i^*(X_{io}^{(k,t)})_\alpha^U+\sum_{k=1}^{K}\sum_{d=1}^{D}f_d^*\hat{z}_{do}^{(k,t-1)^*}}$$

$$(E_o^{(1,t)})_\alpha^L=\frac{\sum_{r\in r^1}u_r^*(Y_{ro}^{(1,t)})_\alpha^L+\sum_{l\in l^1}w_l^*c_{lo}^{(1,t)}+\sum_{d\in d^1}f_d^*z_{do}^{(1,t)^*}}{\sum_{i\in i^1}v_i^*(X_{io}^{(1,t)})_\alpha^U+\sum_{d\in d^1}f_d^*z_{do}^{(1,t-1)^*}}$$

$$(E_o^{(k,t)})_\alpha^L=\frac{\sum_{r\in r^k}u_r^*(Y_{ro}^{(k,t)})_\alpha^L+\sum_{l\in l^k}w_l^*c_{lo}^{(k,t)^*}+\sum_{d\in d^k}f_d^*z_{do}^{(k,t)^*}}{\sum_{i\in i^k}v_i^*(X_{io}^{(k,t)})_\alpha^U+\sum_{d\in d^k}f_d^*z_{do}^{(k,t-1)^*}};\ k=2,\ldots,K-1$$

$$(E_o^{(K,t)})_\alpha^L=\frac{\sum_{r\in r^K}u_r^*(Y_{ro}^{(K,t)})_\alpha^L+\sum_{d\in d^k}f_d^*z_{do}^{(K,t)^*}}{\sum_{i\in i^K}v_i^*(X_{io}^{(K,t)})_\alpha^U+\sum_{l\in l^K}w_l^*c_{lo}^{(K-1,t)^*}+\sum_{d\in d^K}z_{do}^{K(t-1)^*}}$$

Olfat et al. [50] presented an interval type-2 fuzzy dynamic DEA model to deal with uncertainties and measure the performance of DMUs in a dynamic environment. Ebrahimi et al. [41] developed a slacks-based DNDEA model in which different weights are assigned to different divisions, and the divisions are linked through four types of links, namely input link, output link, free link, and fixed or non-discretionary link. All the inputs, outputs, and links are in the form of interval numbers.

10.4.3 Applications of FDDEA

Jafarian-Moghaddam and Ghoseiri [21, 51] proposed fuzzy dynamic DEA in a multi-objective framework and evaluated the performance of 49 railways from all over the world with fuzzy data in a dynamic environment. Kordrostami et al. [31] assessed the efficiency of ten bank areas in Iran by proposing a DNDEA model. Each area comprises three bank branches considered as subunits for three (six-month) periods with interval data for inputs and outputs. Keikha-Javan et al. [32] also used the same data as in Kordrostami et al. [31] to present an application of their model and provided better results than Kordrostami et al. [31]. Yaghoubi et al. [37] and Yaghoubi and Amiri [38] applied a dynamic random fuzzy DEA model on Iranian petroleum company and evaluated the efficiency of five gas stations over two financial periods using the DRF-DEA model and multi-objective stochastic fuzzy DEA (MOFS-DEA) model, respectively. The data used for inputs–outputs were in the form of random triangular fuzzy number with normal distribution, and the efficiency results from both the above-mentioned approaches turned out to be better than the hybrid genetic algorithm proposed by Qin and Liu [52] to deal with fuzzy random inputs–outputs.

Considering the importance of sustainable development, Olfat et al. [26] suggested an extension of DNSBM to calculate the sustainable performance of 28 airports in Iran over two periods. The whole structure is divided into three nodes: (i) airport node, (ii) community node, and (iii) passenger node, and the efficiency is evaluated for each node in different periods as well as the system efficiency in the presence of interval type-2 fuzzy data for inputs and outputs when some of the inputs–outputs are undesirable. Olfat and Pishdar [48] investigated the efficiency of same 28 Iranian airports with the same structure as studied in Olfat et al. [26], but by using both the efficient and inefficient production frontiers, i.e., evaluated the efficiencies from both optimistic and pessimistic viewpoints, whereas, in Olfat et al. [26], efficiencies were evaluated using only optimistic viewpoint and revealed that the former approach exhibits more discrimination power.

Soltanzadeh and Omrani [49] presented a DNDEA model to calculate the efficiency of seven Iranian airlines for the period 2010–12. The network structure of airlines consists of two stages, namely production and consumption, and the data for inputs–outputs belong to the set of triangular fuzzy numbers. In the same manner, Olfat et al. [50] introduced a DNDEA model to assess the performance of the 20 most popular passenger airports in Iran from the viewpoint of sustainability while using interval type-2 fuzzy data for inputs–outputs.

Based on the ideas of Tone and Tsutsui [45] and Wang and Chin [53], Zhou et al. [54] developed a double frontier dynamic network DEA approach for performance evaluation of sustainable supply chains (SSCs) with a network structure of three stages: (i) supplier stage, (ii) manufacturer stage, and (iii) distributor stage. Twenty SSCs are considered for efficiency evaluation (system and period efficiencies) over three periods with interval type-2 fuzzy data for customer satisfaction (desirable output) and environmental pollution (undesirable output). Ebrahimi et al. [41] also evaluated the efficiency of supply chains by using a slacks-based DNDEA model. Thirty Iranian printing supply chains with three divisions (production, assembly, and distribution) for three consecutive periods (2015–2017) are chosen for the case study. The interval overall and period divisional efficiencies are evaluated along with projected values for all the divisions of inefficient DMUs. Hasani and Mokhtari [55] proposed a hybrid fuzzy multi-criteria decision-making (DEA-MCDM) model to evaluate the efficiency of 11 Iranian hospitals with three nodes, namely hospital, community, and patient node. Torabandeh et al. [56] presented a dynamic network DEA model to evaluate and compare the performance of Iran with other countries.

Zhou et al. [25] assessed the efficiency of 20 suppliers for three periods and also set the practical goals (targets) for suppliers by using a goal sequence based on a dynamic DEA model in an uncertain environment. Bansal and Mehra [42] investigated the interval efficiency of 11 Indian airlines over three consecutive periods in the presence of integer and negative data by using dynamic interval DEA. Table 10.1 represents the categorization of publications on applications of FDDEA studies which depicts its implementation in sectors like airlines, supply chains, gas stations, banks, and various other sectors, including railways, oil refineries, bus companies, and hospitals.

10.4.4 Integration of FDDEA with Other Techniques

Khodaparasti and Maleki [57] proposed an integrated approach in a dynamic fuzzy environment by combining a dynamic location model and fuzzy simultaneous DEA model for emergency medical services (EMS). Yaghoubi et al. [37] presented a DRF-DEA model with fuzzy data, which is further converted to a multi-objective programming problem and later on to a single-objective programming problem for performance evaluation. Further, an integrated Monte Carlo simulation and genetic algorithm have been designed to solve the single-objective programming.

Yaghoubi and Amiri [38] proposed a multi-objective stochastic fuzzy DEA (MOFS-DEA) model to evaluate performance in a dynamic environment and designed an integrated meta-heuristic algorithm using imperialist competitive algorithm and Monte Carlo simulation to solve the one objective stochastic model obtained from the initial MOFS-DEA model by using infinite norm approach.

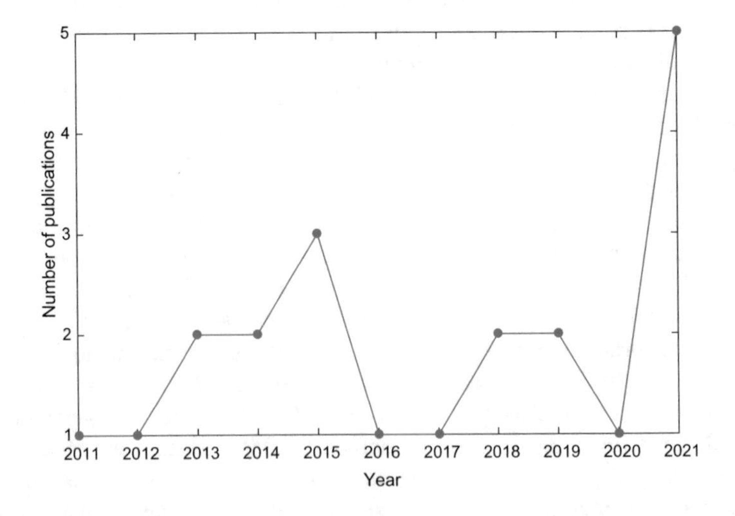

Fig. 10.4 Year-wise publications of fuzzy dynamic data envelopment analysis

Hasani and Mokhtari [55] developed a hybrid fuzzy DEA-MCDM model incorporating fuzzy decision-making trial and evaluation laboratory (DEMATEL) and best-worst method (BWM) to evaluate the system and period efficiencies from interval type-2 fuzzy data. To measure the future as well as past efficiency of suppliers, Nikabadi and Moghaddam [58] developed a hybrid approach by combining an adaptive neuro-fuzzy inference system and dynamic DEA. Figure 10.4 depicts the year-wise publications of FDDEA studies for the periods 2011–2021, from which it can be seen that each year has at least one publication, and the current year 2021 has the highest number of publications on FDDEA in the last decade.

Table 10.1 Publications based on applied study and their characteristics

Area	Study	DMUs	Inputs/outputs/links	Citations
Airports	Olfat et al. [26]	28	Inputs: budget, policy making based on sustainable development concept Outputs: non-aviation income, level of pollution, satisfaction Links: number of aircrafts (takeoff and landing), service quality Carryovers: corporate reputation	63
	Olfat and Pishdar [48]	28	Inputs: policy making based on sustainable development concept, budget Outputs: non-aviation income, pollution levels, satisfaction Links: number of aircrafts (takeoff and landing), service quality, perceived social responsibility Carryovers: corporate reputation	8
	Soltanzadeh and Omrani [49]	7	Inputs: number of employees Outputs: passenger-kilometer performed, passenger ton-kilometer performed Link: number of scheduled flights, available ton-kilometer, available seat-kilometer	25

(continued)

Table 10.1 (continued)

Area	Study	DMUs	Inputs/outputs/links	Citations
	Olfat et al. [50]	20	–	–
	Bansal and Mehra [42]	11	Inputs: operating expenses Outputs: operating revenue, passengers carried per month, pax load factor per month, cargo carried per month Carryovers: losses carried forward after tax, fleet size	–
Supply chains	Zhou et al. [54]	20	Inputs: cost of labor safety, other costs Outputs: degree of environmental pollution Links: value of raw material, value of finished products Carryovers: unrecovered revenue, unpaid cost	24
	Zhou et al. [25]	20	Inputs: technical and financial capability, cost of work safety Outputs: value of raw material, environmental pollution, degree of customer satisfaction Carryovers: accounts receivable, accounts payable	2
	Ebrahimi et al. [41]	30	Inputs: production capacity, planning cost, cardboard and ink cost, electricity cost, machinery cost, labor cost, transportation cost, environmental cost Outputs: label and catalog income, income Links: finished goods, wasted product, recycled waste Carryover: depreciation	–

(continued)

Table 10.1 (continued)

Area	Study	DMUs	Inputs/outputs/links	Citations
Gas stations	Yaghoubi et al. [37]	5	Inputs: employees salaries, operation costs, net profit Outputs: gasoline, net profit Carryover: net profit	3
	Yaghoubi and Amiri [38]	5	Inputs: employees salaries, operation costs, net profit Outputs: gasoline, net profit Carryover: net profit	1
Banks	Kordrostami et al. [31]	30	Inputs: personnel Outputs: usage Carryover: resources	6
	Keikha-Javan et al. [32]	30	Inputs: personnel Outputs: usage Carryover: resources	2
Railways	Jafarian-Moghaddam and Ghoseiri [21]	49	Inputs: length of single, double and electrify track, number of state and private own wagons, fleet size of locomotives, coaches and railcars, and employees Outputs: total train kilometers, gross train tonne kilometers, gross tonne kilometers, gross tonne carried, passengers, passenger kilometers Carryovers: gross tonne kilometers and passenger kilometers	38
Oil refineries	Tavana et al. [59]	9	Inputs: feed, energy consumption, fuel, personal staff, degree of complexity, API Outputs: ratio of light to heavy product, waste (non-permissible CO_2), permissible CO_2	17

(continued)

Table 10.1 (continued)

Area	Study	DMUs	Inputs/outputs/links	Citations
Bus companies	Yen and Chiou [39]	10	Inputs: operating network, number of buses Outputs: operating revenue, number of bus runs, passenger kilometers, passenger satisfaction Carryover: number of buses	1
Hospitals	Hasani and Mokhtari [55]	11	Inputs: policy making based on sustainable concept, budget Outputs: non-healthcare service income, hospital waste, satisfaction Links: social responsibility, population coverage, total bed number, service quality Carryover: hospital reputation	1

10.5 Conclusion

Dynamic DEA with fuzzy set theory is used to measure the inter-temporal efficiency of similar DMUs in an uncertain environment. This study launches a taxonomy and review of recent developments in FDDEA studies in the last decade, and it has been found that FDDEA is still in its initial stage of development. Based on the types of publications used in this paper, FDDEA studies are grouped into four categories, (i) theoretical development of FDDEA models with different fuzzy sets, (ii) FDDEA with network structure, (iii) application of FDDEA, and (iv) integration of FDDEA with other techniques. Figure 10.4 and Table 10.1 clearly depict that FDDEA has been emerging over the years with its concrete applications in various sectors.

References

1. Charnes, A., Cooper, W.W., Rhodes, E.: Measuring the efficiency of decision making units. Eur. J. Oper. Res. **2**, 429–444 (1978)
2. Banker, R.D., Charnes, A., Cooper, W.W.: Some models for estimating technical and scale inefficiencies in data envelopment analysis. Manage. Sci. **30**, 1078–1092 (1984)

3. Cooper, W.W., Seiford, L.M., Tone, K.: Data Envelopment Analysis: A Comprehensive Text with Models, Applications, References and DEA-Solver Software, 2nd edn. Springer, New York (2007)
4. Tone, K.: A slacks-based measure of efficiency in data envelopment analysis. Eur. J. Oper. Res. **130**, 498–509 (2001)
5. Tone, K., Tsutsui, M.: An efficiency measure of goods and bads in DEA and its application to US electric utilities. In: Asia Pacific Productivity Conference (2006)
6. Li, S., Jahanshahloo, G.R., Khodabakhshi, M.: A super-efficiency model for ranking efficient units in data envelopment analysis. Appl. Math. Comput. **184**(2), 638–648 (2007)
7. Kao, C.: Efficiency decomposition for general multi-stage systems in data envelopment analysis. Eur. J. Oper. Res. **232**, 117–124 (2014)
8. Emrouznejad, A., Yang, G.-L.: A survey and analysis of the first 40 years of scholarly literature in DEA: 1978–2016. Soc.-Econ. Plan. Sci. **61**, 4–8 (2018)
9. Contreras, I.: A review of the literature on DEA models under common set of weights. J. Model. Manag. **15**(4), 1277–1300 (2020)
10. Gholizadeh, M.H., Azbari, M.E., Abbasi, R.: Designing dynamic fuzzy data envelopment analysis model for measuring efficiency of the investment corporations in Tehran stock exchange. In: Performance Management and Measurement with Data Envelopment Analysis, p. 96 (2010)
11. Sengupta, J.K.: Dynamics of Data Envelopment Analysis: Theory of Systems Efficiency. Kluwer Academic Publishers, Dordrecht, Netherlands (1995)
12. Sengupta, J.K.: Dynamics data envelopment analysis. Int. J. Syst. Sci. **27**, 277–284 (1996)
13. Sengupta, J.K.: Dynamic aspects of data envelopment analysis. Econ. Notes **25**, 143–164 (1996)
14. Sengupta, J.K.: A dynamic efficiency model using data envelopment analysis. Int. J. Prod. Econ. **62**, 209–218 (1999)
15. Nemoto, J., Goto, M.: Dynamic data envelopment analysis: modeling intertemporal behavior of a firm in the presence of productive inefficiencies. Econ. Lett. **64**(1), 51–56 (1999)
16. Nemoto, J., Goto, M.: Measurement of dynamic efficiency in production: an application of data envelopment analysis to Japanese electric utilities. J. Prod. Anal. **19**(2), 191–210 (2003)
17. Kao, C.: Efficiency decomposition in network data envelopment analysis: a relational model. Eur. J. Oper. Res. **192**(3), 949–962 (2009)
18. Kao, C.: Dynamic data envelopment analysis: a relational analysis. Eur. J. Oper. Res. **227**(2), 325–330 (2013)
19. Puri, J., Yadav, S.P.: A fully fuzzy DEA approach for cost and revenue efficiency measurements in the presence of undesirable outputs and its application to the banking sector in India. Int. J. Fuzzy Syst. **18**(2), 212–226 (2015)
20. Zimmermann, H.J.: Fuzzy Set Theory and Its Applications, 3rd edn. Kluwer-Nijhoff Publishing, Boston (1996)
21. Jafarian-Moghaddam, A.R., Ghoseiri, K.: Fuzzy dynamic multi-objective data envelopment analysis model. Expert Syst. Appl. **38**(1), 850–855 (2011)
22. Soleimani-Damaneh, K.: A fuzzy dynamic DEA model. J. Fuzzy Set Valued Anal. 1–6 (2013)
23. Zadeh, L.A., Fu, K.S., Tanaka, K., Shimura, M.: Fuzzy Sets and Their Applications to Cognitive and Decision Processes. Academic Press, New York (1975)
24. Abdullah, L., Zulkifli, N.: Integration of fuzzy AHP and interval type-2 fuzzy DEMATEL: an application to human resource management. Expert Syst. Appl. **42**(9), 4397–4409 (2015)
25. Zhou, X., Li, L., Wen, H., Tian, X., Wang, S., Lev, B.: Supplier's goal setting considering sustainability: an uncertain dynamic data envelopment analysis based benchmarking model. Inf. Sci. **545**, 44–64 (2021)
26. Olfat, L., Amiri, M., Soufi, J.B., Pishdar, M.: A dynamic network efficiency measurement of airports performance considering sustainable development concept: a fuzzy dynamic network-DEA approach. J. Air Transp. Manag. **57**, 272–290 (2016)
27. Zadeh, L.A.: Fuzzy sets. Inf. Control **8**, 338–353 (1965)
28. Hatami-Marbini, A., Emrouznejad, A., Tavana, M.: A taxonomy and review of the fuzzy data envelopment analysis literature: two decades in the making. Eur. J. Oper. Res. **214**(3), 457–472 (2011)

29. Mariz, F.B., Almeida, M.R., Aloise, D.: A review of dynamic data envelopment analysis: state of the art and applications. Int. Trans. Oper. Res. **25**(2), 469–505 (2017)
30. Chiang, C.I., Tzeng, G.H.: A multiple objective programming approach to data envelopment analysis. In: Shi, Y., Milan, Z. (eds.) New Frontier of Decision Making for the Information Technology Era, pp. 270–285. World Scientific, Singapore (2000)
31. Kordrostami, S., Azmayandeh, O., Bakhoda, Z., Shokri, S.: The new model in interval dynamic network DEA for parallel production systems; an illustration with Iranian banks. Indian J. Sci. Technol. **6**(1), 44–53 (2013)
32. Keikha-Javan, S., Rostamy-Malkhalifeh, M., Payan, A.: The parallel network dynamic DEA model with interval data. J. Data Envel. Anal. Decis. Sci. 1–11 (2014)
33. Ghobadi, S., Jahanshahloo, G.R., Lotfi, F.H., Rostami-Malkhalifeh, M.: Dynamic inverse DEA in the presence of fuzzy data. Adv. Environ. Biol. **8**(24), 139–151 (2014)
34. Jahanshahloo, G.R., Soleimani-Damaneh, M., Reshadi, M.: On Pareto (dynamically) efficient paths. Int. J. Comput. Math. **83**(8–9), 631–635 (2006)
35. Andersen, P., Petersen, N.C.: A procedure for ranking efficient units in data envelopment analysis. Manage. Sci. **39**(10), 1261–1264 (1993)
36. Li, L., Lv, X., Xu, W., Zhang, Z., Rong, X.: Dynamic super-efficiency interval data envelopment analysis. In: 10th International Conference on Computer Science & Education (ICCSE), pp. 213–218. IEEE (2015)
37. Yaghoubi, A., Amiri, M., Safi Samghabadi, A.: A new dynamic random fuzzy DEA model to predict performance of decision making units. J. Optim. Ind. Eng. **9**(20), 75–90 (2016)
38. Yaghoubi, A., Amiri, M.: Designing a new multi-objective fuzzy stochastic DEA model in a dynamic environment to estimate efficiency of decision making units (case study: an Iranian Petroleum Company). J. Ind. Eng. Manag. Stud. **2**(2), 26–42 (2015)
39. Yen, B.T., Chiou, Y.C.: Dynamic fuzzy data envelopment analysis models: case of bus transport performance assessment. RAIRO Oper. Res. **53**(3), 991–1005 (2019)
40. Lan, L.W., Chiou, Y.C., Yen, B.T.: Integrated fuzzy data envelopment analysis to assess transport performance. Transp. A Transp. Sci. **10**(5), 401–419 (2014)
41. Ebrahimi, F., Saen, R.F., Karimi, B.: Assessing the sustainability of supply chains by dynamic network data envelopment analysis: a SCOR-based framework. Environ. Sci. Pollut. Res. 1–29 (2021)
42. Bansal, P., Mehra, A.: Integrated dynamic interval data envelopment analysis in the presence of integer and negative data. J. Ind. Manag. Optim. (2021). https://doi.org/10.3934/jimo.2021023
43. Hashimoto, A., Fukuyama, H., Tone, K.: Dynamic network DEA and an application to Japanese Prefectures. In: Workshop on Dynamic and Network DEA 2013, pp. 39–46 (2013)
44. Avkiran, N.K., McCrystal, A.: Dynamic network range-adjusted measure vs. dynamic network slacks-based measure. J. Oper. Res. Soc. Jpn. **57**(1), 1–14 (2014)
45. Tone, K., Tsutsui, M.: Dynamic DEA with network structure: a slacks-based measure approach. Omega **42**(1), 124–131 (2014)
46. Khalili-Damghani, K., Tavana, M., Santos-Arteaga, F.J., Mohtasham, S.: A dynamic multi-stage data envelopment analysis model with application to energy consumption in the cotton industry. Energy Econ. **51**, 320–328 (2015)
47. Omrani, H., Soltanzadeh, E.: Dynamic DEA models with network structure: an application for Iranian airlines. J. Air Transp. Manag. **57**, 52–61 (2016)
48. Olfat, L., Pishdar, M.: Interval type-2 fuzzy dynamic network data envelopment analysis with undesirable outputs considering double frontiers: an application to Iran airports' sustainability evaluation. Int. J. Ind. Eng. **24**(6) (2017)
49. Soltanzadeh, E., Omrani, H.: Dynamic network data envelopment analysis model with fuzzy inputs and outputs: an application for Iranian airlines. Appl. Soft Comput. **63**, 268–288 (2018)
50. Olfat, L., Amiri, M., BamdadSoufi, J., Pishdar, M.: Developing dynamic network DEA approach and its combination with interval type-2 fuzzy sets theory case of passenger airports' performance based on sustainability principles. J. Prod. Oper. Manag. **9**(2), 23–36 (2018)
51. Jafarian-Moghaddam, A.R., Ghoseiri, K.: Multi-objective data envelopment analysis model in fuzzy dynamic environment with missing values. Int. J. Adv. Manuf. Syst. **61**(5–8), 771–785 (2012)

52. Qin, R., Liu, Y.K.: A new data envelopment analysis model with fuzzy random inputs and outputs. J. Appl. Math. Comput. **33**(1), 327–356 (2010)
53. Wang, Y.M., Chin, K.S.: Fuzzy data envelopment analysis: a fuzzy expected value approach. Expert Syst. Appl. **38**(9), 11678–11685 (2011)
54. Zhou, X., Wang, Y., Chai, J., Wang, L., Wang, S., Lev, B.: Sustainable supply chain evaluation: a dynamic double frontier network DEA model with interval type-2 fuzzy data. Inf. Sci. **504**, 394–421 (2019)
55. Hasani, A.A., Mokhtari, H.: Self-efficiency assessment of sustainable dynamic network healthcare service system under uncertainty: hybrid fuzzy DEA-MCDM method. Sci. Iran. (2020)
56. Torabandeh, M.A., Dorri Nokorani, B., Motameni, A., Rabieh, M.: Comparative-fuzzy analysis of national innovation capability based on results of dynamic network DEA model. J. Ind. Manage. Perspect. **11**(2), 207–246 (2021)
57. Khodaparasti, S., Maleki, H.R.: A new combined dynamic location model for emergency medical services in fuzzy environment. In: 13th Iranian Conference on Fuzzy Systems (IFSC), pp. 1–6. IEEE (2013)
58. Nikabadi, M.S., Moghaddam, H.F.: An integrated approach of adaptive neuro-fuzzy inference system and dynamic data envelopment analysis for supplier selection. Int. J. Math. Oper. **18**(4), 503–527 (2021)
59. Tavana, M., Khalili-Damghani, K., Arteaga, F.J.S., Hosseini, A.: A fuzzy multi-objective multi-period network DEA model for efficiency measurement in oil refineries. Comput. Ind. Eng. **135**, 143–155 (2019)

Chapter 11
Role of Centrality Measures in Link Prediction on Fuzzy Social Networks

Shashank Sheshar Singh, Madhushi Verma, Samya Muhuri, and Divya Srivastava

11.1 Introduction

The rapid growth of online social networks leads to immense application potential such as outbreak detection, viral marketing, information sharing and dissemination, link prediction, and rumor control. The social network analysis becomes a hot topic in the research to tackle these real-world applications by analyzing the relationship between user's [1–3], user's relative importance [4, 5], and investigating network sub-structures [6]. The link prediction is the application of analyzing users and their relationships in the network. Link prediction is the problem of identifying the missing links and future links in the growing networks [1]. Therefore, the centrality measures can be best suited for link prediction by analyzing node and their relationship. In general, social network considers the connection strength based on their direct connections and ignores the indirect impact of users on each other. Although, some studies like the three-degree theory and the six-degree separation suggest that users' impact is limited to their local regions rather than global. However, these studies also consider users' impact on three hops and six hop counts. Fuzzy theory can be more suitable in these scenarios to incorporate imprecision and uncertainty of the user's impact on the network. We have utilized the six-degree separation to investigate an individual influence locally and include the six hop distance between individuals.

The fuzzy social networks consider the fuzzy membership value for each non-existing link along with existing connections by utilizing connectedness and their path [7]. The connectedness measures the affinity of belongingness or closeness to others. We have utilized connectedness with small-world phenomena to explore the

Madhushi Verma, Samya Muhuri and Divya Srivastava are contributed equally to this work.

S. S. Singh (✉) · S. Muhuri
Department of Computer Science and Engineering, Thapar Institute of Engineering and Technology, Patiala, Punjab 147004, India
e-mail: shashank.sheshar@gmail.com

M. Verma · D. Srivastava
School of Computer Science and Engineering, Bennett University, Greater Noida, Uttar Pradesh 201310, India

T. Som et al. (eds.), *Fuzzy, Rough and Intuitionistic Fuzzy Set Approaches for Data Handling*, Forum for Interdisciplinary Mathematics,
https://doi.org/10.1007/978-981-19-8566-9_11

closeness or possibility of knowing each other for predicting future links. Therefore, this chapter utilizes centrality measures and fuzzy theory to predict future links on social networks. The main contribution of the work is as follows.

- The real-world weighted social networks modeled into fuzzy social networks based on six-degree separation and fuzzy theory. The membership function is defined to estimate fuzzy connection strength by associating small-world phenomena.
- The likelihood score computation function is defined based on features like common neighbors, clustering coefficient, and preferential attachments. The node and edge centrality measures are used for similarity index computation.
- The experiments are performed on real-world social networks. The performance of different centrality measures is compared to various link prediction performance matrices on fuzzy social networks. The experimental results validate the utility of centrality measures for the link prediction problem.

The remainder of the chapter is distributed as follows. Section 11.2 elaborates basic concepts like network model, the taxonomy of centrality measures, etc. It also presents an overview of various centrality measures used in this chapter for link prediction. Section 11.3 describes fuzzy social networks along with the link prediction approach using centrality measures. Section 11.4 presents an empirical analysis of various centrality measures corresponding to the link prediction problem. Finally, Sect. 11.5 presents concluding remarks of the chapter along with future possibilities.

11.2 Preliminaries

11.2.1 Social Network

A social network can be represented through an undirected graph $G(V, E)$ where V signifies the set of vertices which in a social network implies people, enterprises, organizations, groups, etc., and E denotes the set of edges. The edges depict the associations, connection, relation, and flows between the nodes. These social networks can also be presented using adjacency matrix which provides an insight about the links between the nodes of the graph by just looking at the matrix. This mathematical representation of social networks is easy to interpret and analyze as the associations between the nodes become clearly visible. The adjacency matrix can be denoted as M consisting of element $m_{u,v} \in (0, 1)$, where

$$m_{u,v} = \begin{cases} 1, & \text{if } (u, v) \in E \\ 0, & \text{otherwise} \end{cases}$$

The degree of each node in a social network is computed by summing up the number of edges incident on a node and can be denoted as $D(u)$. An alternating sequence of

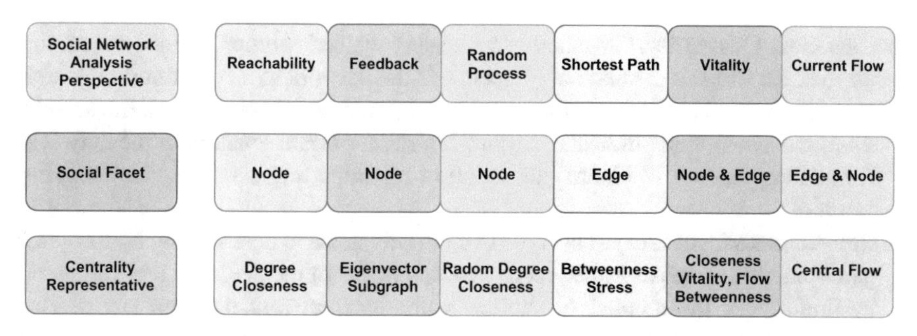

Fig. 11.1 Centrality measures taxonomy [8]

nodes and links where each link lying between two nodes represents the association between them is called a walk. Nodes may be repeated in a walk but a walk with unique representation of nodes and edges is called a path. In a social network, if there is a path between every pair of entities denoted by the nodes, then the network is referred to as connected. The edges in the network can also have weights representing various parameters. Such a network is called weighted. If the edges in the network have direction associated with it describing the direction of flow, then the network is called connected. Neighbors of the node u, represented as $N(u)$ is a set of nodes adjacent to u i.e., $m_{u,v} = 1$.

11.2.2 Centrality Measures

Figure 11.1 presents the taxonomy of the centrality measures [9]. These centrality measures can be node as well as edge centrality.

- **Degree Centrality**. It is a score assigned to the nodes based on the number of edges incident on it. It is easy to compute and is equal to the degree of each node. Degree centrality simply denotes how many connections an entity or component of the network have and higher the score, the more centrally located is the component in the network. In a social network, if a node has 10 connections, then the degree centrality would be 10. Similarly, if the vertex has only one edge incident on it, then the degree centrality would be 1. In this centrality measure, the focus is on the local structure rather than the global perspective and hence the complexity is less. This measure is apt for fast processing applications.
- **Edge Betweenness Centrality**. This type of centrality helps in determining how much a particular link (u, v) exists in-between of a path for a pair of nodes on the network. It is computed by finding out the number of shortest paths in a network between any pair of nodes that cross or pass through this link (u, v) under consideration. A high value of betweenness indicates that the selected link sufficiently influences the flow of information happening in the network as most of exchange that takes places within two nodes passes through (u, v).

- **Closeness Centrality**. Closeness is a measure which is computed by determining the average shortest distance of each node in the network to all other nodes. A high closeness value indicates that the node has large number of close associations, and a small closeness value indicates a more central/important location in the network. It finds application in identifying location for imparting services like shopping complex.
- **Betweenness Centrality**. This type of centrality helps in determining how much a particular node u exists in-between two other nodes of the network. It is computed by finding out the number of shortest paths in a network between any pair of nodes that cross or pass through this node u under consideration. A high value of betweenness indicates that the selected node sufficiently influences the flow of information happening in the network as most of exchange that takes places within two nodes passes through u.
- **PageRank Centrality**. This type of centrality accounts for the direction of the links. A vertex in the network is allotted a score based on the incoming links which also carry weights based on the relative scores of their corresponding source nodes. Hence, a higher PageRank score indicates that the node u under consideration is quite influential in the network and the same is propagated to the other nodes which are connected to u. The importance of a node is computed based on its neighborhood instead of any sort of distance.
- **Harmonic Centrality**. This is a variant of closeness centrality. It can be used in applications where the objective is to decide where the upcoming public services should be located or positioned in the city so that it remains accessible to most of the residents in the city. If the goal is to spread some important message in the social networks, then this method can be used to select the influencing entities which can help in achieving the goal of spreading the information. For a given node u, harmonic centrality is computed as the sum of the inverse of shortest-path distances between all other nodes in the network and u.
- **Load Centrality**. This is an edge-based centrality measure. It is computed as the fraction of all the paths that pass through the edge under consideration (u, v) and have a length less than a defined cut-off value. This centrality can be used for analyzing the flow networks which are functioning well below their capacity constraints.

11.2.3 Link Prediction

In today's era, social networks have become very popular and play an important role in representing the association between the entities or components. The links in the network also signify the behavior of the relationship between the components and on analyzing this behavior of interaction between the components, some useful insights can be generated regarding the properties of the underlying network. Study of the existing connections may also lead to prediction of new linkages which may possibly exist, and this process is defined as link prediction. In link prediction, the task is to analyze the structure and behavior of the existing connections at a particular instant of time and based on that determine the possibility of having links between some other two nodes. Solutions designed for this purpose may lead to a better interpretation of the overall network architecture and behavior, assessment and modeling of the network and classification of the components which are unclassified. In the recent past, this problem has gained a lot of attention from the researchers [10–12]. Initially, link prediction problem was applied in the academic citation networks and social networks by Newman [1] and Liben-Nowell et al. [11], respectively. Few other link prediction approaches include similarity-based methods which gained popularity because of the low computational cost for complex graphs and easier implementation of the algorithms. Out of the several similarity-based methods, in structural similarities, the levels of the structure that include both local and global structures existing in the network are used to calculate the probability score of those links which are not currently existing in the network [13, 14]. Quasi-local similarity method combines the advantages and important aspects of both, i.e., the local and the global indices. In another strategy called the centrality-based link prediction, the likelihood score of those links which may be non-existent at a particular instant is predicted using the centrality index of the existing ones. It includes node as well as edge centrality [9]. The authors of [15] presented a link prediction algorithm considering node and edge relevance on the multiplex social network by exploring distinct relationships. The authors applied network aggregation to capture different relationships. Therefore, single network algorithms can be easily applied to the proposed model. Samanta et al. [16] presented a notion of influence under fuzzy settings. They have developed fuzzy parameters to capture real-world characteristics to measure individual influence.

11.3 Centrality-Based Link Prediction

The growth and development of online social networks lead to the prediction of new nodes and future links. Most of the existing work focuses on analyzing social relationships and topological information to predict future links. This work divides into three steps: (1) fuzzy social network modeling, (2) similarity index computation, and (3) likelihood score computation.

11.3.1 Fuzzy Social Network Modeling

In general, social networks have been considered as a binary relation, i.e., a pair of individuals are either connected or not. The binary relation only focuses on the direct connections and ignores the connection strength. Sometimes, users are not directly connected to each other, but they have some influence on each other based on their similarity and closeness. However, the relationship among the user's in

Table 11.1 Fuzzy connection strength estimation example

		Path specific fuzzy strength										Fuzzy connection strength (unnormalized)									
		A	B	C	D	E	F	G	H	I	J	A	B	C	D	E	F	G	H	I	J
$l = 1$	A	0.00	0.00	0.70	0.00	0.50	0.00	0.60	0.00	0.70	0.50	0.00	0.00	0.70	0.00	0.50	0.00	0.60	0.00	0.70	0.50
	B	0.00	0.00	0.00	0.70	0.00	0.50	0.00	0.40	0.40	0.70	0.00	0.00	0.00	0.70	0.00	0.50	0.00	0.40	0.40	0.70
	C	0.70	0.00	0.00	0.00	0.10	0.00	0.00	0.00	0.80	0.00	0.70	0.00	0.00	0.00	0.10	0.00	0.00	0.00	0.80	0.00
	D	0.00	0.70	0.00	0.00	0.00	0.80	0.00	0.00	0.50	0.00	0.00	0.70	0.00	0.00	0.00	0.80	0.00	0.00	0.50	0.00
	E	0.50	0.00	0.10	0.00	0.00	0.00	0.00	0.00	0.00	0.00	0.50	0.00	0.10	0.00	0.00	0.00	0.00	0.00	0.00	0.00
	F	0.00	0.50	0.00	0.80	0.00	0.00	0.00	0.00	0.00	0.00	0.00	0.50	0.00	0.80	0.00	0.00	0.00	0.00	0.00	0.00
	G	0.60	0.00	0.00	0.00	0.00	0.00	0.00	0.00	0.00	0.40	0.60	0.00	0.00	0.00	0.00	0.00	0.00	0.00	0.00	0.40
	H	0.00	0.40	0.00	0.00	0.00	0.00	0.00	0.00	0.00	0.10	0.00	0.40	0.00	0.00	0.00	0.00	0.00	0.00	0.00	0.10
	I	0.70	0.40	0.80	0.50	0.00	0.00	0.00	0.00	0.00	0.00	0.70	0.40	0.80	0.50	0.00	0.00	0.00	0.00	0.00	0.00
	J	0.50	0.70	0.00	0.00	0.00	0.00	0.40	0.10	0.00	0.00	0.50	0.70	0.00	0.00	0.00	0.00	0.40	0.10	0.00	0.00
$l = 2$	A	0.00	0.25	0.78	0.27	0.44	0.00	0.62	0.04	0.98	0.57	0.00	0.32	1.01	0.35	0.57	0.00	0.80	0.05	1.26	0.74
	B	0.25	0.00	0.25	0.78	0.00	0.82	0.22	0.36	0.58	0.57	0.32	0.00	0.32	1.00	0.00	1.06	0.28	0.47	0.75	0.74
	C	0.78	0.25	0.00	0.31	0.35	0.00	0.33	0.00	1.00	0.27	1.01	0.32	0.00	0.40	0.45	0.00	0.42	0.00	1.29	0.35
	D	0.27	0.78	0.31	0.00	0.00	0.89	0.00	0.22	0.61	0.38	0.35	1.00	0.40	0.00	0.00	1.15	0.00	0.28	0.78	0.49
	E	0.44	0.00	0.35	0.00	0.00	0.00	0.23	0.00	0.17	0.19	0.57	0.00	0.45	0.00	0.00	0.00	0.30	0.00	0.22	0.25
	F	0.00	0.82	0.00	0.89	0.00	0.00	0.00	0.16	0.23	0.27	0.00	1.06	0.00	1.15	0.00	0.00	0.00	0.20	0.30	0.35
	G	0.62	0.22	0.33	0.00	0.23	0.00	0.00	0.03	0.33	0.54	0.80	0.28	0.42	0.00	0.30	0.00	0.00	0.04	0.42	0.70
	H	0.04	0.36	0.00	0.22	0.00	0.16	0.03	0.00	0.12	0.30	0.05	0.47	0.00	0.28	0.00	0.20	0.04	0.00	0.16	0.38
	I	0.98	0.58	1.00	0.61	0.17	0.23	0.33	0.12	0.00	0.25	1.26	0.75	1.29	0.78	0.22	0.30	0.42	0.16	0.00	0.32
	J	0.57	0.57	0.27	0.38	0.19	0.27	0.54	0.30	0.25	0.00	0.74	0.74	0.35	0.49	0.25	0.35	0.70	0.38	0.32	0.00
$l = 3$	A	0.00	0.03	0.04	0.02	0.01	0.00	0.01	0.01	0.03	0.01	0.00	0.35	1.05	0.37	0.58	0.00	0.81	0.06	1.29	0.75
	B	0.03	0.00	0.02	0.04	0.00	0.02	0.01	0.01	0.02	0.00	0.35	0.00	0.34	1.04	0.00	1.08	0.29	0.48	0.77	0.74
	C	0.06	0.03	0.00	0.03	0.02	0.00	0.03	0.00	0.04	0.02	1.07	0.35	0.00	0.43	0.47	0.00	0.45	0.00	1.33	0.37
	D	0.03	0.06	0.03	0.00	0.00	0.04	0.00	0.02	0.03	0.02	0.38	1.06	0.43	0.00	0.00	1.19	0.00	0.30	0.80	0.51
	E	0.02	0.00	0.03	0.00	0.00	0.00	0.02	0.00	0.06	0.02	0.59	0.00	0.48	0.00	0.00	0.00	0.32	0.00	0.28	0.27
	F	0.00	0.06	0.00	0.05	0.00	0.00	0.00	0.02	0.08	0.04	0.00	1.12	0.00	1.20	0.00	0.00	0.00	0.22	0.38	0.39
	G	0.03	0.03	0.04	0.00	0.02	0.00	0.00	0.01	0.05	0.03	0.83	0.31	0.46	0.00	0.32	0.00	0.00	0.05	0.47	0.73
	H	0.01	0.01	0.00	0.02	0.00	0.02	0.01	0.00	0.02	0.02	0.06	0.48	0.00	0.30	0.00	0.22	0.05	0.00	0.18	0.40
	I	0.03	0.02	0.03	0.02	0.03	0.04	0.03	0.01	0.00	0.04	1.29	0.77	1.32	0.80	0.25	0.34	0.45	0.17	0.00	0.36
	J	0.01	0.00	0.02	0.02	0.01	0.02	0.01	0.01	0.04	0.00	0.75	0.74	0.37	0.51	0.26	0.37	0.71	0.39	0.36	0.00

the social network is vague in nature and associated with some terminology such as weakly, strongly, moderately, equally, and extremely connected. Therefore, the fuzzy set theory presented by Zadeh [17, 18] is best suited to incorporate connection strength and dealing imprecision on social networks. There are some studies that incorporate fuzzy theory with social networks to form fuzzy social networks [19]. A fuzzy social network corresponding to a social network $G(V, E)$ can be represented as $\tilde{G}_{\text{FSN}} = (\tilde{V}, \tilde{E})$, where $\tilde{V} = V$ denotes the user's in the network and $\tilde{E} \neq E$ is the fuzzy relationship between individual's which uses membership function to assign relationship strength. If the network is an undirected fuzzy network, then $\tilde{e}_{uv} = \tilde{e}_{vu}$, otherwise $\tilde{e}_{uv} \neq \tilde{e}_{vu}$ in case of directed fuzzy networks. The membership function of connection strength is defined as $\mu(e_{u,v}) = \prod_{i=1}^{i=j} w(e_{v_i, v_{i+1}})$ if there is a path $u - v_1 - v_2 - \cdots - v_j$ exists. Similarly, if there are l paths between u to v then fuzzy connection strength can be computed as $\mu(e_{u,v}) = \sum_{i=1}^{i=l} \mu(e^i_{u,v})$ and $\mu(e_{u,v}) = 0$ when no path exists between u and v.

Some of the studies suggest that an individual influence is limited to its local regions, such as three-degree theory and small-world phenomena. This chapter utilizes a six-degree separation to compute the membership score for a pair of individuals in the fuzzy social network. For example, Table 11.1 presents an influence graph with edge weights are associated to its direct paths, i.e., path length $l = 1$. Then using fuzzy modeling, the social network is converted to fuzzy social networks by considering the indirect path for information spreading and communication. First, the fuzzy strength of each pair is calculated corresponding to path length l, then a cumulative fuzzy strength is calculated by incorporating all the paths less than l. This is because all the paths are independent of each other in information sharing.

11.3.2 Similarity Index Computation

The centrality measures have been used to estimate similarity index between a pair of nodes on fuzzy social networks. We have utilized seven centrality measures for both node and edge similarity in this chapter. The similarity index $I_S(u, v)$ corresponding to an edge (u, v) using edge centrality can be computed as $I_S(u, v) \leftarrow I_C(u, v)$, where $I_C(u, v)$ denotes edge centrality index for edge (u, v). Similarly, the similarity index $I_S(u, v)$ [9] corresponding to an edge (u, v) using node centrality can be computed as $I_S(u, v) \leftarrow \frac{I_C(u)+I_C(v)}{2}$, where $I_C(u)$ is node centrality for node u.

11.3.3 Likelihood Index Computation

Finally, the likelihood score for each non-existing edge (u, v) will be computed to predict missing and future links [9]. The likelihood score $I_L(x, y)$ for a pair of individuals (x, y) can be computed using feature set z as follows.

Table 11.2 Dataset information

Dataset	Nodes	Edges	Avd degree	Radius	Diameter	Density	ASPL	ACC
LESMIS	77	254	6.597	3.000	5.000	0.087	2.607	0.573
CONTACT_DIARY	123	597	9.707	2.000	4.000	0.080	2.047	0.542
FOOTBALL	35	118	6.743	3.000	5.000	0.198	2.062	0.339
WORLD_TRADE	80	875	21.875	2.000	3.000	0.277	1.703	0.752
STARLINKS	113	607	10.743	−1.000	−1.000	0.096	2.127	0.677
SOCIALWORKJ	36	99	5.500	3.000	6.000	0.157	2.389	0.000

$$I_L(x, y) \leftarrow \sum_{z} \frac{I_S(u, z) + I_S(z, v)}{\sum_{w \in N^{\text{out}}(z)} I_S(z, w)}$$

where, z can be computed using common neighbors, clustering coefficient, preferential attachment, etc., feature sets.

11.4 Performance Analysis

This section has performed experiments on six real-world social networks over seven centrality measures to analyze the link prediction performance. Table 11.2 presents the dataset information like size, average degree, radius, diameter, and density. We have evaluated each centrality measure for link prediction on the fuzzy social network over three performance metrics area under the receiver operating characteristics curve (AUC), Balanced Accuracy, and $F1$-Score [12, 20].

11.4.1 AUC

Table 11.3 presents the comparison of different centrality measures performance in terms of AUC metrics over six real-world social networks. The PageRank centrality index performs best among all centrality measures in all the datasets except Starlinks and World_trade datasets. The degree and closeness centrality perform best in Starlinks and World_trad datasets, respectively. The PageRank centrality is approximate to other centrality measures on SocialWorkJ network dataset. The Load centrality performs worst in Lesmis and Contact_diary datasets. The Edge Betweenness centrality performs worst in Football, World_trade, and Starlinks datasets. The degree centrality performs worst in SocialWorkJ dataset.

Table 11.3 AUC results

Dataset	Ratio	Edge betweenness	Closeness	Betweenness	Degree	Load	Harmonic	PageRank
Lesmis	0.1	0.86166	0.91318	0.89102	0.90829	0.88947	0.89051	0.9135
	0.2	0.84341	0.89639	0.88052	0.89103	0.87803	0.89596	0.90494
	0.3	0.82694	0.88606	0.86139	0.87824	0.85575	0.87602	0.87029
	0.4	0.80967	0.85118	0.83806	0.84812	0.83685	0.85048	0.84939
	0.5	0.77652	0.80707	0.80494	0.81114	0.80363	0.81109	0.80799
Contact_diary	0.1	0.88355	0.90382	0.88671	0.90321	0.8766	0.91414	0.90769
	0.2	0.85586	0.87719	0.86337	0.88155	0.85606	0.87915	0.87413
	0.3	0.82168	0.8417	0.82683	0.83868	0.82611	0.84608	0.84367
	0.4	0.78173	0.79079	0.78343	0.79158	0.77508	0.79591	0.79755
	0.5	0.72549	0.73546	0.73152	0.73474	0.72625	0.73092	0.73901
FootballL	0.1	0.6226	0.65395	0.65352	0.66311	0.63973	0.63681	0.66347
	0.2	0.61788	0.63089	0.64875	0.64633	0.63425	0.63381	0.64089
	0.3	0.59977	0.61508	0.63906	0.64033	0.63499	0.62576	0.63009
	0.4	0.59973	0.60918	0.60945	0.62282	0.61088	0.60769	0.61967
	0.5	0.57193	0.592	0.59599	0.59277	0.58988	0.5915	0.59713
World_trade	0.1	0.65015	0.89814	0.77364	0.7504	0.77022	0.82977	0.85136
	0.2	0.63868	0.88158	0.77426	0.78085	0.77655	0.78379	0.84527
	0.3	0.62725	0.86241	0.76874	0.74408	0.7729	0.74805	0.83465
	0.4	0.62062	0.83214	0.77357	0.73559	0.76646	0.71378	0.81829
	0.5	0.60478	0.8014	0.75605	0.73047	0.75147	0.70638	0.79511
Starlinks	0.1	0.87157	0.90925	0.90636	0.91264	0.90061	0.9178	0.90965
	0.2	0.8575	0.89195	0.89135	0.89556	0.88933	0.88993	0.89266
	0.3	0.83613	0.86714	0.86958	0.87401	0.87244	0.86557	0.87348
	0.4	0.80544	0.83528	0.83657	0.84397	0.84175	0.83702	0.84262
	0.5	0.76367	0.79251	0.79443	0.8032	0.80131	0.79372	0.80311
SocialWorkJ	0.1	0.28479	0.28389	0.28513	0.28285	0.28411	0.28208	0.28097
	0.2	0.30186	0.30056	0.29938	0.30014	0.29901	0.29865	0.30211
	0.3	0.32689	0.32325	0.32403	0.32274	0.32481	0.32087	0.32416
	0.4	0.35217	0.34876	0.34806	0.34692	0.34893	0.35212	0.34908
	0.5	0.37974	0.37858	0.37853	0.37734	0.3778	0.37875	0.37835

11.4.2 Balanced Accuracy

Table 11.4 compares different centrality measures performance in terms of Balanced Accuracy metrics over six real-world social networks. The PageRank centrality index performs best among all centrality measures in all the datasets except Football and Starlinks datasets. The degree centrality performs best among all centrality measures in Football and Starlinks datasets. The Edge Betweenness centrality performs worst among all centrality measures in all the datasets except Contact_diary dataset. The Load centrality performs worst in Contact_diary dataset with respect to Balanced Accuracy metrics.

Table 11.4 Balanced accuracy results

Dataset	Ratio	Edge betweenness	Closeness	Betweenness	Degree	Load	Harmonic	PageRank
Lesmis	0.1	0.8111	0.84533	0.83854	0.85086	0.84272	0.83206	0.85443
	0.2	0.79012	0.83755	0.83143	0.84093	0.83087	0.83874	0.85041
	0.3	0.78406	0.83411	0.81257	0.83158	0.81032	0.82633	0.82579
	0.4	0.76778	0.80863	0.79541	0.81094	0.79306	0.80648	0.80955
	0.5	0.74858	0.78126	0.7722	0.78384	0.77391	0.78461	0.78076
Contact_ diary	0.1	0.84621	0.86896	0.82113	0.86676	0.81821	0.87922	0.86983
	0.2	0.82093	0.85351	0.8113	0.85281	0.80452	0.85605	0.84853
	0.3	0.78783	0.8188	0.78379	0.81007	0.78688	0.82383	0.81643
	0.4	0.74839	0.77292	0.75433	0.77388	0.7489	0.77553	0.77924
	0.5	0.70262	0.7178	0.70938	0.71985	0.70624	0.71521	0.72117
FootbalL	0.1	0.5898	0.62446	0.61502	0.64163	0.59195	0.61055	0.63208
	0.2	0.58254	0.61526	0.61159	0.62479	0.60947	0.60748	0.62281
	0.3	0.57246	0.58964	0.60608	0.61371	0.60617	0.60306	0.60004
	0.4	0.5745	0.59125	0.58404	0.60644	0.58779	0.58779	0.59588
	0.5	0.55287	0.58212	0.58099	0.58757	0.58291	0.57748	0.58762
World_ trade	0.1	0.59008	0.80486	0.69138	0.64987	0.68574	0.75672	0.76035
	0.2	0.58461	0.80277	0.69087	0.69504	0.68968	0.69518	0.74716
	0.3	0.58449	0.79074	0.68893	0.67237	0.69059	0.65258	0.73767
	0.4	0.58833	0.76373	0.69277	0.66038	0.68773	0.61692	0.72855
	0.5	0.5786	0.73881	0.68644	0.6575	0.68163	0.61446	0.7177
Starlinks	0.1	0.78696	0.82017	0.82759	0.82897	0.82173	0.82883	0.82981
	0.2	0.77426	0.81366	0.81813	0.82381	0.82042	0.81125	0.81993
	0.3	0.75774	0.79602	0.80289	0.80613	0.8068	0.79259	0.80498
	0.4	0.7349	0.77368	0.77834	0.78205	0.78222	0.77544	0.78471
	0.5	0.70428	0.74713	0.75363	0.76078	0.75776	0.74783	0.7614
SocialWorkJ	0.1	0.35684	0.36306	0.37939	0.38882	0.37753	0.36334	0.37732
	0.2	0.35909	0.36527	0.38113	0.3831	0.38142	0.36586	0.36911
	0.3	0.3672	0.36974	0.37897	0.38665	0.38134	0.37018	0.37879
	0.4	0.37553	0.36432	0.37736	0.38673	0.37629	0.36645	0.37511
	0.5	0.39301	0.37942	0.38337	0.38633	0.38322	0.379	0.38102

11.4.3 F1-Score

Table 11.5 compares different centrality measures performance in terms of $F1$-Score metrics over six real-world social networks. The Betweenness centrality performs best among all centrality measures in Contact_diary and Starlinks datasets. The Load centrality performs best in the Lesmis dataset with respect to $F1$-Score metrics. The PageRank centrality index performs best among all centrality measures in Contact_diary dataset. The Edge Betweenness centrality performs worst among all centrality measures in all the datasets except Lesmis and Contact_diary datasets.

Table 11.5 $F1$-score results

Dataset	Ratio	Edge betweenness	Closeness	Betweenness	Degree	Load	Harmonic	PageRank
Lesmis	0.1	0.09351	0.08035	0.09586	0.08797	0.09722	0.07997	0.08842
	0.2	0.16613	0.15711	0.17016	0.16403	0.17167	0.15628	0.16624
	0.3	0.23396	0.22658	0.22489	0.2392	0.22632	0.21847	0.22434
	0.4	0.28577	0.25998	0.25645	0.28296	0.25485	0.26073	0.26481
	0.5	0.33848	0.31873	0.3104	0.32916	0.31277	0.31828	0.31105
Contact_diary	0.1	0.09206	0.09007	0.10495	0.09102	0.10364	0.09158	0.08973
	0.2	0.15971	0.173	0.18131	0.17442	0.17889	0.17385	0.17024
	0.3	0.19742	0.22632	0.22476	0.21573	0.22994	0.22788	0.21861
	0.4	0.22328	0.25143	0.24944	0.25075	0.2431	0.25226	0.25429
	0.5	0.22888	0.24135	0.24237	0.24638	0.24343	0.23789	0.24399
FootbalL	0.1	0.06891	0.07969	0.07925	0.08779	0.07227	0.07543	0.08404
	0.2	0.12123	0.13995	0.14123	0.14924	0.14036	0.13599	0.14752
	0.3	0.16052	0.17297	0.18949	0.19416	0.18976	0.18351	0.18291
	0.4	0.19925	0.21601	0.21193	0.23363	0.21471	0.21259	0.22152
	0.5	0.20566	0.23518	0.23585	0.24167	0.2376	0.23048	0.24194
World_trade	0.1	0.11471	0.18751	0.16825	0.16453	0.1649	0.20593	0.21883
	0.2	0.18239	0.32934	0.28151	0.30719	0.27933	0.29781	0.34552
	0.3	0.23361	0.42666	0.35712	0.36303	0.35978	0.33009	0.43226
	0.4	0.27673	0.4714	0.41787	0.39515	0.40944	0.33005	0.4776
	0.5	0.29675	0.49515	0.44806	0.42381	0.44164	0.35403	0.50107
Starlinks	0.1	0.06294	0.06914	0.07786	0.07574	0.0765	0.0706	0.07499
	0.2	0.11819	0.13376	0.1469	0.14919	0.14807	0.13287	0.1457
	0.3	0.16475	0.18771	0.20622	0.20209	0.20574	0.18502	0.19852
	0.4	0.20119	0.22633	0.24311	0.24863	0.24627	0.22691	0.2461
	0.5	0.22546	0.25758	0.27438	0.28571	0.27803	0.25833	0.27788
SocialWorkJ	0.1	0.11418	0.16322	0.1651	0.17954	0.15493	0.17253	0.17624
	0.2	0.18563	0.23695	0.24948	0.26	0.24824	0.23877	0.24704
	0.3	0.21859	0.2858	0.30321	0.30405	0.30334	0.27501	0.29962
	0.4	0.23041	0.29035	0.31039	0.32152	0.31105	0.292	0.31374
	0.5	0.23091	0.27878	0.30443	0.30886	0.30543	0.28098	0.30879

Some conclusions can also be made from the experiments about the correlation between network properties and the performance of centrality measures corresponding to link prediction. Datasets with clustering coefficient and Gini Coefficient have a positive relationship with performance, while average shortest path (ASP) and diameter negatively correlate with performance. This is because diameter and ASP reflect how easy to communicate with each other, while the clustering coefficient explores the mutual connectedness between individuals.

11.5 Conclusion and Future Directions

This chapter presents the study of link prediction over various centrality measures on fuzzy social networks. Firstly, we model online social networks into the fuzzy system based on path information and connectedness, such as three-degree theory and six-degree separation phenomena. Then similarity indices of all the existing connections have been computed. Next, the likelihood index of each non-existing relationship has been estimated based on some feature sets such as common neighbors, preferential attachment, and clustering coefficient. Finally, missing links are predicted using supervised learning. The experimental results are obtained on real-world social networks to analyze the performance of different centrality measures over link prediction problems under fuzzy settings. This work only considers the fuzzy connection strength between peers as it only considers centrality measures. New fuzzy parameters, such as incorporating properties of the associated network, the weighting factor of different relationships, structure-specific, etc., and centrality indices, can effectively solve real-world problems. Many future directions can be possible with the incorporation of information dissemination, contextual features, multiple and dynamic networks, rough set theory and granular computing, game theory, etc., to simulate more realistic settings.

References

1. Newman, M.E.: Clustering and preferential attachment in growing networks. Phys. Rev. E Stat. Nonlinear Soft Matter Phys. **64**(2 Pt 2), 025102 (2001)
2. Singh, S.S., Srivastva, D., Verma, M., Singh, J.: Influence maximization frameworks, performance, challenges and directions on social network: a theoretical study. J. King Saud Univ. Comput. Inf. Sci. (2021). https://doi.org/10.1016/j.jksuci.2021.08.009
3. Singh, S.S., Mishra, S., Kumar, A., Biswas, B.: CLP-ID: community-based link prediction using information diffusion. Inf. Sci. **514**, 402–433 (2020). https://doi.org/10.1016/j.ins.2019.11.026
4. Singh, S.S., Singh, K., Kumar, A., Biswas, B.: ACO-IM: maximizing influence in social networks using ant colony optimization. Soft Comput., 1–23 (2019). https://doi.org/10.1007/s00500-019-04533-y
5. Singh, S.S., Kumar, A., Singh, K., Biswas, B.: IM-SSO: maximizing influence in social networks using social spider optimization. Concurr. Comput. Pract. Exp. **32**(2), 5421 (2020) https://arxiv.org/abs/, https://onlinelibrary.wiley.com/doi/pdf/10.1002/cpe.5421, https://doi.org/10.1002/cpe.5421.e5421cpe.5421
6. Biswas, A., Biswas, B.: FuzAg: fuzzy agglomerative community detection by exploring the notion of self-membership. IEEE Trans. Fuzzy Syst. **26**(5), 2568–2577 (2018). https://doi.org/10.1109/TFUZZ.2018.2795569
7. Singh, S.S., Srivastva, D., Kumar, A., Srivastava, V.: FLP-ID: fuzzy-based link prediction in multiplex social networks using information diffusion perspective. Knowl. Based Syst. **248**, 108821 (2022). https://doi.org/10.1016/j.knosys.2022.108821
8. Koschützki, D., Lehmann, K.A., Peeters, L., Richter, S., Tenfelde-Podehl, D., Zlotowski, O.: In: Brandes, U., Erlebach, T. (eds.) Centrality indices, pp. 16–61. Springer, Berlin, Heidelberg (2005). https://doi.org/10.1007/978-3-540-31955-9_3

9. Singh, S.S., Mishra, S., Kumar, A., Biswas, B.: In: Biswas, A., Patgiri, R., Biswas, B. (eds.) Link prediction on social networks based on centrality measures, pp. 71–89. Springer, Singapore (2022). https://doi.org/10.1007/978-981-16-3398-0_4
10. Kumar, A., Singh, S.S., Singh, K., Biswas, B.: Link prediction techniques, applications, and performance: a survey. Phys. A Stat. Mech. Appl., 124289 (2020). https://doi.org/10.1016/j.physa.2020.124289
11. Liben-Nowell, D., Kleinberg, J.: The link prediction problem for social networks. In: Proceedings of the Twelfth International Conference on Information and Knowledge Management. CIKM'03, pp. 556–559. ACM, New York, NY, USA (2003). https://doi.org/10.1145/956863.956972
12. Kumar, A., Mishra, S., Singh, S.S., Singh, K., Biswas, B.: Link prediction in complex networks based on significance of higher-order path index (SHOPI). Phys. A Stat. Mech. Appl., 123790 (2019). https://doi.org/10.1016/j.physa.2019.123790
13. Jaccard, P.: Distribution de la flore alpine dans le bassin des dranses et dans quelques régions voisines. Bull. Soc. Vaudoise Sci. Nat. **37**, 241–272 (1901)
14. Cannistraci, C.V., Alanis-Lobato, G., Ravasi, T.: From link-prediction in brain connectomes and protein interactomes to the local-community-paradigm in complex networks. Sci. Rep. **3**, 1613 (2013). https://doi.org/10.1038/srep01613
15. Mishra, S., Singh, S.S., Kumar, A., Biswas, B.: MNERLP-MUL: merged node and edge relevance based link prediction in multiplex networks. J. Comput. Sci. **60**, 101606 (2022). https://doi.org/10.1016/j.jocs.2022.101606
16. Samanta, S., Dubey, V.K., Sarkar, B.: Measure of influences in social networks. Appl. Soft Comput. **99**, 106858 (2021). https://doi.org/10.1016/j.asoc.2020.106858
17. Zadeh, L.A.: Fuzzy sets. Inf. Control **8**(3), 338–353 (1965). https://doi.org/10.1016/S0019-9958(65)90241-X
18. Zadeh, L.: A fuzzy-algorithmic approach to the definition of complex or imprecise concepts. Int. J. Hum. Comput. Stud. Int. J. Man Mach. Stud. **8**, 249–291 (1976)
19. Ren-jie, H.: On the definition and property analysis of fuzzy social network based on fuzzy graph. J. Guangdong Univ. Technol. (2012)
20. Manning, C.D., Raghavan, P., Schütze, H.: Introduction to Information Retrieval. Cambridge University Press, New York, NY, USA (2008)

Chapter 12
Interval Solutions of Fractional Integro-differential Equations by Using Modified Adomian Decomposition Method

Kasimala Narasimha Rao and S. Chakraverty

12.1 Introduction

In this chapter, we consider CFVFIDE as [1–3]:

$$^{c}D^{\alpha}v(x) = a(x)v(x) + g(x) + \int_{0}^{x} K_1(x,s)M_1(v(s))\mathrm{d}s + \int_{0}^{1} K_2(x,s)M_2(v(s))\mathrm{d}s \tag{12.1.1}$$

with the initial condition,

$$v(0) = v_0, \tag{12.1.2}$$

where $^{c}D^{\alpha}$ is Caputo's fractional derivatives (CFD), $0 < \alpha \leq 1$, and $v : J \rightarrow R$, where $J = [0, 1]$ is the continuous function (CF) which has to be determined, $g : J \rightarrow R$ and $K_i : J \times J \rightarrow R$, $i = 1, 2$ are CF's $M_i : R \rightarrow R$, $i = 1, 2$ are Lipchitz continuous functions. In Eqs. (12.1.1) and (12.1.2), to prove existence and uniqueness result proved [3].

K. N. Rao (✉) · S. Chakraverty
Department of Mathematics, National Institute of Technology Rourkela, Rourkela 769008, India
e-mail: knarasimharao2222@gmail.com

T. Som et al. (eds.), *Fuzzy, Rough and Intuitionistic Fuzzy Set Approaches for Data Handling*, Forum for Interdisciplinary Mathematics,
https://doi.org/10.1007/978-981-19-8566-9_12

The great mathematician Abel was the first person to give an application of fractional derivatives (FD) in 1823 [4], who applied fractional calculus (FC) in the solutions of an IE that arises in the formulations of the Tautochrone problem. The fractional integro-differential equations have attracted much more interest from mathematicians and physicists, which provides efficiency for the description of many practical dynamical problems arising in engineering and scientific disciplines such as physics, biology, electrochemistry, chemistry, economy, electromagnetic, control theory, and viscoelasticity [1, 5–12].

The idea of ADM is originally emerged in a pioneering paper by Adomian [13]. Researchers who made the most significant contributions in the applications and developments of ADM are Rach [14], Wazwaz [15], Abbaoui [16], among others. The MADM was introduced by Wazwaz [15].

The main objective of this chapter is to obtain the approximate solution of the CFVFIDE in an uncertain environment by using the MADM. The uncertainty has been taken here as an interval in the initial condition which is transformed to parametric form to solve the CFVFIDE in different cases.

The structure of this chapter is divided into few sections, starting with an introduction in Sect. 12.1, followed by some preliminaries related to basic definitions of FC as Riemann–Liouville fractional derivatives (RLFD's), CD's, some basic interval arithmetic, and integro-differential equation in Sect. 12.2. MADM is constructed for solving the CFVFIDE in Sect. 12.3. Examples are illustrated in Sect. 12.4. Lastly, the conclusion is drawn in Sect. 12.5.

12.2 Preliminaries

Some of the needful basic definitions are as follows:

Definition 2.1 (*Riemann–Liouville Fractional Integral* (*RLFI*) [17, 18]) The RLFI of an order $\alpha > 0$ of a function f is defined as

$$J^{\alpha} f(X) = \frac{1}{\Gamma(\alpha)} \int_{0}^{x} (x-s)^{\alpha-1} f(s)\mathrm{d}s, \quad x > 0,\ \alpha \in R^{+},$$

$$J^{0} f(X) = f(x),$$

Here, R^{+} is called positive real numbers.

Definition 2.2 (*Caputo Fractional Derivative* (*CFD*) [17]) The CFD of $f(x)$ is defined as

$${}^{c}D_{x}^{\alpha} f(x) = J^{m-\alpha} D^{m} f(x)$$

$$= \left\{ \frac{1}{\Gamma(m-\alpha)} \int_0^x (x-s)^{m-\alpha-1} \frac{\mathrm{d}^m f(s)}{\mathrm{d}s^m} \mathrm{d}s,\ m-1<\alpha<m \right\}$$

$$= \frac{\mathrm{d}^m f(x)}{\mathrm{d}x^m}, \quad \alpha = m,\ m \in N,$$

where, α is called the parameter and order of the FD.

In the present chapter, α is considered as real and positive.

Some of the needful axioms are [3]:

I. $J^\alpha J^\upsilon f = J^{\alpha+\upsilon} f, \alpha, \upsilon > 0.$
II. $J^\alpha x^\beta = \frac{\Gamma(\beta+1)}{\Gamma(\beta+\alpha+1)} x^{\beta+\alpha}$
III. $D^\alpha x^\beta = \frac{\Gamma(\beta+1)}{\Gamma(\beta-\alpha+1)} x^{\beta-\alpha}, \alpha > 0, \beta > -1, x > 0.$
IV. $J^\alpha D^\upsilon f(x) = f(x) - \sum_{k=0}^{m-1} f^{(k)}(0^+) \frac{x^k}{k!}, x > 0, m-1 < \alpha \le m.$

Definition 2.3 (*Riemann–Liouville Fractional Derivative* (*RLFD*) [17]) The RLFD of order $\alpha > 0$ is defined as

$$D^\alpha f(x) = D^m J^{m-\alpha} f(x), \quad m-1 < \alpha \le m,\ m \in N$$

Definition 2.4 (*Interval Arithmetic* (*IA*) [19]) Usually, an interval is defined as $\tilde{\delta} = [\underline{\delta}, \overline{\delta}]$, where $\underline{\delta}, \overline{\delta} \in \tilde{\delta}$ which are the lower and upper bounds of the interval $\tilde{\delta}$. Any two intervals $\tilde{\delta}$ and $\tilde{\eta}$ are considered to be equal if their corresponding bounds are equal.

r-cut [19].

In interval analysis, interval uncertainty expresses uncertain data in terms of closed intervals in the real line R such that

$$\tilde{\delta} = [\underline{\delta}, \overline{\delta}] = \{\delta \in R / \underline{\delta} \le \delta \le \overline{\delta}\},$$

where $\underline{\delta}, \overline{\delta} \in R$ are the lower and upper bounds of $\tilde{\delta}$, respectively.

Now r-cut is defined as

$$\tilde{\delta} = \underline{\delta} + (\overline{\delta} - \underline{\delta})r, \quad \text{where } r \text{ is any parameter and } r \in [0, 1].$$

Definition 2.5 (*Integro-Differential Equations* (*IDE*) [20]) An IE equation which involves both integrals and derivatives then it is called an IDE. For example,

$$\psi'(x) + \int_{x_0}^{x} f(t, \psi(s))\mathrm{d}s = g(x, \psi(x)), \quad \psi(x_0) = \psi_0,\ x_0 \ge 0$$

12.3 Modified Adomian Decomposition Method (MADM) [3]

Let us consider the CFVFIDE as

$$^{c}D^{\alpha}v(x) = a(x)v(x) + g(x) + \int_{0}^{x} K_1(x,s)M_1(v(s))\mathrm{d}s + \int_{0}^{1} K_2(x,s)M_2(v(s))\mathrm{d}s. \tag{12.3.1}$$

Applying J^{α} on both sides of Eq. (12.3.1), we have

$$v(x) = v_0 + J^{\alpha}(a(x)v(x) + g(x) + \int_{0}^{1} K_1(x,s)M_1(v(s)\mathrm{d}s$$
$$+ \int_{0}^{1} K_2(x,s)M_2(v(s))\mathrm{d}s$$

Adomain's method $v(x)$ is described in the series form

$$v = \sum_{n=0}^{\infty} v_n \tag{12.3.2}$$

and the nonlinear function M decomposed as

$$M_1 = \sum_{n=0}^{\infty} D_n, \quad M_2 = \sum_{n=0}^{\infty} E_n \tag{12.3.3}$$

where D_n, E_n are called Adomian polynomial's given by

$$D_n = \frac{1}{n!}\left[\frac{\mathrm{d}^n}{\mathrm{d}\varphi^n}\left(M_1 \sum_{i=0}^{n} \varphi^i v_i\right)\right]_{\varphi=0}$$
$$E_n = \frac{1}{n!}\left[\frac{\mathrm{d}^n}{\mathrm{d}\varphi^n}\left(M_2 \sum_{i=0}^{n} \varphi^i v_i\right)\right]_{\varphi=0}$$

The Adomian polynomials are written as,

$$D_0 = M_1(v_0),$$
$$D_1 = v_1 M_1^{'}(v_0),$$

$$D_2 = v_2 M_1'(v_0) + \frac{1}{2} v_1^2 M_1''(v_0),$$

$$D_3 = v_3 M_1'(v_0) + v_1 v_2 M_1''(v_0) + \frac{1}{3} v_1^3 M_1'''(v_0),$$

$$\vdots$$

$$\vdots$$

and

$$E_0 = M_2(v_0),$$

$$E_1 = v_1 M_2'(v_0),$$

$$E_2 = v_2 M_2'(v_0) + \frac{1}{2} v_1^2 M_2''(v_0),$$

$$E_3 = v_3 M_2'(v_0) + v_1 v_2 M_2''(v_0) + \frac{1}{3} v_1^3 M_2'''(v_0),$$

$$\vdots$$

$$\vdots$$

The components $v_0, v_1, v_2 \ldots$ are determined recursively by

$$v_0(x) = v(0) + J^\alpha(g(x)),$$

$$\vdots$$

$$v_{k+1}(x) = J^\alpha(a(x)v_k(x)) + J^\alpha\left(\int_0^x K_1(x,s)D_k \mathrm{d}s + \int_0^1 K_2(x,s)E_k \mathrm{d}s\right).$$

Having defined the components $v_0, v_1, v_2, \ldots$. the solution 'v' in a series form defined by (12.3.2) follows immediately. It is important to note that the MADM proposes that v_0 is defined by the initial conditions and the function $g(x)$ is as described above. The other approximations, namely $v_1, v_2, \ldots,$ are derived recurrently.

In MADM method, we assume the function $J^\alpha g(x) = W(x)$ can be divided into two parts, namely $W_1(x)$ and $W_2(x)$. That is,

$$W(x) = W_1(x) + W_2(x). \tag{12.3.4}$$

We apply this decomposition when the function $W(x)$ consists of many parts that can be decomposed into two different parts [8, 13, 15, 16]. In this case, $W(x)$ is usually a summation of a polynomial and trigonometric or transcendental functions. A proper choice for the part $W_1(x)$ is important and $W_2(x)$ consists of the remaining terms of $W(x)$. In comparison standard decomposition method, the MADM minimizes the size of calculations and the cost of computational operations in the algorithm.

Now, MADM procedure is,

$$v_0(x) = v(0) + W_1(x) \tag{12.3.5}$$

$$v_1(x) = W_2(x) + J^{\alpha}(a(x)v_0(x)) + J^{\alpha}\left(\int_0^x K_1(x,s)D_0\mathrm{d}s + \int_0^1 K_2(x,s)E_0\mathrm{d}s\right)$$

$$\vdots \tag{12.3.6}$$

$$v_{k+1}(x) = J^{\alpha}(a(x)v_k(x)) + J^{\alpha}\left(\int_0^x K_1(x,s)D_k\mathrm{d}s + \int_0^1 K_2(x,s)E_k\mathrm{d}s\right), \quad k \geq 1 \tag{12.3.7}$$

12.4 Illustrative Examples

In this section, we present the analytical technique based on MADM to solve CFVFID equations.

Example 4.1 Consider the CFVFIDE of the form [2]:

$${}^{c}D^{0.5}[v(x)] = \frac{x^{0.5}}{\Gamma(1.5)} - \frac{x^2}{2} - \frac{x^2 v(x)}{3} + \int_0^x s v(s)\mathrm{d}s + \int_0^1 x^2 v(s)\mathrm{d}s \tag{12.4.1.1}$$

with the initial condition

$$\tilde{v}(0) = [0, 0.5] \tag{12.4.1.2}$$

using the r-cut approach, the initial condition becomes

$$\begin{aligned} &v(0) = a + (b-a)r, \text{ here } a = 0,\ b = 0.5, \text{ and } r = \text{parameter} \\ &v(0) = 0.5r \end{aligned} \tag{12.4.1.3}$$

By using operator $J^{0.5}$ on Eq. (12.4.1.1), we have

$$v(x) = v(0) + J^{0.5}\left[\frac{x^{0.5}}{\Gamma(1.5)} - \frac{x^2}{2} - \frac{x^2 v(x)}{3} + \int_0^x s v(s)\mathrm{d}s + \int_0^1 x^2 v(s)\mathrm{d}s\right]$$

$$v(x) = 0.5r + J^{0.5}\left[\frac{x^{0.5}}{\Gamma(1.5)} - \frac{x^2}{2}\right] + J^{0.5}\left[-\frac{x^2 v(x)}{3}\right] + J^{0.5}\left[\int_0^x s v(s)\mathrm{d}s + \int_0^1 x^2 v(s)\mathrm{d}s\right] \quad (12.4.1.4)$$

from Eq. (12.4.1.1), we observe $g(x) = \frac{x^{0.5}}{\Gamma(1.5)} - \frac{x^2}{2}$. Suppose $W(x) = J^{0.5}g(x)$ from Eq. (12.4.1.4), we have

$$\begin{aligned} W(x) = J^{0.5}g(x) &= J^{0.5}\left[\frac{x^{0.5}}{\Gamma(1.5)} - \frac{x^2}{2}\right] \\ &= \frac{1}{\Gamma(1.5)\Gamma(0.5)}\int_0^x \frac{s^{0.5}}{(x-s)^{1-0.5}}\mathrm{d}s - \frac{1}{2\Gamma(0.5)}\int_0^x \frac{s^2}{(x-s)^{1-0.5}}\mathrm{d}s \\ &= \frac{1}{\Gamma(1.5)\Gamma(0.5)}\int_0^x \frac{s^{0.5}}{(x-s)^{0.5}}\mathrm{d}s - \frac{1}{2\Gamma(0.5)}\int_0^x \frac{s^2}{(x-s)^{0.5}}\mathrm{d}s \\ &= \frac{1}{\Gamma(1.5)\Gamma(0.5)}\int_0^1 \tau^{0.5}(1-\tau)^{-0.5} x\,\mathrm{d}\tau - \frac{1}{2\Gamma(0.5)}\int_0^1 \tau^2(1-\tau)^{-0.5} x^{2.5}\mathrm{d}\tau \\ &= \frac{x}{\Gamma(1.5)\Gamma(0.5)}\beta(0.5, 1.5) - \frac{x^{2.5}}{2\Gamma(0.5)}\beta(0.5, 3) \\ &= x - \frac{x^{2.5}}{\Gamma(3.5)} \end{aligned} \quad (12.4.1.5)$$

By using MADM,

$$W(x) = W_1(x) + W_2(x) = J^{0.5}g(x) = J^{0.5}\left[\frac{x^{0.5}}{\Gamma(1.5)} - \frac{x^2}{2}\right] = x - \frac{x^{2.5}}{\Gamma(3.5)} \quad (12.4.1.6)$$

The MADM recursive relations are,
$v_0(x) = v(0) + W_1(x)$, from Eqs. (12.4.1.3), (12.4.1.5) and (12.4.1.6), we get

$$v_0(x) = 0.5r + x \quad (12.4.1.7)$$

$$v_1(x) = W_2(x) + J^{0.5}(f(x)v_0(x)) + J^{0.5}\left(\int_0^x K_1(x,s)D_0(s)\mathrm{d}s + \int_0^1 K_2(x,s)E_0(s)\mathrm{d}s\right)$$

$$= -\frac{x^{2.5}}{\Gamma(3.5)} + J^{0.5}\left[-\frac{x^2 v_0(x)}{3}\right] + J^{0.5}\left[\int_0^x s D_0(s)\mathrm{d}s + \int_0^1 x^2 E_0(s)\mathrm{d}s\right]$$

$$= -\frac{x^{2.5}}{\Gamma(3.5)} + J^{0.5}\left(-\frac{x^2}{3}(0.5r + x)\right)$$

$$+ J^{0.5}\left(\int_0^x s(0.5r + s)\mathrm{d}s\right) + J^{0.5}\left(\int_0^1 x^2(0.5r + s)\mathrm{d}s\right)$$

$$= -\frac{x^{2.5}}{\Gamma(3.5)} + J^{0.5}\left(\frac{7(0.5r)x^2}{6} + \frac{x^2}{2}\right)$$

$$v_1(x) = 0.3510512964 r x^{2.5} \tag{12.4.1.8}$$

Substitute Eq. (12.4.1.8) in Eq. (12.3.7), then we get

$$v_2(x) = -0.01701388888 r x^5 + 0.06036098582 r x^{2.5} \tag{12.4.1.9}$$

using Eqs. (12.4.1.9) and (12.3.7), we have

$$\begin{aligned} v_3(x) &= 0.001163806421 r x^{7.5} - 0.002925427473 r x^5 + 0.008672179333 r x^{2.5} \\ &\vdots \\ &\vdots \end{aligned} \tag{12.4.1.10}$$

The solution is $v(x) = v_0(x) + v_1(x) + v_2(x) + v_3(x) + \cdots$
By using Eqs. (12.4.1.7)–(12.4.1.10), the obtained solution is

$$\begin{aligned} v(x) &= 0.5r + x + 0.3510512964 r x^{2.5} - 0.01701388888 r x^5 \\ &\quad + 0.06036098582 r x^{2.5} + 0.001163806421 r x^{7.5} - 0.002925427473 r x^5 \\ &\quad + 0.008672179333 r x^{2.5} + \cdots \end{aligned}$$

$$\begin{aligned} v(x) &= 0.5r + x + 0.4200844615 r x^{2.5} - 0.01993931635 r x^5 \\ &\quad + 0.001163806421 r x^{7.5} + \cdots \end{aligned} \tag{12.4.1.11}$$

From Eq. (12.4.1.11), if $r = 1$

$$\begin{aligned} v(x) &= 0.5 + x + 0.4200844615 x^{2.5} - 0.01993931635 x^5 \\ &\quad + 0.001163806421 x^{7.5} + \cdots \end{aligned} \tag{12.4.1.12}$$

By using Eq. (12.4.1.11), if $r = 0$

$$v(x) = x \tag{12.4.1.13}$$

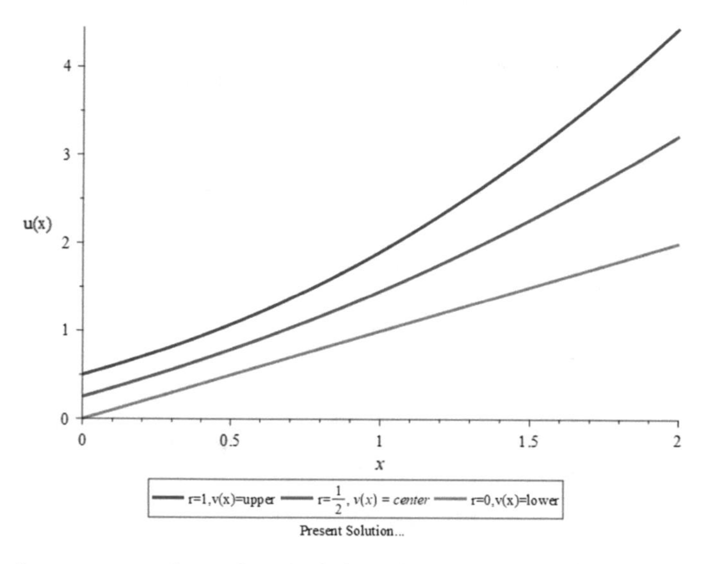

Fig. 12.1 Lower, center, and upper bound solutions of Example 4.1

From Eq. (12.4.1.11), if $r = \frac{1}{2}$ (Fig. 12.1)

$$\begin{aligned} v(x) &= 0.2500000000 + x + 0.2100422308x^{2.5} - 0.009969658175x^{5} \\ &\quad + 0.0005819032105x^{7.5} + \cdots \end{aligned} \tag{12.4.1.14}$$

Problem 4.2 Here, the following form of CFVFIDE is considered [2]:

$$^{c}D^{0.75}[v(x)] = \frac{x^{0.25}}{\Gamma(1.25)} - \frac{x^2}{2} - \frac{x^2 v(x)}{3} + \int_0^x s v(s)\mathrm{d}s + \int_0^1 x^2 v(s)\mathrm{d}s \tag{12.4.2.1}$$

with the initial condition

$$\tilde{v}(0) = [0, 0.5] \tag{12.4.2.2}$$

using the r-cut approach, the initial condition becomes

$$\begin{aligned} &v(0) = a + (b-a)r, \text{ here } a = 0, b = 0.5, \text{ and } r = \text{parameter} \\ &v(0) = 0.5r \end{aligned} \tag{12.4.2.3}$$

Now, applying $J^{0.75}$ operator on Eq. (12.4.2.1), we have

$$v(x) = v(0) + J^{0.75}\left[\frac{x^{0.75}}{\Gamma(1.25)} - \frac{x^2}{2} - \frac{x^2 v(x)}{3} + \int_0^x s v(s)\mathrm{d}s + \int_0^1 x^2 v(s)\mathrm{d}s\right]$$

$$v(x) = 0.5r + J^{0.75}\left[\frac{x^{0.75}}{\Gamma(1.25)} - \frac{x^2}{2}\right] + J^{0.75}\left[-\frac{x^2 v(x)}{3}\right]$$

$$+ J^{0.75}\left[\int_0^x s v(s)\mathrm{d}s + \int_0^1 x^2 v(s)\mathrm{d}s\right] g(x)$$

$$= \frac{x^{0.75}}{\Gamma(1.25)} - \frac{x^2}{2} \tag{12.4.2.4}$$

from Eq. (12.4.2.1) we observe $g(x) = \frac{x^{0.5}}{\Gamma(1.5)} - \frac{x^2}{2}$. Suppose $R(x) = J^{0.75} g(x)$ from Eq. (12.4.2.4), we have

$$W(x) = J^{0.75} g(x) = J^{0.75}\left[\frac{x^{0.25}}{\Gamma(1.25)} - \frac{x^2}{2}\right]$$

$$= \frac{1}{\Gamma(1.25)\Gamma(0.75)} \int_0^x \frac{s^{0.25}}{(x-s)^{1-0.75}}\mathrm{d}s - \frac{1}{2\Gamma(0.75)} \int_0^x \frac{s^2}{(x-s)^{1-0.75}}\mathrm{d}s$$

$$= \frac{1}{\Gamma(1.25)\Gamma(0.75)} \int_0^1 \tau^{0.25}(1-\tau)^{-0.25} x\,\mathrm{d}\tau$$

$$- \frac{1}{2\Gamma(0.75)} \int_0^1 \tau^2 (1-\tau)^{-0.25} x^{2.75}\mathrm{d}\tau$$

$$= \frac{x}{\Gamma(1.25)\Gamma(0.75)}\beta(0.75, 1.25) - \frac{x^{2.75}}{2\Gamma(0.75)}\beta(0.75, 3)$$

$$= x - \frac{x^{2.75}}{\Gamma(3.75)} \tag{12.4.2.5}$$

Now using MADM is,

$$W(x) = W_1(x) + W_2(x) = J^{0.75} g(x) = J^{0.75}\left[\frac{x^{0.25}}{\Gamma(1.25)} - \frac{x^2}{2}\right] = x - \frac{x^{2.75}}{\Gamma(3.75)} \tag{12.4.2.6}$$

The MADM recursive relations are,
$v_0(x) = v(0) + W_1(x)$, from Eqs. (12.4.2.3), (12.4.2.5), and (12.4.2.6), we have

$$v_0(x) = 0.5r + x \tag{12.4.2.7}$$

$$
\begin{aligned}
v_1(x) &= W_2(x) + J^{0.75}(f(x)v_0(x)) \\
&\quad + J^{0.75}\left(\int_0^x K_1(x,s)D_0(s)\mathrm{d}s + \int_0^1 K_2(x,s)E_0(s)\mathrm{d}s\right) \\
&= -\frac{x^{2.75}}{\Gamma(3.75))} + J^{0.75}\left[-\frac{x^2 v_0(x)}{3}\right] + J^{0.75}\left[\int_0^x sD_0(s)\mathrm{d}s + \int_0^1 x^2 E_0(s)\mathrm{d}s\right] \\
&= -\frac{x^{2.75}}{\Gamma(3.75))} + J^{0.75}\left(-\frac{(0.5r)x^2}{3} - \frac{x^3}{3}\right) + J^{0.75}\left(\frac{(0.5r)x^2}{2} + \frac{x^3}{3}\right) \\
&\quad + J^{0.75}\left((0.5r)x^2 + \frac{x^2}{2}\right) \\
v_1(x) &= 0.2637733945rx^{2.75}
\end{aligned}
\tag{12.4.2.8}
$$

Substitute Eq. (12.4.2.8) in Eq. (12.3.7), then we get

$$
v_2(x) = -0.008864931729rx^{5.50} + 0.03180635595rx^{2.75} \tag{12.4.2.9}
$$

From Eqs. (12.4.2.9) and (12.3.7), then we obtained

$$
\begin{aligned}
v_3(x) &= 0.0003600661419rx^{8.25} - 0.001068952290rx^{5.50} + 0.003218574687rx^{2.75} \\
&\vdots \\
&\vdots
\end{aligned}
\tag{12.4.2.10}
$$

The solution is $v(x) = v_0(x) + v_1(x) + v_2(x) + v_3(x) + \cdots$
By using Eqs. (12.4.2.7)–(12.4.2.10), then we obtained a solution that is

$$
\begin{aligned}
v(x) &= 0.5r + x + 0.2637733945rx^{2.75} - 0.008864931729rx^{5.50} \\
&\quad + 0.03180635595rx^{2.75} + 0.0003600661419rx^{8.25} - 0.001068952290rx^{5.50} \\
&\quad + 0.003218574687rx^{2.75} + \cdots \\
v(x) &= 0.5r + x + 0.2987983252rx^{2.75} - 0.009933884019rx^{5.50} \\
&\quad + 0.0003600661419rx^{8.25} + \cdots
\end{aligned}
\tag{12.4.2.11}
$$

From Eq. (12.4.2.11), if $r = 1$

$$
\begin{aligned}
v(x) &= 0.5 + x + 0.2987983252x^{2.75} - 0.009933884019x^{5.50} \\
&\quad + 0.0003600661419x^{8.25} + \cdots
\end{aligned}
\tag{12.4.2.12}
$$

By using Eq. (12.4.2.11), if $r = 0$

$$
v(x) = x \tag{12.4.2.13}
$$

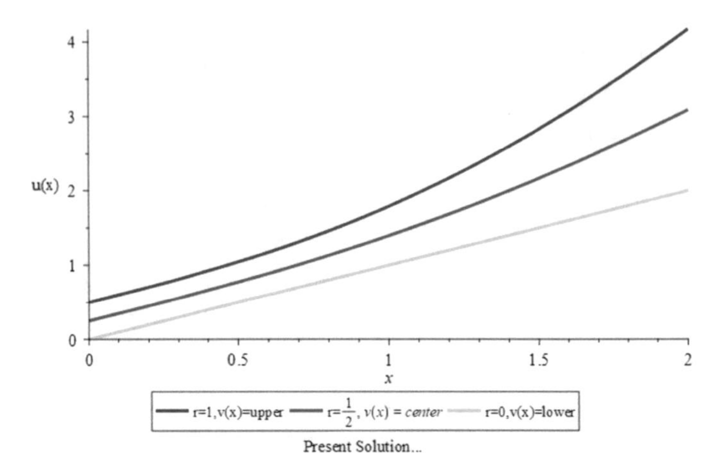

Fig. 12.2 Lower, center, and upper bound solutions of Example 4.2

From (12.4.2.11), if $r = \frac{1}{2}$ (Fig. 12.2)

$$v(x) = 0.2500000000 + x + 0.1493991626x^{2.75} - 0.004966942010x^{5.50} + 0.0001800330710x^{8.25} + \cdots \quad (12.4.2.14)$$

12.5 Conclusions

The behavior of the approximate solution of the CFVFID equation in an uncertain environment was successfully examined in this chapter using MADM. This approach converts the interval form of the initial condition to the parametric form, which is then utilized to solve the integral problem. Corresponding results are presented, and validation was accomplished by effectively comparing different cases.

References

1. Alkan, S., Hatipoglu, V.: Approximate solutions of Volterra-Fredholm integro-differential equations of fractional order. Tbilisi Math. J. **10**(2), 1–13 (2017)
2. Hamoud, A.A, Ghadle, K.P., Bani Issa, M.S.H., Giniswamy, G.: Existence and uniqueness theorem for fractional Volterra-Fredholm integro-differential equations. Int. J. Appl. Math. **31**(3), 333–348 (2018)
3. Hamoud, A.A, Ghadle, K.P., Atshan, S.M.: The approximate solutions of fractional integro-differential equations by using modified Adomian decomposition method. Khayyam J. Math. **5**(1), 21–39 (2019). https://doi.org/10.22034/kjm.2018.73593

4. Abel, N.: Solutions de quelques problemes a laide dintegrales definites, pp. 16–18. Christiania Grondahl, Norway (1881)
5. Hamoud, A.A., Azez, A.D., Ghadle, K.P.: A study of some iterative methods for solving fuzzy Volterra-Fredholm integral equations, Indonesian. J. Electr. Eng. Comput. Sci. **11**(3), 1228–1235 (2018)
6. Hamoud, A.A., Ghadle, K.P.: The reliable modified of Laplace Adomian decomposition method to solve nonlinear interval Volterra-Fredholm integral equation. Korean J. Math. **25**(3), 323–334 (2017)
7. Hamoud, A.A., Ghadle, K.P.: The combined modified Laplace with Adomian decomposition method for solving the nonlinear Volterra-Fredholm integro-differential equations. J. Korean Soc. Ind. Appl. Math. **21**, 17–28 (2017)
8. Hamoud, A.A., Ghadle, K.P.: Modified Adomian decomposition method for solving fuzzy Volterra-Fredholm integral equations. J. Indian Math. Soc. **85**(1–2), 52–69 (2018)
9. Hamoud, A.A., Ghadle, K.P.: Existence and uniqueness of solutions for fractional mixed Volterra-Fredholm integro-differential equations. Indian J. Math. **60**(3) (2018) (to appear)
10. Ma, X., Huang, C.: Numerical solutions of fractional integro-differential equations by a hybrid collocation method. Appl. Math. Comput. **219**(12), 6750–6760 (2013)
11. Mittal, R., Nigam, R.: Solution of fractional integro-differential equations by Adomian decomposition method. Int. J. Appl. Mech. **4**, 87–94 (2008)
12. Yang, C., Hou, J.: Numerical solutions of integro-differential equations of fractional order by Laplace decomposition method. Wseas Trans. Math. **12**(12), 1173–1183 (2013)
13. Adomian, G.: A review of the decomposition method in applied mathematics. J. Math. Anal. Appl. **135**(2), 501–544 (1988)
14. Rach, R.: On the Adomian decomposition method and comparisons with Picard's method. J. Math. Anal. Appl. **128**(2), 480–483 (1987)
15. Wazwaz, A.M.: A reliable modification of Adomian decomposition method. Appl. Math. Comput. **102**, 77–86 (1992)
16. Abbaoui, K., Cherruault, Y.: Convergence of Adomian's method applied to nonlinear equations. Math. Comput. Model. **20**(9), 69–73 (1994)
17. Kilbas, A., Srivastava, H., Trujillo, J.: Theory and applications of fractional differential equations. In: North-Holland Mathematics Studies. Elsevier, Amsterdam (2006)
18. Podlubny, I.: Fractional Differential Equations: Mathematics in Science and Engineering, vol. 198 (1999)
19. Chakraverty, S., Mahato, N.R., Karunakar, P., Rao, T.D.: Advanced Numerical and Semi-Analytical Methods for Differential Equations. Wiley (2019)
20. Kanwal, R.P.: Linear Integral Equations: Theory and Technique. Academic Press Inc. (1971)

Chapter 13
Generalized Hukuhara Subdifferentiability for Convex Interval-Valued Functions and Its Applications in Nonsmooth Interval Optimization

Krishan Kumar, Anshika, and Debdas Ghosh

13.1 Introduction

Commonly, optimization problems are used to deal with deterministic values. So that one can find precise solutions. However, there are many mathematical or computer models of some deterministic real-world phenomena in which uncertainty appears. We cannot handle this uncertainty with the exact solution. Therefore, to tackle this kind of imprecise, interval analysis was introduced. Keeping this practical importance in mind, optimization problems having interval coefficient of objective functions, namely interval optimization problems (IOPs), have been a significant research topic in the past two decades. In 1966, Moore introduced interval arithmetic to deal with intervals [1]. Subsequently, there were few improvements on this arithmetic proposed in [2]. However, H-difference in [2] was further found to be pretty much restrictive [3]. Stefaninni and Bede [3] introduced a concept known as generalized Hukuhara difference (gH-difference), which has been extensively used in interval analysis.

One can note that intervals are not linearly ordered. Therefore, the ordering of intervals plays vital role in the study of interval arithmetic. Ishibuchi and Tanaka [4] observed many partial ordering structures and their solutions for IOPs. After that, many partial ordering relations are studied with their respective solutions concepts for IOPs [5–8]. Recently, Ghosh et al. [9] studied the variable ordering relations for intervals and their application to IOPs.

K. Kumar · Anshika · D. Ghosh (✉)
Department of Mathematical Sciences, Indian Institute of Technology (Banaras Hindu University), Varanasi 221005, Uttar Pradesh, India
e-mail: debdas.mat@iitbhu.ac.in

K. Kumar
e-mail: krishankumar.rs.mat19@itbhu.ac.in

Anshika
e-mail: anshika.rs.mat19@itbhu.ac.in

T. Som et al. (eds.), *Fuzzy, Rough and Intuitionistic Fuzzy Set Approaches for Data Handling*, Forum for Interdisciplinary Mathematics,
https://doi.org/10.1007/978-981-19-8566-9_13

Calculus plays a significant role to observe the characteristics of an IVF. Initially, to develop the calculus for IVFs, the articles [10, 11] introduced the concept of H-derivative for diverse mathematical analysis. However, this H-differentiability was restrictive [8] because if $\mathbf{T}(x) = \mathbf{A} \odot t(x)$, where $\mathbf{A}$ is a compact interval and $t(x)$ is a real-valued function, then $\mathbf{T}$ is not H-differentiable in case of $t'(x) < 0$ [10]. To remove this deficiency, the concepts of gH-derivative for IVF have been introduced by Stefanini and Bede [3]. After the definition of gH-derivative, Ghosh [12] presented the notions of gH-partial derivative and gH-gradient for IVF. Ghosh et al. [13] have proposed the concepts of gH-directional derivative, gH-Gâteaux derivative, and gH-Fréchet derivative of IVFs. Also , in [14], a new concept of gH-differentiability that is equipped with a linearity concept of IVFs has been illustrated. After that, using the special product, the concept of gH-subdifferentiability for non gH-differentiable IVF was introduced by Kumar et al. [15]. Many researchers have extended this calculus of IVFs for instance [7, 15–18] and their reference in.

In the direction of solving IOPs, Wu [5] presented KKT optimality conditions along with two solution concepts of IOPs. In the continuation, Bhurjee and Panda [7] introduced efficient solution and a methodology to study the existence of solution of an IOP. A linear bilevel program in which the coefficients of both the objective functions are intervals is addressed by Calvate and Galê [19]. Subsequently, Chalco-Cano et al. [8] considered two types of order relations and KKT conditions for each relation as well. Also, Osuna-Gomez et al. [20] gave necessary and sufficient conditions for gH-differentiable IVF. Ghosh et al. [21] using gH-differentiability of IVFs, derived KKT conditions and duality theories for IOPs. Recently, Kumar et al. [15] have given a new concept weak sharp minima to solve nonsmooth IOPs using gH-subgradient of an IVF. Apart from this, many researchers have investigated solution concepts of IOPs. They have developed the theories and techniques to find efficient solutions to the IOPs (see [12, 16, 22, 23] and their references). From the literature of IVFs and IOPs, it can be observed that the concept of gH-subgradient and gH-subdifferential is recently introduced [15] but the compactness of the gH-subdifferential set is not studied. For this chapter, we have two major contributions: first, we show that if $\mathbf{T}$ is gH-continuous convex IVF, then its gH-subdifferential set is compact on dom$(\mathbf{T})$. Second, we present two optimality conditions using gH-subgradient and gH-subdifferential in obtaining the efficient solution of nonsmooth IOPs.

We have presented our work in the following manner. In Sect. 13.2, basic terminologies and definition on intervals are provided. In the same section, several important concepts such as special product, the dominance of intervals followed by convexity and calculus of IVFs, are presented briefly. Next, the concept of the support function of a subset of a set of intervals, the convex combination of intervals, and the convex hull of a set of intervals are defined. In Sect. 13.3, the compactness of a gH-subdifferential set of gH-continuous convex IVF is presented based on the notion of gH-subdifferential set and on some of its properties. Thereafter, an interrelation of the support function of a set of intervals is observed with the gH-subdifferential set of convex IVF. Next, in the same section, we provide the supremum rule of gH-subdifferential calculus for convex IVFs. Further, two optimality conditions to find an efficient point of IOPs are given in Sect. 13.4 along with an example that exemplify that the proposed conditions are necessary but not sufficient. Finally, the last section is concerned with the conclusion and future scopes.

13.2 Preliminaries and Terminologies

In this section, we start with a few notations. After that, fundamental operations of intervals are given. Subsequently, two important concepts special product and dominance of intervals are illustrated that are used throughout the chapter. Next, we present the necessary calculus of IVFs. Lastly, a convex combination of intervals and a convex hull of a set of intervals are defined.

Following notations are used throughout the chapter:

- $\mathbb{R}$ and $\mathbb{R}^+$ denote the set of real numbers and the set of nonnegative real numbers, respectively
- $\|\cdot\|$ denotes the Euclidean norm and $\langle \cdot, \cdot \rangle$ denotes the standard inner product on $\mathbb{R}^n$
- $I(\mathbb{R})$ represents the set of all closed and bounded intervals
- Bold capital letters refer to the elements of $I(\mathbb{R})$
- Bold capital letters with a cap refers to the elements of $I(\mathbb{R})^n$
- $\overline{I(\mathbb{R})} = I(\mathbb{R}) \cup \{-\infty, +\infty\}$
- $\mathbb{B} = \{x \in \mathbb{R}^n : \|x\| \leq 1\}$ denotes the closed unit ball in $\mathbb{R}^n$
- $\mathbf{0}$ represents the interval $[0, 0]$.

13.2.1 Fundamental Operations on Intervals

Consider two intervals $\mathbf{P} = [\underline{p}, \overline{p}]$ and $\mathbf{Q} = \left[\underline{q}, \overline{q}\right]$. Then, the addition and the difference of two intervals are defined by

$$\mathbf{P} \oplus \mathbf{Q} = \left[\underline{p} + \underline{q},\ \overline{p} + \overline{q}\right],\ \mathbf{P} \ominus \mathbf{Q} = \left[\underline{p} - \overline{q},\ \overline{p} - \underline{q}\right],$$

respectively.

Similarly, the product of an interval $\mathbf{P}$ with a real number α is defined by

$$\alpha \odot \mathbf{P} = \mathbf{P} \odot \alpha = \begin{cases} [\alpha \underline{p},\ \alpha \overline{p}], & \text{if } \alpha \geq 0 \\ [\alpha \overline{p},\ \alpha \underline{p}], & \text{if } \alpha < 0. \end{cases}$$

The norm [1] of an interval $\mathbf{P} = [\underline{p}, \overline{p}] \in I(\mathbb{R})$ and an interval vector $\widehat{\mathbf{P}} = (\mathbf{P}_1, \mathbf{P}_2, \ldots, \mathbf{P}_n)^\top \in I(\mathbb{R})^n$ is defined by

$$\|\mathbf{P}\|_{I(\mathbb{R})} = \max\{|\underline{p}|, |\overline{p}|\} \text{ and } \|\widehat{\mathbf{P}}\|_{I(\mathbb{R})^n} = \sum_{i=1}^{n} \|\mathbf{P}_i\|_{I(\mathbb{R})},$$

respectively.

It is to note that a real number p, or more appropriately the singleton set $\{p\}$, can be represented by the interval $[p, p]$. In this case interval $\mathbf{P} = [p, p]$ is called degenerate interval.

Definition 1 (*gH-difference of intervals* [3]). Let $\mathbf{P}, \mathbf{Q} \in I(\mathbb{R})$ such that $\mathbf{P} = [\underline{p}, \overline{p}]$ and $\mathbf{Q} = [\underline{q}, \overline{q}]$. Then, the gH-difference between $\mathbf{P}$ and $\mathbf{Q}$, denoted by $\mathbf{P} \ominus_{gH} \mathbf{Q}$, is defined by

$$\mathbf{P} \ominus_{gH} \mathbf{Q} = \left[\min\{\underline{p} - \underline{q}, \overline{p} - \overline{q}\},\ \max\{\underline{p} - \underline{q}, \overline{p} - \overline{q}\}\right].$$

It can be seen that if $\mathbf{C} \in I(\mathbb{R})$ is gH-difference between $\mathbf{P}$ and $\mathbf{Q}$, then

$$\mathbf{P} = \mathbf{Q} \oplus \mathbf{C} \text{ or } \mathbf{Q} = \mathbf{P} \ominus \mathbf{C},$$

and $\mathbf{P} \ominus_{gH} \mathbf{P} = \mathbf{0}$.

Definition 2 (*Special product* [15]). For an $x = (x_1, x_2, \ldots, x_n)^\top \in \mathbb{R}^n$ and a vector of intervals $\widehat{\mathbf{P}} = (\mathbf{P}_1, \mathbf{P}_2, \ldots, \mathbf{P}_n)^\top \in I(\mathbb{R})^n$ with $\mathbf{P}_i = [\underline{p}_i, \overline{p}_i]$ for each $i = 1, 2, \ldots, n$, the special product between x and $\widehat{\mathbf{P}}$, denoted by $x^\top \odot \widehat{\mathbf{P}}$, is given by

$$x^\top \odot \widehat{\mathbf{P}} = \left[\min\left\{\sum_{i=1}^{n} x_i \underline{p}_i, \sum_{i=1}^{n} x_i \overline{p}_i\right\}, \max\left\{\sum_{i=1}^{n} x_i \underline{p}_i, \sum_{i=1}^{n} x_i \overline{p}_i\right\}\right].$$

Remark 1 It is easy to see that if all the components of $\widehat{\mathbf{P}}$ are degenerate intervals, i.e., $\widehat{\mathbf{P}} \in \mathbb{R}^n$, then the special product $x^\top \odot \widehat{\mathbf{P}}$ reduces to the standard inner product of $x \in \mathbb{R}^n$ and $\widehat{\mathbf{P}}$.

Definition 3 (*Dominance of intervals* [6]). Let $\mathbf{P} = [\underline{p}, \overline{p}]$ and $\mathbf{Q} = [\underline{q}, \overline{q}]$ be two elements in $I(\mathbb{R})$.

(i) $\mathbf{Q}$ is said to be dominated by $\mathbf{P}$ if $\underline{p} \leq \underline{q}$ and $\overline{p} \leq \overline{q}$, and denoted by $\mathbf{P} \preceq \mathbf{Q}$;
(ii) $\mathbf{Q}$ is said to be strictly dominated by $\mathbf{P}$ if $\mathbf{P} \preceq \mathbf{Q}$ and $\mathbf{P} \neq \mathbf{Q}$, and denoted by $\mathbf{P} \prec \mathbf{Q}$. Equivalently, $\mathbf{P} \prec \mathbf{Q}$ if and only if any of the following holds: '$\underline{p} < \underline{q}$ and $\overline{p} \leq \overline{q}$' or '$\underline{p} \leq \underline{q}$ and $\overline{p} < \overline{q}$' or '$\underline{p} < \underline{q}$ and $\overline{p} < \overline{q}$';
(iii) if neither $\mathbf{P} \preceq \mathbf{Q}$ nor $\mathbf{Q} \preceq \mathbf{P}$, we say that none of $\mathbf{P}$ and $\mathbf{Q}$ dominates the other, or $\mathbf{P}$ and $\mathbf{Q}$ are not comparable. Equivalently, $\mathbf{P}$ and $\mathbf{Q}$ are not comparable if either '$\underline{p} < \underline{q}$ and $\overline{p} > \overline{q}$' or '$\underline{p} > \underline{q}$ and $\overline{p} < \overline{q}$'.

Lemma 1 (See [1]). *For $\boldsymbol{P}, \boldsymbol{Q} \in I(\mathbb{R})$ and $x \in \mathbb{R}$,*

$$x \odot (\boldsymbol{P} \oplus \boldsymbol{Q}) = x \odot \boldsymbol{P} \oplus x \odot \boldsymbol{Q}.$$

13.2.2 IVF and Its Calculus

Throughout this chapter, X is nonempty and X' is nonempty convex subsets of $\mathbb{R}^n$, unless mentioned otherwise.

Definition 4 (*gH-continuous IVF* [12]). Let $\mathbf{T} : X \to I(\mathbb{R})$ be an IVF on X. Let $\bar{x} \in X$, and $h \in \mathbb{R}^n$ such that $\bar{x} + h \in X$. The IVF $\mathbf{T}$ is said to be gH-continuous at $\bar{x}$ if

$$\lim_{\|h\|\to 0} \big(\mathbf{T}(\bar{x} + h) \ominus_{gH} \mathbf{T}(\bar{x})\big) = \mathbf{0}.$$

Definition 5 (*gH-derivative* [3]). The gH-derivative of an IVF $\mathbf{T} : \mathbb{R} \to I(\mathbb{R})$ at $\bar{x} \in \mathbb{R}$ is defined by

$$\mathbf{T}'(\bar{x}) = \lim_{h\to 0} \frac{1}{h} \odot (\mathbf{T}(\bar{x} + h) \ominus_{gH} \mathbf{T}(\bar{x})), \text{ provided the limit exists.}$$

Definition 6 (gH-partial derivative [12]). Let $\bar{x} = (\bar{x}_1, \bar{x}_2, \ldots, \bar{x}_n)^\top$ be a point of X. For a given $i \in \{1, 2, \ldots, n\}$, we define a function $\mathbf{H}_j$ by

$$\mathbf{H}_j(x_j) = \mathbf{T}(\bar{x}_1, \bar{x}_2, \ldots, \bar{x}_{j-1}, x_j, \bar{x}_{j+1}, \ldots, \bar{x}_n).$$

If the generalized derivative of $\mathbf{H}_j$ exists at $\bar{x}_j$, then the jth *gH-partial derivative* of $\mathbf{T}$ at $\bar{x}$, denoted as $\boldsymbol{D}_j\mathbf{T}(\bar{x})$, is defined as

$$\boldsymbol{D}_j\mathbf{T}(\bar{x}) = \mathbf{H}'_j(\bar{x}_j) \text{ for all } j = 1, 2, \ldots, n.$$

Definition 7 (gH-gradient [12]). Let X be a subset of $\mathbb{R}^n$, then the *gH-gradient* of an IVF $\mathbf{T}$ at a point $\bar{x} \in X$, denoted by $\nabla\mathbf{T}(\bar{x}) \in I(\mathbb{R})^n$, defined by

$$\nabla\mathbf{T}(\bar{x}) = (D_1\mathbf{T}(\bar{x}), D_2\mathbf{T}(\bar{x}), \ldots, D_n\mathbf{T}(\bar{x}))^\top.$$

Definition 8 (*Convex IVF* [5]). Let $\mathbf{T} : X' \to I(\mathbb{R})$ be an IVF. The IVF $\mathbf{T}$ is said to be convex on X' if for any $x_1, x_2 \in X'$,

$$\mathbf{T}(\lambda_1 x_1 + \lambda_2 x_2) \preceq \lambda_1 \odot \mathbf{T}(x_1) \oplus \lambda_2 \odot \mathbf{T}(x_2) \text{ for all } \lambda_1, \lambda_2 \in [0, 1] \text{ with } \lambda_1 + \lambda_2 = 1.$$

Lemma 2 (See [5]). *Let $\boldsymbol{T} : X' \to I(\mathbb{R})$ be an IVF such that $\boldsymbol{T}(x) = [\underline{T}(x), \overline{T}(x)]$ for all $x \in X'$, where $\underline{T}$ and $\overline{T}$ are real-valued functions defined on X'. Then, $\boldsymbol{T}$ is convex on X' if and only if $\underline{T}$ and $\overline{T}$ are convex on X'.*

Definition 9 (*gH-directional derivative* [13]). Let $\mathbf{T}$ be an IVF on X. Let $\bar{x} \in X$ and $d \in \mathbb{R}^n$. Then, $\mathbf{T}$ has gH-directional derivative at $\bar{x}$ in the direction d, if the limit

$$\lim_{\lambda\to 0+} \frac{1}{\lambda} \odot \big(\mathbf{T}(\bar{x} + \lambda d) \ominus_{gH} \mathbf{T}(\bar{x})\big)$$

exists, and it is denoted by $\mathbf{T}_{\mathscr{D}}(\bar{x})(d)$.

Definition 10 (*gH-differentiability* [12]). An IVF $\mathbf{T}$ is said to be gH-differentiable at $\bar{x} \in X$ if there exist two IVFs $\mathbf{E}(\mathbf{T}(\bar{x}); h)$ and $\mathbf{L}_{\bar{x}} : \mathbb{R}^n \to I(\mathbb{R})$ such that

$$\mathbf{T}(\bar{x} + h) \ominus_{gH} \mathbf{T}(\bar{x}) = \mathbf{L}_{\bar{x}}(h) \oplus \|h\| \odot \mathbf{E}(\mathbf{T}(\bar{x}); h)$$

for $\|h\| < \delta$ for some $\delta > 0$, where $\lim_{\|h\| \to 0} \mathbf{E}(\mathbf{T}(\bar{x}); h) = \mathbf{0}$ and $\mathbf{L}_{\bar{x}}$ is such a function that satisfies

(i) $\mathbf{L}_{\bar{x}}(x + y) = \mathbf{L}_{\bar{x}}(x) \oplus \mathbf{L}_{\bar{x}}(y)$ for all $x, y \in X$, and
(ii) $\mathbf{L}_{\bar{x}}(cx) = c \odot \mathbf{L}_{\bar{x}}(x)$ for all $c \in \mathbb{R}$ and $x \in X$.

Remark 2 (*See* [8]). Let $\mathbf{T} : X \to I(\mathbb{R})$ be an IVF such that $\mathbf{T}(x) = [\underline{T}(x), \overline{T}(x)]$ for all $x \in X$, where $\underline{T}$ and $\overline{T}$ are real-valued functions defined on X. Then, the gH-derivative of $\mathbf{T}$ at $\bar{x} \in X$ exists if the derivatives of $\underline{T}$ and $\overline{T}$ at $\bar{x}$ exist and

$$\mathbf{T}'(\bar{x}) = \left[\min\left\{\underline{T}'(\bar{x}), \overline{T}'(\bar{x})\right\}, \max\left\{\underline{T}'(\bar{x}), \overline{T}'(\bar{x})\right\}\right].$$

Definition 11 (*Proper IVF*). An extended IVF $\mathbf{T} : X \to \overline{I(\mathbb{R})}$ is called a proper IVF if there exists $\bar{x} \in X$ such that $\mathbf{T}(\bar{x}) \prec [+\infty, +\infty]$ and $[-\infty, -\infty] \prec \mathbf{T}(x)$ for all $x \in X$.

Definition 12 (*Effective domain of IVF*). The effective domain of an extended IVF $\mathbf{T} : X \to \overline{I(\mathbb{R})}$ is the collection of all such points at which $\mathbf{T}$ is finite. It is denoted by dom$(\mathbf{T})$, i.e.,

$$\text{dom}(\mathbf{T}) = \left\{x \in X : \|\mathbf{T}(x)\|_{I(\mathbb{R})} < \infty\right\}.$$

Definition 13 (*Indicator IVF*). Let S be a nonempty subset of $\mathbb{R}^n$. Then, the indicator function $\boldsymbol{\delta}_S : \mathbb{R}^n \to \overline{I(\mathbb{R})}$ of S is defined by

$$\boldsymbol{\delta}_S(s) = \begin{cases} \mathbf{0} & \text{if } s \in S \\ +\infty & \text{if } s \notin S. \end{cases}$$

Definition 14 (*Convex combination of intervals*). Let $\widehat{\mathbf{P}}$ be an interval in $I(\mathbb{R})^n$. Then, $\widehat{\mathbf{P}}$ is said to be a convex combination of intervals $\widehat{\mathbf{P}}_1, \widehat{\mathbf{P}}_2, \ldots, \widehat{\mathbf{P}}_m \in I(\mathbb{R})^n$ if

$$\widehat{\mathbf{P}} = \bigoplus_{i=1}^{m} \lambda_i \odot \widehat{\mathbf{P}}_i \text{ with } \lambda_i \geq 0 \text{ and } \sum_{i=1}^{m} \lambda_i = 1.$$

Definition 15 (*Convex hull of a set of intervals*). Let $\mathbf{S}$ be a nonempty subset of $I(\mathbb{R})^n$. Then, the convex hull of $\mathbf{S}$, denoted by $\mathbf{co}(\mathbf{S})$, is defined by

$$\mathbf{co}(\mathbf{S}) = \left\{ \widehat{\mathbf{P}} \in I(\mathbb{R})^n : \widehat{\mathbf{P}} = \bigoplus_{i=1}^{m} \lambda_i \odot \widehat{\mathbf{P}}_i,\ \widehat{\mathbf{P}}_i \in \mathbf{S} \text{ with } \lambda_i \geq 0 \text{ and } \sum_{i=1}^{m} \lambda_i = 1 \right\}.$$

Remark 3 It is to be noted that if $\widehat{\mathbf{P}} = \bigoplus_{i=1}^{m} \lambda_i \odot \widehat{\mathbf{P}}_i$, where $\widehat{\mathbf{P}} = (\mathbf{P}_1, \mathbf{P}_2, \ldots, \mathbf{P}_n)^\top \in I(\mathbb{R})^n$ and each $\widehat{\mathbf{P}}_i = (\mathbf{P}_{i1}, \mathbf{P}_{i2}, \ldots, \mathbf{P}_{in})^\top \in I(\mathbb{R})^n$, then

$$\mathbf{P}_k = [\underline{p}_k, \overline{p}_k] = \oplus_{k=1}^{m} \lambda_k \odot \mathbf{P}_{ik} \implies \underline{p}_k = \sum_{k=1}^{m} \lambda_k \underline{p}_{ik},\ \overline{p}_k = \sum_{k=1}^{m} \lambda_k \overline{p}_{ik}.$$

Definition 16 (*Supremum of a subset of* $\overline{I(\mathbb{R})}$ [16]). Let $\mathbf{S} \subseteq \overline{I(\mathbb{R})}$. If an interval $\bar{\mathbf{P}} \in I(\mathbb{R})$ is satisfying $\mathbf{Q} \preceq \bar{\mathbf{P}}$ for all $\mathbf{Q}$ in $\mathbf{S}$. Then, $\bar{\mathbf{P}}$ is an upper bound of $\mathbf{S}$. Moreover, if an upper bound $\bar{\mathbf{P}}$ of $\mathbf{S}$ satisfy $\bar{\mathbf{P}} \preceq \mathbf{C}$ for all upper bounds $\mathbf{C}$ of $\mathbf{S}$ in $I(\mathbb{R})$, then $\bar{\mathbf{P}}$ is called the supremum of $\mathbf{S}$. We denote the supremum of $\mathbf{S}$ by $\sup \mathbf{S}$.

Remark 4 (*See* [16]). Let Λ be an index set. For any $\mathbf{S} = \left\{[a_\alpha, b_\alpha] \in \overline{I(\mathbb{R})} : \alpha \in \Lambda\right\}$, we have $\sup \mathbf{S} = \left[\sup_{\alpha \in \Lambda} a_\alpha,\ \sup_{\alpha \in \Lambda} b_\alpha\right]$.

Definition 17 (*Support function of a subset of* $I(\mathbb{R})^n$ [15]). Let $\mathbf{S}$ be a nonempty subset of $I(\mathbb{R})^n$. Then, the support function of $\mathbf{S}$ at $x \in \mathbb{R}^n$, denoted by $\psi^*_{\mathbf{S}}(x)$, is defined by

$$\psi^*_{\mathbf{S}}(x) = \sup_{\widehat{\mathbf{P}} \in \mathbf{S}} x^\top \odot \widehat{\mathbf{P}}.$$

Definition 18 (*Supremum of an IVF* [16]). Let S be a nonempty subset of X and $\mathbf{T} : S \to \overline{I(\mathbb{R})}$ be an extended IVF. Then, the supremum of $\mathbf{T}$ denoted as $\sup_{x \in S} \mathbf{T}(x)$ is equal to the supremum of range set of $\mathbf{T}$, i.e.,

$$\sup_{x \in S} \mathbf{T}(x) = \sup\{\mathbf{T}(x) : x \in S\}.$$

Lemma 3 (See [16]). *Let* $\mathbf{T}_1$ *and* $\mathbf{T}_2$ *be two proper extended IVFs, and* S *be a nonempty subset of* X. *Then,*

$$\sup_{x \in S}\{\mathbf{T}_1(x) \oplus \mathbf{T}_2(x)\} \preceq \sup_{x \in S} \mathbf{T}_1(x) \oplus \sup_{x \in S} \mathbf{T}_2(x).$$

13.3 gH-Subdifferential for Convex IVFs

This section provides the notion of gH-subdifferential set of convex IVFs and its properties. Closedness, convexity, nonemptyness, etc. are given in [15]. Using these properties, we present compactness of gH-subdifferential set of gH-continuous

IVF. Few interrelations between support function of a set of intervals and gH-subdifferential set of convex IVF are also established. At the end of this section, the supremum rule of gH-subdifferential calculus for IVFs is presented.

Definition 19 (*gH-subdifferentiability* [15]). Let $\mathbf{T} : X' \to \overline{I(\mathbb{R})}$ be a proper convex IVF. Then, gH-subdifferential of $\mathbf{T}$ at $\bar{x} \in \text{dom}(\mathbf{T})$, denoted by $\boldsymbol{\partial}\mathbf{T}(\bar{x})$, is defined by

$$\boldsymbol{\partial}\mathbf{T}(\bar{x}) = \left\{\widehat{\mathbf{G}} \in I(\mathbb{R})^n : (x - \bar{x})^\top \odot \widehat{\mathbf{G}} \preceq \mathbf{T}(x) \ominus_{gH} \mathbf{T}(\bar{x}) \text{ for all } x \in \mathbb{R}^n\right\}. \quad (13.1)$$

The elements of (13.1) are known as gH-subgradients of $\mathbf{T}$ at $\bar{x}$. Further, if $\boldsymbol{\partial}\mathbf{T}(\bar{x}) \neq \emptyset$, we say that $\mathbf{T}$ is gH-subdifferentiable at $\bar{x}$.

Example 1 Consider an IVF $\mathbf{T} : X \to \overline{I(\mathbb{R})}$ be defined by $\mathbf{T}(x) = \boldsymbol{\delta}_S(x)$, where $\boldsymbol{\delta}_S(x)$ is indicator IVF of S, defined in Definition 13. If $\widehat{\mathbf{G}}$ is a gH-subgradient of $\mathbf{T}$ at any $\bar{x} \in S$, then for all $x \in X$, we have

$$\begin{aligned} & (x - \bar{x})^\top \odot \widehat{\mathbf{G}} \preceq \boldsymbol{\delta}_S(x) \ominus_{gH} \boldsymbol{\delta}_S(\bar{x}) \\ \Longleftrightarrow\ & (x - \bar{x})^\top \odot \widehat{\mathbf{G}} \preceq \mathbf{0}. \end{aligned}$$

Hence, the gH-subdifferential set of $\boldsymbol{\delta}_S(x)$ at $\bar{x}$ is

$$\{\widehat{\mathbf{G}} \in I(\mathbb{R})^n : (x - \bar{x})^\top \odot \widehat{\mathbf{G}} \preceq \mathbf{0} \text{ for all } x \in X\}. \quad (13.2)$$

Remark 5 It is noteworthy to see that if $\mathbf{T}(x) = \boldsymbol{\delta}_S(x)$ in Example 1 is real-valued, then $\widehat{\mathbf{G}} \in \mathbb{R}^n$. In this case, the gH-subdifferential set of $\boldsymbol{\delta}_S(x)$ at $\bar{x}$ is normal cone of S at $\bar{x}$.

Definition 20 (*Non-decreasing IVF*). Let $\mathbf{T} : X \to I(\mathbb{R})$ be an IVF. Then, $\mathbf{T}$ is a non-decreasing IVF if for any $x, y \in \mathbb{R}^n$ such that

$$\begin{aligned} & x_i \leq y_i, \text{ for each } i = 1, 2, \ldots, n \\ \Longrightarrow\ & \mathbf{T}(x) \preceq \mathbf{T}(y). \end{aligned}$$

Theorem 4 *Let $\boldsymbol{T} : X' \to \overline{I(\mathbb{R})}$ be a proper non-decreasing convex IVF. Then, for any $\bar{x} \in \text{int}(\text{dom}(\boldsymbol{T}))$, $\boldsymbol{\partial T}(\bar{x}) \subseteq I(\mathbb{R}_+)^n$.*

Proof Since $\mathbf{T}$ is a non-decreasing convex IVF. Therefore, for any $e_i = (0, \ldots 0, 1, 0, \ldots, 0)^\top \in \mathbb{R}^n$ with 1 at the ith place, we have

$$\begin{aligned} & \mathbf{T}(\bar{x} - e_i) \preceq \mathbf{T}(\bar{x}) \\ \Longrightarrow\ & \mathbf{T}(\bar{x} - e_i) \ominus_{gH} \mathbf{T}(\bar{x}) \preceq \mathbf{0}. \end{aligned}$$

From Remark 6, $\boldsymbol{\partial}\mathbf{T}(\bar{x})$ is nonempty. Thus, there exists $\widehat{\mathbf{G}} \in \boldsymbol{\partial}\mathbf{T}(\bar{x})$ such that

$$\begin{aligned} & (-e_i)^\top \odot \widehat{\mathbf{G}} \preceq \mathbf{T}(\bar{x} - e_i) \ominus_{gH} \mathbf{T}(\bar{x}) \preceq \mathbf{0} \\ \implies & (-e_i)^\top \odot \widehat{\mathbf{G}} \preceq \mathbf{0} \\ \implies & \mathbf{0} \preceq \mathbf{G}_i. \end{aligned}$$

Since i is arbitrary, therefore $\mathbf{0} \preceq \mathbf{G}_i$ for each $i = 1, 2, \ldots, n$. Thus, $\boldsymbol{\partial}\mathbf{T}(\bar{x}) \subseteq I(\mathbb{R}_+)^n$. □

Theorem 5 *Let $\boldsymbol{T} : X' \to I(\mathbb{R})$ be a convex IVF. Then, for any $x \in dom(\boldsymbol{T})$ and $\lambda \geq 0$,*

$$\boldsymbol{\partial}(\lambda \odot \boldsymbol{T})(x) = \lambda \odot \boldsymbol{\partial}\boldsymbol{T}(x),$$

where $dom(\lambda \odot \boldsymbol{T}) = dom(\boldsymbol{T})$.

Proof Let $\widehat{\mathbf{G}} \in \partial\mathbf{T}(x)$. Then, for any $y \in \text{dom}(\mathbf{T})$,

$$\begin{aligned} & (y - x)^\top \odot \widehat{\mathbf{G}} \preceq \mathbf{T}(y) \ominus_{gH} \mathbf{T}(x) \\ \iff & \lambda \odot ((y - x)^\top \odot \widehat{\mathbf{G}}) \preceq \lambda \odot \big(\mathbf{T}(y) \ominus_{gH} \mathbf{T}(x)\big) \text{ for } \lambda \geq 0 \\ \iff & (y - x)^\top \odot (\lambda \odot \widehat{\mathbf{G}}) \preceq \lambda \odot (\mathbf{T}(y) \ominus_{gH} \mathbf{T}(x)) \\ \iff & (y - x)^\top \odot (\lambda \odot \widehat{\mathbf{G}}) \preceq (\lambda \odot \mathbf{T})(y) \ominus_{gH} (\lambda \odot \mathbf{T})(x) \\ \iff & \lambda \odot \widehat{\mathbf{G}} \in \boldsymbol{\partial}(\lambda \odot \mathbf{T})(x). \end{aligned}$$

Hence, the result follows. □

Next, we introduce a lemma that is used to prove Theorem 9.

Lemma 6 *Let $x \in \mathbb{B}$ and $\widehat{\boldsymbol{P}} = (\boldsymbol{P}_1, \boldsymbol{P}_2, \ldots, \boldsymbol{P}_n)^\top \in I(\mathbb{R})^n$ with $\boldsymbol{P}_i = [\underline{p}_i, \overline{p}_i]$. Suppose there exists an $M > 0$ such that $x^\top \odot \widehat{\boldsymbol{P}} \preceq M$. Then,*

$$\|\widehat{\boldsymbol{P}}\|_{I(\mathbb{R})^n} \leq M', \text{ where } M' = nM.$$

Proof We have

$$\begin{aligned} & x^\top \odot \widehat{\mathbf{P}} = \left[\min\left\{\sum_{i=1}^{n} x_i \underline{p}_i, \sum_{i=1}^{n} x_i \overline{p}_i\right\}, \max\left\{\sum_{i=1}^{n} x_i \underline{p}_i, \sum_{i=1}^{n} x_i \overline{p}_i\right\}\right] \preceq M \\ \implies & \sum_{i=1}^{n} x_i \underline{p}_i \leq M \text{ and } \sum_{i=1}^{n} x_i \overline{p}_i \leq M. \end{aligned}$$

Take $\sum_{i=1}^{n} x_i \underline{p}_i \leq M$. Then, by Remark 1, we have

$$\langle x, \underline{p} \rangle \leq M, \text{ where } \underline{p} = (\underline{p}_1, \underline{p}_2, \ldots, \underline{p}_n)^\top \in \mathbb{R}^n. \tag{13.3}$$

If $\underline{p} \neq 0$, then choosing $x = \frac{\underline{p}}{\|\underline{p}\|}$, (13.3) gives

$$\left\langle \frac{\underline{p}}{\|\underline{p}\|}, \underline{p} \right\rangle \leq M$$
$$\implies \|\underline{p}\| \leq M, \text{ where } \underline{p} = (\underline{p}_1, \underline{p}_2, \ldots, \underline{p}_n) \in \mathbb{R}^n$$
$$\implies |\underline{p}_i| \leq M \text{ for each } i = 1, 2, \ldots, n.$$

Similarly, when we take $\sum_{i=1}^{n} x_i \overline{p}_i \leq M$, we get $|\overline{p}_i| \leq M$ for each $i = 1, 2, \ldots, n$. Therefore, we have

$$\|\mathbf{P}_i\|_{I(\mathbb{R})} = \max\{|\underline{p}_i|, |\overline{p}_i|\} \leq M \text{ for each } i = 1, 2, \ldots, n$$
$$\implies \|\widehat{\mathbf{P}}\|_{I(\mathbb{R})^n} = \sum_{i=1}^{n} \|\mathbf{P}_i\|_{I(\mathbb{R})} \leq M', \text{ where } M' = nM.$$

□

Theorem 7 (See [15]). *Let $\boldsymbol{T} : X' \to \overline{I(\mathbb{R})}$ be a proper convex IVF. Then, for any $\bar{x} \in \text{dom}(\boldsymbol{T})$, $\boldsymbol{\partial T}(\bar{x})$ is closed and convex.*

Theorem 8 (See [15]). *Let $\boldsymbol{T} : X' \to \overline{I(\mathbb{R})}$ be a proper convex IVF and $\bar{x} \in \text{int}(\text{dom}(\boldsymbol{T}))$. Then, the gH-subdifferential set of $\boldsymbol{T}$ at $\bar{x}$ is bounded.*

By Theorems 7 and 8, it can be seen that for a proper convex IVF $\mathbf{T}$, the gH-subdifferential set $\boldsymbol{\partial}\mathbf{T}(\bar{x})$ is compact for any $\bar{x} \in \text{int}(\text{dom}(\mathbf{T}))$. However, if $\mathbf{T}$ is gH-continuous on $\text{dom}(\mathbf{T})$, then $\boldsymbol{\partial}\mathbf{T}(\bar{x})$ is compact for any $\bar{x} \in \text{dom}(\mathbf{T})$. In the next result, we prove this fact.

Theorem 9 *Let $\boldsymbol{T} : X' \to \overline{I(\mathbb{R})}$ be a proper convex IVF. Let $\boldsymbol{T}$ is gH-continuous on $\text{dom}(\boldsymbol{T})$. Then, for any $\bar{x} \in \text{dom}(\boldsymbol{T})$, $\boldsymbol{\partial T}(\bar{x})$ is compact.*

Proof Since $\mathbf{T}$ is gH-continuous at $\bar{x} \in \text{dom}(\mathbf{T})$, it cannot be unbounded in the neighborhood of $\bar{x}$. Thus, there exist $\epsilon > 0$ and $M \geq 0$ such that

$$\mathbf{T}(\bar{x} + \epsilon z) \preceq M \text{ for all } z \in \mathbb{B}. \tag{13.4}$$

Let $\widehat{\mathbf{G}} \in \boldsymbol{\partial}\mathbf{T}(\bar{x})$. Then, by Definition 19, we have

$$(x - \bar{x})^\top \odot \widehat{\mathbf{G}} \preceq \mathbf{T}(x) \ominus_{gH} \mathbf{T}(\bar{x}) \text{ for all } x \in \mathbb{R}^n. \tag{13.5}$$

Take $x = \bar{x} + \epsilon z$, where $z \in \mathbb{B}$ in (13.5). Then,

$$
\begin{aligned}
& \epsilon z^\top \odot \widehat{\mathbf{G}} \preceq \mathbf{T}(\bar{x}+\epsilon z) \ominus_{gH} \mathbf{T}(\bar{x}) \\
\Longrightarrow\ & \epsilon z^\top \odot \widehat{\mathbf{G}} \preceq M \ominus_{gH} \mathbf{T}(\bar{x}) \text{ using (13.4)} \\
\Longrightarrow\ & \epsilon z^\top \odot \widehat{\mathbf{G}} \preceq M \oplus \mathbf{T}(\bar{x}) \\
\Longrightarrow\ & z^\top \odot \widehat{\mathbf{G}} \preceq \frac{1}{\epsilon}(M \oplus N), \text{ where } \|\mathbf{T}(\bar{x})\|_{I(\mathbb{R})} \leq N \\
\Longrightarrow\ & z^\top \odot \widehat{\mathbf{G}} \preceq N', \text{ where } N' = \frac{1}{\epsilon}(M \oplus N) \\
\Longrightarrow\ & \|\widehat{\mathbf{G}}\|_{I(\mathbb{R})^n} \leq N'' \text{ using Lemma 6, where } N'' = nN'.
\end{aligned}
$$

Since $\widehat{\mathbf{G}} \in \boldsymbol{\partial}\mathbf{T}(\bar{x})$ is arbitrary chosen, therefore $\boldsymbol{\partial}\mathbf{T}(\bar{x})$ is bounded for any $\bar{x} \in \mathrm{dom}(\mathbf{T})$. Hence, along with Theorem 7, we have the required result. □

Example 2 Consider $\mathbf{F} : \mathbb{R} \to I(\mathbb{R})$ be a convex IVF such that $\mathbf{F}(x) = |x| \odot \mathbf{A}$, where $\mathbf{0} \preceq \mathbf{A}$. Let us check gH-subdifferentiability of $\mathbf{F}$ at 0.

$$
\begin{aligned}
\boldsymbol{\partial}\mathbf{F}(0) &= \left\{\mathbf{G} \in I(\mathbb{R}) : (x-0) \odot \mathbf{G} \preceq \mathbf{F}(x) \ominus_{gH} \mathbf{F}(0) \text{ for all } x \in \mathbb{R}\right\} \\
&= \{\mathbf{G} \in I(\mathbb{R}) : x \odot \mathbf{G} \preceq |x| \odot \mathbf{A} \text{ for all } x \in \mathbb{R}\} \qquad (13.6)
\end{aligned}
$$

- Case 1. $x \leq 0$. In this case, for all $x \in \mathbb{R}$, (13.6) gives,

$$x \odot \mathbf{G} \preceq (-x) \odot \mathbf{A} \implies (-1) \odot \mathbf{A} \preceq \mathbf{G}.$$

- Case 2. $x > 0$. In this case, for all $x \in \mathbb{R}$, (13.6) gives,

$$x \odot \mathbf{G} \preceq x \odot \mathbf{A} \implies \mathbf{G} \preceq \mathbf{A}.$$

Hence, from Case 1 and Case 2, we have $\boldsymbol{\partial}\mathbf{F}(0) = \{\mathbf{G} \in I(\mathbb{R}) : (-1) \odot \mathbf{A} \preceq \mathbf{G} \preceq \mathbf{A}\}$.

Theorem 10 (See [15]). *Let $\boldsymbol{T} : X' \to I(\mathbb{R})$ be a convex and gH-differentiable IVF at $\bar{x} \in X'$. Then, $\boldsymbol{T}$ has gH-directional derivative at $\bar{x}$ for every direction $h \in \mathbb{R}^n$ and*

$$\boldsymbol{\partial}\boldsymbol{T}(\bar{x}) = \{\nabla \boldsymbol{T}(\bar{x})\},$$

where $\nabla \boldsymbol{T}(\bar{x})$ is gH-gradient of $\boldsymbol{T}$.

Lemma 11 (See [15]). *Let $\boldsymbol{T} : X' \to \overline{I(\mathbb{R})}$ be a proper convex IVF with $\boldsymbol{T}(x) = [\underline{T}(x), \overline{T}(x)]$, where $\underline{T}$, $\overline{T} : X' \to \overline{\mathbb{R}}$ are extended real-valued functions. Then, the subdifferential set of $\boldsymbol{T}$ at $\bar{x} \in \mathrm{int}(\mathrm{dom}(\boldsymbol{T}))$ can be obtained by the subdifferential sets of $\underline{T}$ and $\overline{T}$ at $\bar{x}$ and vice-versa.*

Remark 6 (See [15]). By Lemma 11, it is easy to note that for any proper convex IVF $\mathbf{T}(x) = [\underline{T}(x), \overline{T}(x)]$ and $\bar{x} \in \mathrm{int}(\mathrm{dom}(\mathbf{T}))$, $\boldsymbol{\partial}\mathbf{T}(\bar{x})$ is nonempty.

Lemma 12 (See [15]). *Let $\mathbf{S}_1$ and $\mathbf{S}_2$ be two nonempty subsets of $I(\mathbb{R})^n$ such that $\mathbf{S}_1 \subseteq \mathbf{S}_2$. Then, for any $x \in X$,*

$$\boldsymbol{\psi}^*_{\mathbf{S}_1}(x) \ \preceq \ \boldsymbol{\psi}^*_{\mathbf{S}_2}(x).$$

Theorem 13 *Let* **S** *be a nonempty subset of* $I(\mathbb{R})^n$. *Then, for any* $x \in X$,

$$\boldsymbol{\psi}^*_{\mathbf{S}}(x) = \boldsymbol{\psi}^*_{\boldsymbol{co(S)}}(x),$$

where ***co(S)*** *is the convex hull of* **S**.

Proof Since for any set **S**, $\mathbf{S} \subseteq \mathbf{co}(\mathbf{S})$. Thus, from Lemma 12, we have

$$\boldsymbol{\psi}^*_{\mathbf{S}}(x) \preceq \boldsymbol{\psi}^*_{\mathbf{co(S)}}(x). \tag{13.7}$$

Now to prove the reverse inequality, assume any $\widehat{\mathbf{Z}} \in \mathbf{co}(\mathbf{S})$. Then, from Definition 15, it can be written in the convex combination of elements of **S**:

$$\widehat{\mathbf{Z}} = \bigoplus_{i=1}^{m} \lambda_i \odot \widehat{\mathbf{G}}_i, \ \widehat{\mathbf{G}}_i \in \mathbf{S}, \ \lambda_i \geq 0 \text{ with } \sum_{i=1}^{m} \lambda_i = 1.$$

From the above relation, we see that for any $x \in X$,

$$x^\top \odot \widehat{\mathbf{Z}} = x^\top \odot \bigoplus_{i=1}^{m} \lambda_i \odot \widehat{\mathbf{G}}_i$$
$$\Longrightarrow \sup_{\widehat{\mathbf{Z}} \in \mathbf{coS}} x^\top \odot \widehat{\mathbf{Z}} = \sup_{\widehat{\mathbf{Z}} \in \mathbf{co(S)}} \left(x^\top \odot \bigoplus_{i=1}^{m} \lambda_i \odot \widehat{\mathbf{G}}_i \right).$$

From Lemmas 1 and 3, we have

$$\begin{aligned}
&\sup_{\widehat{\mathbf{Z}} \in \mathbf{co(S)}} \ x^\top \odot \widehat{\mathbf{Z}} \\
&\preceq \sup_{\widehat{\mathbf{G}}_1 \in \mathbf{S}} x^\top \odot \lambda_1 \odot \widehat{\mathbf{G}}_1 \oplus \sup_{\widehat{\mathbf{G}}_2 \in \mathbf{S}} x^\top \odot \lambda_2 \odot \widehat{\mathbf{G}}_2 \oplus \cdots \oplus \sup_{\widehat{\mathbf{G}}_m \in \mathbf{S}} x^\top \odot \lambda_m \odot \widehat{\mathbf{G}}_m \\
&= \lambda_1 \odot \sup_{\widehat{\mathbf{G}}_1 \in \mathbf{S}} x^\top \odot \widehat{\mathbf{G}}_1 \oplus \lambda_2 \odot \sup_{\widehat{\mathbf{G}}_2 \in \mathbf{S}} x^\top \odot \widehat{\mathbf{G}}_2 \oplus \cdots \oplus \lambda_m \odot \sup_{\widehat{\mathbf{G}}_m \in \mathbf{S}} x^\top \odot \widehat{\mathbf{G}}_m \\
&= \lambda_1 \odot \boldsymbol{\psi}^*_{\mathbf{S}}(x) \oplus \lambda_2 \odot \boldsymbol{\psi}^*_{\mathbf{S}}(x) \oplus \cdots \oplus \lambda_m \odot \boldsymbol{\psi}^*_{\mathbf{S}}(x) \\
&= \bigoplus_{i=1}^{m} \lambda_i \odot \boldsymbol{\psi}^*_{\mathbf{S}}(x) \\
&= \boldsymbol{\psi}^*_{\mathbf{S}}(x).
\end{aligned}$$

As $\widehat{\mathbf{Z}}$ is arbitrary, this will hold for every $\widehat{\mathbf{Z}} \in \mathbf{co}(\mathbf{S})$. Hence,

$$\boldsymbol{\psi}^*_{\mathbf{co(S)}}(x) \ \preceq \ \boldsymbol{\psi}^*_{\mathbf{S}}(x). \tag{13.8}$$

In view of (13.7) and (13.8), we have

$$\boldsymbol{\psi}^*_{\mathbf{S}}(x) = \boldsymbol{\psi}^*_{\mathbf{co(S)}}(x).$$

□

Theorem 14 (See [15]). *Let $\boldsymbol{T} : X' \to \overline{I(\mathbb{R})}$ be a proper convex IVF with $\boldsymbol{T}(x) = [\underline{T}(x), \overline{T}(x)]$, where $\underline{T}, \overline{T} : X' \to \overline{\mathbb{R}}$ are extended real-valued functions. Then, at any $\bar{x} \in \text{int}(\text{dom}\boldsymbol{T})$,*

$$\boldsymbol{T}_{\mathscr{D}}(\bar{x})(h) = \boldsymbol{\psi}^*_{\partial \boldsymbol{T}(\bar{x})}(h) \text{ for all } h \in \mathbb{R}^n \text{ such that } \bar{x} + h \in X',$$

where $\boldsymbol{T}_{\mathscr{D}}(\bar{x})(h)$ is the gH-directional derivative of $\boldsymbol{T}$ at $\bar{x}$ in the direction of h.

Remark 7 From Theorem 10, for a gH-differentiable IVF, one can see how the gH-subgradient and the gH-gradient are related to each other. Similarly, a relation between gH-directional derivative and gH-gradient for a gH-differentiable IVF can be establish by Theorems 10 and 14 which is $\mathbf{T}_{\mathscr{D}}(\bar{x})(h) = h^\top \odot \nabla \mathbf{T}(\bar{x})$.

Theorem 15 (See [17]) *Let Λ be any finite set of indices. For each $i \in \Lambda$, let $\boldsymbol{T}_i : X' \to \overline{I(\mathbb{R})}$ be convex and gH-continuous IVF such that $\boldsymbol{T}_{i_{\mathscr{D}}}(\bar{x})(d)$ exists for any $\bar{x} \in X'$. Define*

$$\boldsymbol{T}(x) = \sup_{i \in \Lambda} \boldsymbol{T}_i(x).$$

Then, for any $\bar{x} \in X'$ and $h \in \mathbb{R}^n$ such that $\bar{x} + h \in X'$,

$$\boldsymbol{T}_{\mathscr{D}}(\bar{x})(h) = \sup_{i \in I(\bar{x})} \boldsymbol{T}_{i_{\mathscr{D}}}(\bar{x})(h), \text{ where } I(\bar{x}) = \{i \in \Lambda : \boldsymbol{T}_i(\bar{x}) = \boldsymbol{T}(\bar{x})\}.$$

Theorem 16 (Supremum rule for gH-subdifferential calculus of IVFs) *Let Λ be any finite set of indices. For each $i \in \Lambda$, let $\boldsymbol{T}_i : X' \to \overline{I(\mathbb{R})}$ be a proper convex IVF which is gH-continuous on X'. Define*

$$\boldsymbol{T}(x) = \sup_{i \in \Lambda} \boldsymbol{T}_i(x).$$

Then, for any $\bar{x} \in \bigcap_{i \in \Lambda} \text{int}(\text{dom}(\boldsymbol{T}_i))$, and $h \in \mathbb{R}^n$ such that $\bar{x} + h \in X'$,

$$\boldsymbol{\psi}^*_{\partial \boldsymbol{T}(\bar{x})}(h) = \boldsymbol{\psi}^*_{\boldsymbol{S}}(h),$$

where $\boldsymbol{S} = \boldsymbol{co}\left(\bigcup_{i \in I(\bar{x})} \partial \boldsymbol{T}_i\right)$ and $I(\bar{x}) = \{i \in \Lambda : \boldsymbol{T}_i(\bar{x}) = \boldsymbol{T}(\bar{x})\}$.

Proof From Theorem 15, the gH-directional derivative of supremum of convex IVFs is given by

$$\mathbf{T}_{\mathscr{D}}(\bar{x})(h) = \sup_{i \in I(\bar{x})} \mathbf{T}_{i_{\mathscr{D}}}(\bar{x})(h). \tag{13.9}$$

For simplicity, assume that $\Lambda = \{1, 2, \ldots, m\}$ and $I(\bar{x}) = \{1, 2, \ldots, k\}$ for some $k \in \{1, 2, \ldots, m\}$. Now in view of Theorem 14, we see that for any $\bar{x} \in \bigcap_{i=1}^{m} \text{int}(\text{dom}(\mathbf{T}_i))$ and $h \in \mathbb{R}^n$ such that $\bar{x} + h \in X'$,

$$\mathbf{T}_{\mathscr{D}}(\bar{x})(h) = \boldsymbol{\psi}^*_{\boldsymbol{\partial}\mathbf{T}(\bar{x})}(h) = \sup\left\{h^\top \odot \widehat{\mathbf{G}} : \widehat{\mathbf{G}} \in \boldsymbol{\partial}\mathbf{T}(\bar{x})\right\}. \tag{13.10}$$

From (13.10), we observe that for each $i = 1, 2, \ldots, m$ and corresponding to each $\mathbf{T}_i$ there exists $\widehat{\mathbf{G}}_i = (\mathbf{G}_{i1}, \mathbf{G}_{i2}, \ldots, \mathbf{G}_{in})^\top \in I(\mathbb{R})^n$ with $\mathbf{G}_{ij} = [\underline{g}_{ij}, \overline{g}_{ij}]$ such that

$$\begin{aligned}
&\mathbf{T}_{i_{\mathscr{D}}}(\bar{x})(h)\\
&= \sup_{\widehat{\mathbf{G}}_i \in \boldsymbol{\partial}\mathbf{T}_i(\bar{x})} \left\{h^\top \odot \widehat{\mathbf{G}}_i\right\}\\
&= \sup_{\substack{\underline{g}_i \in \partial \underline{T}_i\\ \overline{g}_i \in \partial \overline{T}_i}} \left[\min\left\{\sum_{j=1}^{n} h_j \underline{g}_{ij}, \sum_{j=1}^{n} h_j \overline{g}_{ij}\right\}, \max\left\{\sum_{j=1}^{n} h_j \underline{g}_{ij}, \sum_{j=1}^{n} h_j \overline{g}_{ij}\right\}\right],
\end{aligned} \tag{13.11}$$

where $\underline{g}_i = (\underline{g}_{i1}, \underline{g}_{i2}, \ldots, \underline{g}_{in})^\top$, $\overline{g}_i = (\underline{g}_{i1}, \underline{g}_{i2}, \ldots, \underline{g}_{in})^\top \in \mathbb{R}^n$. We now consider the following two cases.

- Case 1. Let $\min\left\{\sum_{j=1}^{n} h_j \underline{g}_{ij}, \sum_{j=1}^{n} h_j \overline{g}_{ij}\right\} = \sum_{j=1}^{n} h_j \underline{g}_{ij}$ and $\max\left\{\sum_{j=1}^{n} h_j \underline{g}_{ij}, \sum_{j=1}^{n} h_j \overline{g}_{ij}\right\} = \sum_{j=1}^{n} h_j \overline{g}_{ij}$. Therefore, from (13.9) and (13.11), we get

$$\begin{aligned}
\mathbf{T}_{\mathscr{D}}(\bar{x})(h) &= \sup_{i \in I(\bar{x})} \left[\sup_{\underline{g}_i \in \partial \underline{\mathbf{T}}_i} \left(\sum_{j=1}^{n} h_j \underline{g}_{ij}\right), \sup_{\overline{g}_i \in \partial \overline{T}_i} \left(\sum_{j=1}^{n} h_j \overline{g}_{ij}\right)\right]\\
&= \sup_{i \in I(\bar{x})} \left[\sup_{\underline{g}_i \in \partial \underline{T}_i} \langle \underline{g}_i, h\rangle, \sup_{\overline{g}_i \in \partial \overline{T}_i} \langle \overline{g}_i, h\rangle\right] \text{Remark 1}\\
&= \left[\sup_{i \in I(\bar{x})} \left(\sup_{\underline{g}_i \in \partial \underline{\mathbf{T}}_i} \langle \underline{g}_i, h\rangle\right), \sup_{i \in I(\bar{x})} \left(\sup_{\overline{g}_i \in \partial \overline{T}_i} \langle \overline{g}_i, h\rangle\right)\right].
\end{aligned} \tag{13.12}$$

Using the fact that for any finite numbers $a_1, a_2, \ldots, a_k \in \mathbb{R}$,

$$\sup\{a_1, a_2, \ldots, a_k\} = \sup_{\lambda \in \Delta_k} \sum_{i=1}^{k} \lambda_i a_i.$$

On applying above relation in (13.12), we get

$$\mathbf{T}_{\mathscr{D}}(\bar{x})(h)$$

$$= \left[\sup_{i \in I(\bar{x})} \left(\sup_{\underline{g}_i \in \partial \underline{T}_i} \langle \underline{g}_i, h \rangle \right), \sup_{i \in I(\bar{x})} \left(\sup_{\overline{g}_i \in \partial \overline{T}_i} \langle \overline{g}_i, h \rangle \right) \right]$$

$$= \left[\sup_{\lambda \in \Delta_k} \left(\sum_{i=1}^{k} \lambda_i \left(\sup_{\underline{g}_i \in \partial \underline{T}_i} \langle \underline{g}_i, h \rangle \right) \right), \sup_{\lambda \in \Delta_k} \left(\sum_{i=1}^{k} \left(\sup_{\overline{g}_i \in \partial \overline{T}_i} \langle \overline{g}_i, h \rangle \right) \right) \right]$$

$$= \left[\sup_{\substack{\underline{g}_i \in \partial \underline{T}_i \\ \lambda \in \Delta_k}} \left\langle \sum_{i=1}^{k} \lambda_i \underline{g}_i, h \right\rangle, \sup_{\substack{\overline{g}_i \in \partial \overline{T}_i \\ \lambda \in \Delta_k}} \left\langle \sum_{i=1}^{k} \lambda_i \overline{g}_i, h \right\rangle \right]$$

$$= \sup_{\substack{\underline{g}_i \in \partial \underline{T}_i \\ \overline{g}_i \in \partial \overline{T}_i}} \left[\left\langle \sum_{i=1}^{k} \lambda_i \underline{g}_i, h \right\rangle, \left\langle \sum_{i=1}^{k} \lambda_i \overline{g}_i, h \right\rangle \right], \ \lambda \in \Delta_k.$$

Therefore, from Definition 15 and Remark 3, we get

$$\begin{aligned} \mathbf{T}_{\mathscr{D}}(\bar{x})(h) &= \sup \left\{ h^\top \odot \widehat{\mathbf{G}} : \ \widehat{\mathbf{G}} \in \mathbf{co} \left(\bigcup_{i=1}^{k} \partial \mathbf{T}_i \right) \right\} \\ &= \boldsymbol{\psi}^*_{\mathbf{S}}(h), \ \text{ where } \mathbf{S} = \mathbf{co} \left(\bigcup_{i=1}^{k} \partial \mathbf{T}_i \right). \end{aligned} \tag{13.13}$$

Thus, using (13.13) and (13.10), we get

$$\boldsymbol{\psi}^*_{\partial \mathbf{T}(\bar{x})}(h) = \boldsymbol{\psi}^*_{\mathbf{S}}(h), \ \text{ where } \mathbf{S} = \ \mathbf{co} \left(\bigcup_{i=1}^{k} \partial \mathbf{T}_i \right).$$

- Case 2. Let $\min \left\{ \sum_{j=1}^{n} h_j \underline{g}_{ij}, \sum_{j=1}^{n} h_j \overline{g}_{ij} \right\} = \sum_{j=1}^{n} h_j \overline{g}_{ij}$ and $\max \left\{ \sum_{j=1}^{n} h_j \underline{g}_{ij}, \sum_{j=1}^{n} h_j \overline{g}_{ij} \right\} = \sum_{j=1}^{n} h_j \underline{g}_{ij}$. Proof contains similar steps as in Case 1.

From Case 1 and Case 2, it is clear that for any $\bar{x} \in \bigcap_{i=1}^{m} \text{int}(\text{dom}(\mathbf{T}_i))$ and $h \in \mathbb{R}^n$ such that $\bar{x} + h \in X'$,

$$\boldsymbol{\psi}^*_{\partial \mathbf{T}(\bar{x})}(h) = \boldsymbol{\psi}^*_{\mathbf{S}}(h), \ \text{ where } \mathbf{S} = \mathbf{co} \left(\bigcup_{i=1}^{k} \partial \mathbf{T}_i \right).$$

□

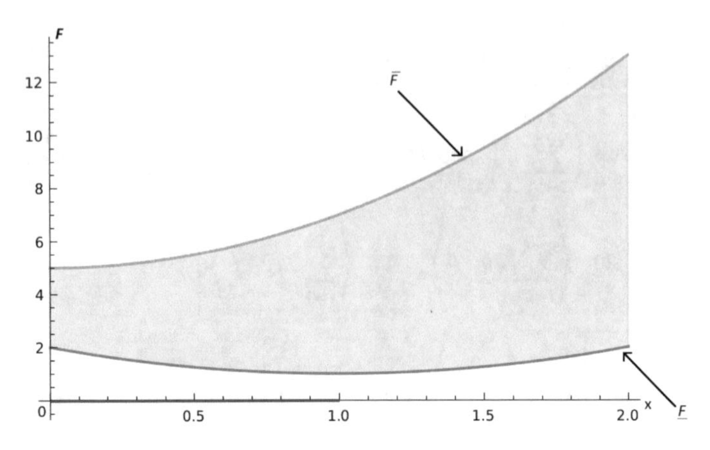

Fig. 13.1 IVF **T** and efficient solution of IOP (13.15)

13.4 Application on Nonsmooth Interval Optimization

This section has two optimality conditions (Theorems 17 and 18) to find the efficient point of IOP (13.14) and an example that illustrates these conditions are necessary but not sufficient.

Definition 21 (*Efficient solution (ES)* [7]). Let $\mathbf{T} : X \to I(\mathbb{R})$ be an IVF. A point $\bar{x} \in X$ is called an ES of the IOP

$$\min_{x \in X} \mathbf{T}(x), \tag{13.14}$$

if there does not exist any $x \in X$ such that $\mathbf{T}(x) \prec \mathbf{T}(\bar{x})$.

Definition 22 (*Weak efficient solution (WES)* [17]). A point $\bar{x} \in X$ is said to be a WES of the IOP (13.14), if $\mathbf{T}(\bar{x}) \preceq \mathbf{T}(x)$ for all $x \in X$.

Theorem 17 *Let $\boldsymbol{T} : X' \to I(\mathbb{R})$ be a convex IVF. If $\bar{\boldsymbol{0}} \in \boldsymbol{\partial T}(\bar{x})$ for some $\bar{x} \in X'$, where $\bar{\boldsymbol{0}} = (\boldsymbol{0}, \boldsymbol{0}, \ldots, \boldsymbol{0})^\top \in I(\mathbb{R})^n$, then $\bar{x}$ is an ES of (13.14).*

Proof Let $\bar{\mathbf{0}} \in \boldsymbol{\partial}\mathbf{T}(\bar{x})$. Thus, by Definition 19, we have

$$\begin{aligned} & (x - \bar{x})^\top \odot \bar{\mathbf{0}} \preceq \mathbf{T}(x) \ominus_{gH} \mathbf{T}(\bar{x}) \text{ for all } x \in X' \\ \Longrightarrow\ & \mathbf{0} \preceq \mathbf{T}(x) \ominus_{gH} \mathbf{T}(\bar{x}) \\ \Longrightarrow\ & \mathbf{T}(\bar{x}) \preceq \mathbf{T}(x) \text{ for all } x \in X'. \end{aligned}$$

Hence, $\bar{x}$ is an ES of (13.14). □

Remark 8 The converse of Theorem 17 is not true. For instance, consider the following IOP:

$$\min_{x \in X} \mathbf{T}(x) = [1, 2] \odot x^2 \oplus [-2, 0] \odot x \oplus [2, 5], \tag{13.15}$$

where $X = [0, 2]$. Note here that $\underline{T}(x) = x^2 - 2x + 2$ and $\overline{T}(x) = 2x^2 + 5$ are the lower and upper functions of $\mathbf{T}$, for all $x \in X$. Clearly, $\underline{T}$ and $\overline{T}$ are convex and differentiable real-valued functions. Therefore, due to Lemma 2 and Remark 2, $\mathbf{T}$ is also convex and gH-differentiable. Hence,

$$\begin{aligned} \partial \mathbf{T}(x) &= \{\nabla \mathbf{T}(x)\} \text{ for all } x \in X \text{ by Theorem 10} \\ &= [2, 4] \odot x \oplus [-2, 0] \text{ for all } x \in X. \end{aligned}$$

We draw the graph of IVF $\mathbf{T}$, which is shown by the gray region in Fig. 13.1. It is easy to observe that $\bar{x} \in [0, 1]$ is an ES of (13.15), which is shown by red bold line on x-axis.

Note that for each $\bar{x} \in [0, 1]$,

$$\nabla \mathbf{T}(x) = [2x - 2, 4x] \neq \mathbf{0}.$$

Hence, $\mathbf{0} \notin \partial \mathbf{T}(\bar{x})$.

Remark 9 In Remark 8, we have seen that the converse of Theorem 17 is not true in general. However, the converse can be made true if we take WES defined in Definition 22 instead of ES of IOP (13.14). The reason is as follows.

Let $\bar{x}$ be a WES of IOP (13.14). Then, for all $x \in X'$

$$\begin{aligned} & \mathbf{T}(\bar{x}) \preceq \mathbf{T}(x) \\ \iff & \mathbf{0} \preceq \mathbf{T}(x) \ominus_{gH} \mathbf{T}(\bar{x}) \\ \iff & (x - \bar{x})^\top \odot \bar{\mathbf{0}} \preceq \mathbf{T}(x) \ominus_{gH} \mathbf{T}(\bar{x}) \text{ where } \bar{\mathbf{0}} = (\mathbf{0}, \mathbf{0}, \ldots, \mathbf{0})^\top \in I(\mathbb{R})^n \\ \iff & \bar{\mathbf{0}} \in \partial \mathbf{T}(\bar{x}). \end{aligned}$$

Theorem 18 *Let $\boldsymbol{T} : X' \to I(\mathbb{R})$ be a convex IVF. Let there exists $\widehat{\mathbf{G}} \in \partial \boldsymbol{T}(\bar{x})$ for some $\bar{x} \in X'$, for which there does not exist any $x \in X'$, such that*

$$(x - \bar{x})^\top \odot \widehat{\mathbf{G}} \prec \boldsymbol{0}. \tag{13.16}$$

Then, $\bar{x}$ is an ES of the IOP (13.14).

Proof Let $\widehat{\mathbf{G}} \in \partial \mathbf{T}(\bar{x})$ for some $\bar{x} \in X'$, which satisfies (13.16). Therefore, there does not exist any $x \in X'$ such that

$$(x - \bar{x})^\top \odot \widehat{\mathbf{G}} \prec \mathbf{0}.$$

Thus, due to Definition 19, there does not exist any $x \in X'$ for which

$$\mathbf{T}(x) \ominus_{gH} \mathbf{T}(\bar{x}) \prec \mathbf{0}.$$

Hence, there does not exist any $x \in X'$ such that $\mathbf{T}(x) \prec \mathbf{T}(\bar{x})$, and therefore, $\bar{x}$ is an ES of (13.14), by Definition 21. □

Remark 10 The converse of the Theorem 18 is not true. For instance, consider IOP (13.15) of Remark 8, for which $\bar{x} \in [0, 1]$ is the ES. Thus, at $\bar{x} = 0$, we have

$$(x - \bar{x})^\top \odot \widehat{\mathbf{G}} = (x - \bar{x}) \odot \nabla\mathbf{T}(\bar{x}) = x \odot [-2, 0].$$

Therefore, $(x - \bar{x})^\top \odot \widehat{\mathbf{G}} = x \odot [-2, 0] \prec \mathbf{0}$ for all $x \in (0, 2] \subseteq X$.

13.5 Conclusion and Future Scopes

In this chapter, calculus of gH-subdifferential set of convex IVFs has been extended. With the help of existing properties of gH-subdifferential set, compactness of gH-subdifferential set of gH-continuous IVF (Theorem 9) and supremum rule for IVFs (Theorem 16) have been found. To prove these two theorems, related concepts are also given (Definitions 14, 15 and Lemma 6). Thereafter, two necessary optimality conditions (Theorems 17 and 18) are given with appropriate example (Remark 8).

In future, we have a few following directions to work on.

Problem 1. Although we have studied many properties and characterizations of the gH-subdifferential set, we could not provide gH-subdifferential sum rule for convex IVFs. We shall try to make a conclusion on it. Because once we get this sum rule, we can use the proposed study of gH-subdifferential set and optimality conditions to solve constrained IOPs. The reason is as follows.
Let $\mathbf{T} : X' \to I(\mathbb{R})$ be a proper convex IVF. Let S be a nonempty subset of X'. Consider a constrained IOP

$$\min_{x \in S} \mathbf{T}(x). \tag{13.17}$$

Note that IOP (13.17) can be converted into unconstrained IOP

$$\min_{x \in X'} \mathbf{T}_o(x), \tag{13.18}$$

where $\mathbf{T}_o : X' \to I(\mathbb{R})$ such that

$$\mathbf{T}_o(x) = \mathbf{T}(x) \oplus \boldsymbol{\delta}_S(x) \text{ for all } x \in X'.$$

Therefore, we have

$$\boldsymbol{\partial}\mathbf{T}_o(x) = \boldsymbol{\partial}(\mathbf{T}(x) \oplus \boldsymbol{\delta}_S(x)).$$

Thus, optimality conditions given in Theorems 17 and 18 can be applied on $\mathbf{T}_o$ to find the efficient point of IOP (13.17) with the help of gH-subdifferential sum rule.

Problem 2. In future, we shall also try to solve the following IOP (13.19) using the Sup-IVF approach, in which we shall use presented Supremum rule for IVF Theorem 16. The idea is as follows. Let $\mathbf{T}$ and $\mathbf{H}_i$ be convex IVFs on X'. Consider an IOP

$$\min_{x \in S} \mathbf{T}(x), \tag{13.19}$$

where $S = \{x \in X' : \mathbf{H}_i(x) \preceq \mathbf{0}\}$. We say $\bar{x} \in S$ is an ES of (13.19), if it is an ES of the following IOP (13.20) as well.

$$\min_{x \in X} \mathbf{T}_o(x), \tag{13.20}$$

where $\mathbf{T}_o(x) = \sup\{\mathbf{T}(x) \ominus_{gH} \mathbf{T}(\bar{x}), \mathbf{H}_1(x), \mathbf{H}_2(x), \ldots, \mathbf{H}_n(x)\}$. Now we can find the gH-subdifferential set of $\mathbf{T}_o$ at $\bar{x}$ using supremum rule for IVF Theorem 16. Thereafter, we can use our optimality condition Theorem 17, i.e.,

$$\mathbf{0} \in \partial \mathbf{T}_o(\bar{x})$$

to investigate ES of IOP (13.20). Using this idea, we shall try also to give an attempt to derive KKT conditions for IOP (13.19).

Another future direction is to propose a gH-subgradient method and its convergence to solve the unconstrained IOPs. We further plan to use the gH-subgradient method for maximum margin structured learning [24] under interval data uncertainty.

References

1. Moore, R.E.: Interval Analysis. Prentice-Hall, Englewood Cliffs, EJ (1966)
2. Hukuhara, M.: Intégration des applications mesurables dont la valeur est un compact convexe. Funkcialaj Ekvacioj **10**(3), 205–223 (1967)
3. Stefanini, L., Bede, B.: Generalized Hukuhara differentiability of interval-valued functions and interval differential equations. Nonlinear Anal. Theory Methods Appl. **71**(3–4), 1311–1328 (2009)
4. Ishibuchi, H., Tanaka, H.: Multiobjective programming in optimization of the interval objective function. Eur. J. Oper. Res. **48**(2), 219–225 (1990)
5. Wu, H.-C.: The Karush–Kuhn–Tucker optimality conditions in an optimization problem with interval-valued objective function. Eur. J. Oper. Res. **176**(1), 46–59 (2007)
6. Wu, H.-C.: Wolfe duality for interval-valued optimization. J. Optim. Theory Appl. **138**(3), 497–509 (2008)
7. Bhurjee, A.K., Panda, G.: Efficient solution of interval optimization problem. Math. Methods Oper. Res. **76**(3), 273–288 (2012)
8. Chalco-Cano, Y., Lodwick, W.A., Rufián-Lizana, A.: Optimality conditions of type KKT for optimization problem with interval-valued objective function via generalized derivative. Fuzzy Optim. Decis. Making **12**(3), 305–322 (2013)
9. Ghosh, D., Debnath, A.K., Pedrycz, W.: A variable and a fixed ordering of intervals and their application in optimization with interval-valued functions. Int. J. Approx. Reas. **121**, 187–205 (2020)

10. Bede, B., Gal, S.G.: Generalizations of the differentiability of fuzzy-number-valued functions with applications to fuzzy differential equations. Fuzzy Sets Syst. **151**(3), 581–599 (2005)
11. Chalco-Cano, Y., Roman-Flores, H.: On new solutions of fuzzy differential equations. Chaos Solitons Fractals **38**(1), 112–119 (2008)
12. Ghosh, D.: Newton method to obtain efficient solutions of the optimization problems with interval-valued objective functions. J. Appl. Math. Comput. **53**(1–2), 709–731 (2017)
13. Ghosh, D., Chauhan, R.S., Mesiar, R., Debnath, A.K.: Generalized Hukuhara Gâteaux and Fréchet derivatives of interval-valued functions and their application in optimization with interval-valued functions. Inf. Sci. **510**, 317–340 (2020)
14. Ghosh, D., Debnath, A.K., Chauhan, R.S., Castillo, O.: Generalized-Hukuhara-gradient efficient-direction method to solve optimization problems with interval-valued functions and its application in least-squares problems. Int. J. Fuzzy Syst. 1–26 (2021)
15. Kumar, K., Ghosh, D., Kumar, G.: Weak sharp minima for interval-valued functions and its primal-dual characterizations using generalized Hukuhara subdifferentiability. Soft. Comput. **26**(19), 10253-10273 (2021)
16. Kumar, G., Ghosh, D.: Ekeland's variational principle for interval-valued functions. arXiv preprint arXiv:2104.11167 (2021)
17. Ghosh, D., Chauhan, R.S., Mesiar, R.: Generalized-Hukuhara subdifferential analysis and its application in nonconvex composite optimization problems with interval-valued functions. arXiv preprint arXiv:2109.14586 (2021)
18. Chauhan, R.S., Ghosh, D., Ramik, J., Debnath, A.K.: Generalized Hukuhara-Clarke derivative of interval-valued functions and its properties. Soft Comput. **25**(23), 14629-14643 (2020). arXiv preprint arXiv:2010.16182
19. Calvete, H.I., Galé, C.: Linear bilevel programming with interval coefficients. J. Comput. Appl. Math. **236**(15), 3751–3762 (2012)
20. Osuna-Gómez, R., Chalco-Cano, Y., Hernández-Jiménez, B., Ruiz-Garzón, G.: Optimality conditions for generalized differentiable interval-valued functions. Inf. Sci. **321**, 136–146 (2015)
21. Ghosh, D., Singh, A., Shukla, K., Manchanda, K.: Extended Karush-Kuhn-Tucker condition for constrained interval optimization problems and its application in support vector machines. Inf. Sci. **504**, 276–292 (2019)
22. Bhurjee, A.K., Padhan, S.K.: Optimality conditions and duality results for non-differentiable interval optimization problems. J. Appl. Math. Comput. **50**(1), 59–71 (2016)
23. Ghosh, D., Ghosh, D., Bhuiya, S.K., Patra, L.K.: A saddle point characterization of efficient solutions for interval optimization problems. J. Appl. Math. Comput. **58**(1), 193–217 (2018)
24. Ratliff, N., Bagnell, J.A., Zinkevich, M.: Subgradient methods for maximum margin structured learning. In: ICML Workshop on Learning in Structured Output Spaces, vol. 46 (2006). Citeseer

Chapter 14
Rule-Based Classifiers for Identifying Fake Reviews in E-commerce: A Deep Learning System

Saleh Nagi Alsubari, Sachin N. Deshmukh, Theyazn H. H. Aldhyani, Abdullah H. Al Nefaie, and Melfi Alrasheedi

14.1 Introduction

Opinion mining systems are broadly utilized in diverse e-commerce platforms, which include Taobao, eBay, and Amazon. These sorts of systems gather personal feedback from clients and provide reliable material regarding the vender to assist capability purchasers create their buying decisions [10, 12, 34]. 91% of respondents specified that they might consider customer critiques before buying online. Excellent critiques and high product scores will convince purchasers to buy, whereas the poor ones will decrease shoppers' intention to purchase [7]. Further, poor opinions have a greater impact on clients' buy choices than positive evaluations [31].

The value of Internet customer feedback gives robust motivation for assessment handling, which leads to large numbers of phony satisfactory evaluations [15]. These overviews can benefit various accounts. For instance, vendors masked as purchasers post online critiques. Internet-based company Amazon carelessly disclosed the identification of a reviewer who turned out to be the writer of the e-book [36]. Merchants hire human workers to post fake positive assessments to inflate their shops' standings [39]. Subsequently, gathering dataset included 2.14 million reviewers and 58 million reviews from Amazon's website, and Jindal and Liu [16] introduced a study for

S. N. Alsubari · S. N. Deshmukh
Department of Computer Science and Information Technology, Dr. Babasaheb Ambedkar, Marathwada University, Aurangabad, India

T. H. H. Aldhyani
Applied College in Abqaiq, King Faisal University, 31982, Al-Ahsa, Saudi Arabia
e-mail: taldhyani@kfu.edu.sa

A. H. Al Nefaie · M. Alrasheedi (✉)
Department of Quantitaive Methods, School of Business, King Faisal University, Al-Ahsa 31982, Saudi Arabia
e-mail: malrasheedy@kfu.edu.sa

T. Som et al. (eds.), *Fuzzy, Rough and Intuitionistic Fuzzy Set Approaches for Data Handling*, Forum for Interdisciplinary Mathematics,
https://doi.org/10.1007/978-981-19-8566-9_14

opinion spam analysis for detecting manipulation of opinions. For example, reviews of text that were posted by the same reviewer for exceptional goods are identical or very similar to opinions that were written by different reviewers for the identical product or exclusive merchandise. This leads to the assumption that an individual reviewer has submitted many evaluations of specific products. Both are considered as single user or enumerated as different consumer's names. Glaringly, those opinions are fake [9, 37].

Online product feedback (reviews) can be defined as evaluations given by customers after purchasing the products from e-commerce websites. The existence of manipulation of reviews in online purchasing websites will extend an effect on the opinion mining systems performance. According to previous research, 80% of customers have the authenticity of online reviews [39]. Deceptive evaluations substantially misguide online purchasers. Hence, the detection of fake online reviews/opinions has been a hot challenge topic of theoretical and practical research. The contributions of this research work are the following:

1. Analyzing and detecting online deceptive reviews in electronic product reviews on Amazon and Yelp platforms.
2. Proposing an enhanced framework for online fake/deceptive feature reviews.
3. Proposing novel features such as authenticity and analytical thinking that are used to differentiate between online fake and genuine reviews.
4. Proposing mathematical equations for some used feature.
5. Labeling the Amazon reviews dataset used, based on extracted linguistic features from the review text.
6. Comparing and analyzing the performance of the Recurrent Neural Network (Bidirectional Long Short Memory) model on Amazon and Yelp datasets.

14.2 Related Work

Previous research has attempted to figure out which online opinions in e-commerce websites are fake or true, to identify fake online evaluations of products, searching to learn the qualities of fake opinions or fraud reviewers after filtering them. Furthermore, there are ongoing attempts by fraudsters to find ways to prevent false evaluations from being detected [21]. Previous research pointed out that language used in the writing of fake opinions has simply been like that used in truthful opinions [30]. Consequently, an identification and analysis of the fake reviews and deceptive reviewers is a challenging task. Therefore, this paper focuses on detecting such reviews and reviewers and proposes a model based on deep learning, and comparing it to a machine leaning model in order to observe which model can provide more accuracy. Normally, purchasers do not lie without a reason. As a result, the real source of fake opinion is the sellers. This take seems to be the advice of the overview and the reviewer and has initially evolved from the evaluate manipulators and dealers, to set up a data mining model that can be used to analyze the intention of sellers' evaluation manipulation and recognize whether or not sellers have an ability to control online evaluation. This will furnish a brand new standpoint for fake evaluation identification.

Prior researchers have suggested becoming aware of attributes to distinguish between real and fake reviews. They encompass the textual content of the review, evaluation duration, rating, emotion, clarity, subjectivity, style of writing, and product-rating characteristics [8, 30]. Other research work focused on the distinction between a reviewer's rating and the average rating of the product and the score differences between reviews of numerous products written by identical reviewers [38]. Further, Deng et al. suggested 11 linguistic fraud indications for fake reviews detection and divided them into three categories: word occurrence, information richness, and accounts integrity of the reviewer [35]. However, fake reviews posted by fraud individuals are designed to mimic authentic reviews to prevent them from being recognized. Studies have shown that some fake remarks are similar in language to real reviews [11]. Consequently, it is a challenging task to pick out fake reviews by the customer manually. Every other indication for identifying fake reviews is based entirely on the reviewers' behavior traits, such as the number of reviews published by the same reviewers every day, the time between the first and the last review from the same reviewer, the share of the first review on the merchandise website, all the reviews published by means of a reviewer, the range of votes obtained, and the provision of video facts [17, 29]. Zhang et al. have carried out research in figuring out fake evaluations via combining statistics of overview textual content and reviewer activities [40]. However, there is difficulty detecting fake reviews using opinion mining system due to the fact that evaluation manipulation is covert and invisible, and only the evaluation companies or the operators genuinely recognize whether an assessment is fake or real [24]. So far, the problem of fake opinion on the Internet has still not been addressed because there is no effective machine learning-based system providing great accuracy for detecting fake reviews. Most previous researchers created synthetic review datasets to train a machine learning classifier [33].

Fake overview detection is much like crime detection. Similarly to amassing clues on the scene of a criminal offense, police should additionally determine who has a motive for the crime. Stimulated by this similarity, this paper focuses on analyzing the sellers' evaluation manipulation behavior to assist in fake reviews detection. Few research works concentrated on reviewing manipulation behavior. It has been identified that supplier reputation growth had been a worthwhile hidden commercial enterprise. Mayzlin [25] mentioned that the cost of a review manipulation process can determine the quantity of a manipulation system. As an example, branded chain resorts are much less likely to engage in evaluation manipulation than premium accommodations, because their name would be jeopardized if manipulation activity was exposed publicly. Gao and Liu [18] advanced an evaluation manipulation proxy to identify what type of dealers may also engage in evaluating manipulation based on the discretionary accrual primarily based on earning of control framework. Most marketplaces give preference to well-evaluated goods, potentially rewarding companies that pay for false reviews. Vast amounts of positive reviews motivate buyers to make a purchase and boost manufacturers' finances, while negative reviews allow customers to search for alternatives, leading to financial losses. It can also be difficult to identify and differentiate fake reviews from trusted ones because of the number of reviews released online and the skills of review fraudsters. Moreover, detecting and removing such reviews from review websites and product recommendation systems is important for businesses and consumers [26]. The increase of spam review threats has actually accelerated, since anyone can just write and share spam reviews online with no restrictions. Some manufactures may hire persons for their time and services to compose fake reviews in order to popularize their products. These persons are known as spammers. Fake reviews are usually published to earn profit, as well as to promote online services or products. This case is called Spamming Review [3, 28]. According to the existing studies, there is no effective method of discriminating between the features of truthful and false reviews. Getting a credible and comprehensive review website is the main objective of deceptive reviews detection methods that filter the text content from deceptive and unwanted reviews. "Credibility" is of great importance principally for applications of opinion mining. Credibility includes how stable the credibility of the intended system is, so the deceptive reviews identification methods are important for deleting and filtering deceptive reviews in the online e-commerce websites.

14.3 Materials and Methods

Materials and methods explain the details of the recommended methodology for analyzing and detection of fake opinions in product reviews of e-commerce platforms. Figure 14.1 illustrates the steps which are applied in this methodology.

Our proposed methodology consists of eight phases, which are the e-commerce website, dataset collection, dataset labeling, preprocessing, feature extraction (TF-IDF/Word2Vec), supervised machine learning/deep learning hybrid neural network methods, assessment metrics, and results. The steps of this methodology are explained below.

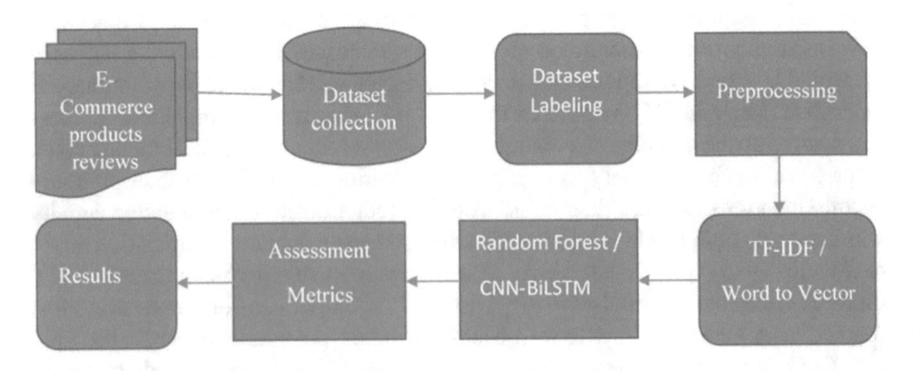

Fig. 14.1 Framework for the used methodology

Table 14.1 Distribution of the used dataset

Product name	Number of reviews
Phones	12,100
Batteries	8400
Chargers	5400
Headphones	4570

14.3.1 E-commerce Product Reviews

Recently, e-commerce businesses have increasingly sprung up all over the world. Almost every business platform functions as an online marketplace in some way. The demand for e-commerce has grown rapidly to new heights in recent years, due to easy Internet access and available advanced technologies. There are variety of elements to determine the popularity of e-commerce companies and their reputation such as credibility, product quality, and easy access websites. Product reviews, on the other hand, are a critical factor in improving an e-commerce store's credibility, standards, and assessment. Product reviews give an e-commerce business one of the most useful resources for detecting issues with products and analyzing the feeling of consumers 143 [1]. For example, customer reviews of the product represent the satisfaction of customers when purchased.

14.3.2 Dataset Collection

This is the most important phase in our methodology. We have scrapped raw product reviews of mobile phones and their accessories from Amazon.com, which is one of the largest online e-commerce websites usually accessed for selling and purchasing online consumer products. The scrapping process has been completed with the help of a web scrapper application that is programmed and developed using Python library Scrapy [19]. The collected unlabeled reviews consist of 30,470 distributed over four different categories of products collected to perform our experiments. The meta-features of the dataset include reviewer ID, reviewer name, review title, product name, rating value, verified purchase (TRUE or FALSE), and the review text. Table 14.1 shows the distribution reviews per product category.

14.3.3 Dataset Labeling

Data labeling can be defined as the process of identifying raw data samples by adding one or more significant and useful labels for all samples of the dataset. Labels are usually used to bring a context for machine learning models, which can have the ability to learn from them. The labels utilized to discriminate the features of the dataset should be informative, distinguished, and distinct, to lead to the creation of a sophisticated algorithm. A correctly labeled dataset usually presents ground truth for a machine learning algorithm that can verify its forecasts for accuracy. As the collected dataset is unlabeled reviews, we aim to abstract important features from the text of the product review that has been given by the reviewer and combine these features with meta-features for labeling the review as fake or truthful. This can be obtained by extracting and computing various features from the dataset such as the following.

A. **The authenticity of a written review text**

Authenticity is adopted to calculate and measure the personalized and trustworthiness of the writer from his/her written text. Once individuals speak and write about their experience, attitudes, and feeling authentically, they become more friendly, modest, and vulnerable [32]. We have applied the authenticity feature which is one of the LIWC outputs dimensions for computing and analyzing a score of truthfulness for each review text that is given by the reviewer. Furthermore, it provides useful ideas to discriminate between fake and truthful reviews. LIWC is Linguistic Inquiry and Word Count, It is text exploration computerized software mostly implemented in the field of natural language processing. However, it has 90 output dimensions. We have selected an authenticity dimension of LIWC because it is appropriate to explore and identify the deception markers in the text. The equation for calculation of authenticity is given below.

$$A(s) = \sum \text{FPS} + \text{TPS} + \text{TPP} + \text{exclusive} - \text{Negemo} - \text{Motion} \tag{14.1}$$

where $A(s)$ represents an authenticity score of a review text that is given by reviewer (person, customer). FPS, TPS, and TPP denote a frequency and overall percentage of First Person Singular, Third Person Singular, and Third Person Plural pronouns as well as percentage of negative emotion, exclusive, differ, and motion words in the text review. This feature has value in a range from 1 to 100.

Newman et al. [32] have used LIWC dictionary for predicting lying words and deception from linguistic styles present in the text's contents. They discovered that the truthful content is described by the use of fewer self-references (pronouns), fewer exclusive words (e.g., except, but, without), fewer negative words, and fewer motion verbs, whereas the fake content is characterized by the utilization of more self-references, more negative sentiment words (e.g., hate, anger, enemy), more exclusive words, and more motion verbs (walk, move, go).

B. **Verified purchasing**

This feature is a meta-feature of dataset and indicates whether a person has purchased the product or not from the targeted website. It has two parameters which are TRUE (the product has indeed been bought by the customer and his/her review is verified) and FALSE (the product has not been purchased by the customer and his/her review has not been verified).

C. **Sentiment deviation**

Sentiment analysis is extensively applied to analyze the individuals' material like reviews and survey answers, online and social media data, and healthcare resources, for statements that vary from marketing to customer service to clinical medicine. Sentiment analysis is also called opinion mining, a process of calculating a sentiment score of a written text. It is a subdivision of text categorization that is a type of research applied to examine the feeling of customers, filtering the opinions of customers toward products and exploring the thoughts of consumers and their emotions toward specific entities along with their features conveyed in the given text [22]. In this section, we calculate the sentiment scores for review text and its title using the equation that is presented as follows:

$$S = \frac{P(ws) - N(ws)}{P(ws) + N(ws)} \tag{14.2}$$

where S is sentiment score of the review text, $P(ws)$ denotes positive words score, and $N\ (ws)$ represents negative words score. With the help of the above equation, we can calculate the polarity score of the reviewed text and its title and establish whether it is positive or negative or neutral. As a reviewer always writes a review when buying a product, he/she also gives for his/her review a title. Sentiment deviation can be defined as the process of a sentiment score of the reviewed text alongside its title not matching, meaning that the review text has positive polarity and its title has a negative one. In other words, this review or opinion that is written by the reviewer is not truthful and expresses a fake assessment of the product.

D. **Rating value**

This is also one of the meta-features of the dataset used. It is an evaluation value given by the customer or shopper while purchasing the product from a specific e-commerce website. The satisfaction of the consumer for the online product or service is represented with this value which has a range from 1 (worst rating) to 5 (the best rating).

14.3.3.1 Labeling Using Rules-Based Method

In this section, we label the reviews of dataset based on mentioned above features. The approach used for labeling process is the rule-based approach. Here we have set of features with their values, each feature should have parameters, and threshold values will be used as a measurement to distinguish between fake reviews and truthful ones. In the applied method, groups of rules that are based on the linguistic and behavioral features such as authenticity, verified purchasing percentage, sentiment deviation, and rating value could be created. Furthermore, one threshold is assigned for authenticity feature, and for rest features, different parameters are identified. For each review, an authenticity is calculated and associated with the distinct threshold value, which is $X = 50$. On the foundation of these assessments beside the sentiment deviation, rating value, and verified purchasing, the review is classified as a fake review or as a truthful one. Equation (14.3) displays how the review is labeled as fake or a truthful.

$$R_L = \begin{cases} \text{Review}_{\text{Fake}}, & \text{if AU} < X \text{ and VP} = \text{False and} \\ & \text{SD} = \text{True and RV} = 1 \text{ or } 5 \\ \text{Review}_{\text{Truthful}}, & \text{if AU} \geq X \text{ and VP} = \text{True and} \\ & \text{SD} = \text{false and RV} = 2 \text{ or } 3 \text{ or } 4 \end{cases} \quad (14.3)$$

where

R_L: Review label
AU: Authenticity of review text
VP: Verified purchasing
SD: Sentiment deviation
R: Rating value.

14.3.4 Preprocessing

Before applying the representation and transformation methods on the dataset, the preprocessing step is implemented for data cleaning and removing of noise. The key point of text preprocessing is to convert and express the review text in an effective form, to which text analysis methods can be applied. However, the dataset should be subjected to certain cleansing steps such as the following:

- Removal of punctuation, which is a process of eliminating punctuation marks from the review's text like, ? !:;,".
- Stop words removal, which refers to cleaning certain words from the dataset; for example, "the", "a", "an", "in" are removed from text.
- Stripping unnecessary words and characters from the whole data.
- Tokenization: It is the process of splitting each sentence of review text into individual elements such as words, keywords, phrases, and pieces of information.
- Padding sequences: We implement deep learning neural network as one classification method for identifying the fake and truthful reviews, so that all sequences of reviews texts have to possess equal real-valued vectors. This has been completed using post-padding sequence method.

14.3.5 TF-IDF and Word to Vector Methods

This section presents two transformation methods which are used to map each word of review text into machine-readable form. Particularly, these methods are adopted to transform training and testing data into numerical form in order to be fed to the machine learning algorithm. In this research work, hybrid deep learning neural network and supervised machine learning techniques are implemented for analyzing and classification of the fake reviews and truthful ones; therefore, two dissimilar word representation approaches are utilized like TF-IDF and word to vector.

14.3.5.1 TF-IDF Method

The full form of TF-IDF is term frequency-inverse document frequency. It is a word representation technique, which is widely used in natural language processing domain for transforming a list of textual documents to a matrix conformation. Every single document is then transformed to a row of the TF-IDF vector, and each word can be represented in a real-valued vector [2]. Furthermore, unique words present in a document can be observed as the number of nonzero values in the vector, and similar words, as it is also measured as a feature extraction method that can be applied in the text mining systems. TF-IDF is a numerical method adopted to measure how significant a term or word is to a single document. There are two parts contained in TF-IDF that are term frequency; this is utilized to compute a recurrence of specified words in the document to determine the similarity between documents. The equation for TF is given below.

$$\mathrm{TF}(w)_d = \frac{n_w(d)}{|d|} \tag{14.4}$$

Set D point to a set of documents and d is a document where $d \in D$. A document can be defined as a group of sentences which are contained sets of words w. Put n_w (d) to symbolize the numbers of recurrent words w obtainable in document d. Consequently, the volume of document d can be expressed in an Eq. (14.5) as follows.

$$|d| = \sum_{w \in d} n_w(d) \tag{14.5}$$

The frequency with which a word is observed in a single document is computed using Eq. (14.5). The second part of TF-IDF is inverse document frequency (IDF), which is always adopted to calculate the total number of documents in the corpus divided by the number of documents in which that specified word has appeared. The equation for calculating the IDF is given below.

$$\mathrm{IDF}(w)_d = 1 + \log\left(\frac{|D|}{|\{d : D|w \in d\}|}\right) \tag{14.6}$$

Hence, figuring the TF-IDF for word w associated with document d and corpus D can be achieved by the following equation.

$$\text{TF.IDF} = \text{TF}(w)_d \times \text{IDF}(w)_D \tag{14.7}$$

Generally, The TF-IDF method helps the classifier-based model determine which keywords are more or less frequent in the document.

14.3.5.2 Word2Vect Method

The Word2Vect is word embeddings technique [5], and it has capability of capturing and understanding semantics, meaning, and context in vectors representations of text data [27]. This technique was developed by Mikolov et al. [20] in 2013 for word mapping comprising the implication and context of words in a document. There are two types of methods of Word2Vec that are continuous, bag-of-word (CBOW), and skip-gram. The similarity among words is computed through a cosine similarity of word vectors. However, it is used to map each word of review text into n-dimensional vector. Specifically, we utilized this method while applying hybrid deep learning neural network for classification review text into fake or truthful review. For that purpose, the word to vector method is adopted to convert the words of review text into n-dimensional vectors of numerical values called word embeddings.

14.3.6 Classification Techniques

After conducting preprocess steps on whole dataset, the next step is the classification of product reviews. For this purpose, we have implemented two different supervised learning techniques that are Random Forest (RF) and Convolutional Neural Network combined with Bidirectional Long Short-Term Memory (CNN-BiLSTM).

14.3.6.1 Fake Review Detection-Based Random Forest Model

Random Forest (RF) is one of the broadly applied techniques in supervised machine learning applications [4, 6, 23]. RF, as the name suggests, contains a forest of trees. While evaluating textual data using RF, numerous decision trees can aid in constructing best decision for sample classification. After performing preprocess steps on the reviews of the dataset and gained TF-IDF features, Random Forest classifier is applied to classify products reviews into fake and truthful reviews. Each tree in the Random Forest algorithm works in the same policy of decision tree algorithm. Taking a decision is based on polls of a minor decision tree that will decide a class label by mainstream votes. RF is named the “divide and conquer method”, and hence, it utilizes a small number of the weak learners to make strong learner. Each single tree in the RF classifier has a root node made of N data points or samples. Each node t in the tree also comprises N_t dataset features and positioned a split S_t for generating two sub-nodes that are t_L (left node) and t_R (right node). For computing and deciding the best split of the dataset features which have the highest information, an impurity measure is calculated using Gini index function that is given below in Eq. (14.8)

$$\text{Gini} = 1 - \sum_{i=1}^{C} (P_i)^2 \tag{14.8}$$

where P_i is the likelihood of occurrence of the data features existing and perceived in the dataset and C is the denoted number of classes in the dataset.

14.3.6.2 Fake Reviews Detection-Based CNN-BiLSTM Model

This proposed method applies and assists in an evaluation of hybrid convolution neural network combined with Bidirectional Long Short-Term Memory (CNN-BiLSTM) to distinguish and classify the review text containing content with fake linguistic indications. In order to achieve robust performance, the deep learning-based hybrid neural network model is implemented for detecting fake reviews using Amazon reviews dataset. Figure 14.2 demonstrates the construction of the CNN-BiLSTM model.

As shown in Fig. 14.2, the structure of CNN-BiLSTM consists of different layers such as embedding layer, convolutional layer, bidirectional LSTM layers, and sigmoid layer.

- Embedding Layer

The review text is comprised of sentences that contain a list of words and are labeled as $X_1, X_2, X_3, \ldots, X_t$ as cited in Fig. 14.2, and every single word is assigned a specific index integer number. The function of this layer is to create word embeddings for each word of review content included in the used dataset. However, Word2Vec, as one of a word embeddings techniques, is implemented to map or transform each word in training and testing data into real-valued vector. On the other hand, a group of words are created as a features set of the dataset and converted into numerical form. This task is termed word embeddings. In this model, an embedding layer is structured through three modules: maximum features, embedding dimension, and input sequence length. Maximum features are defined as the most recurrent words present in review text. Embedding dimension determines the dimensions vector of each word of the review text. Finally, the input sequence length is the extreme length of the review text. Further, word embeddings as a matrix of sequences are input to the succeeding layer. Equation (14.9) shows embedding matrix.

$$E(w) = R^{V \times D} \tag{14.9}$$

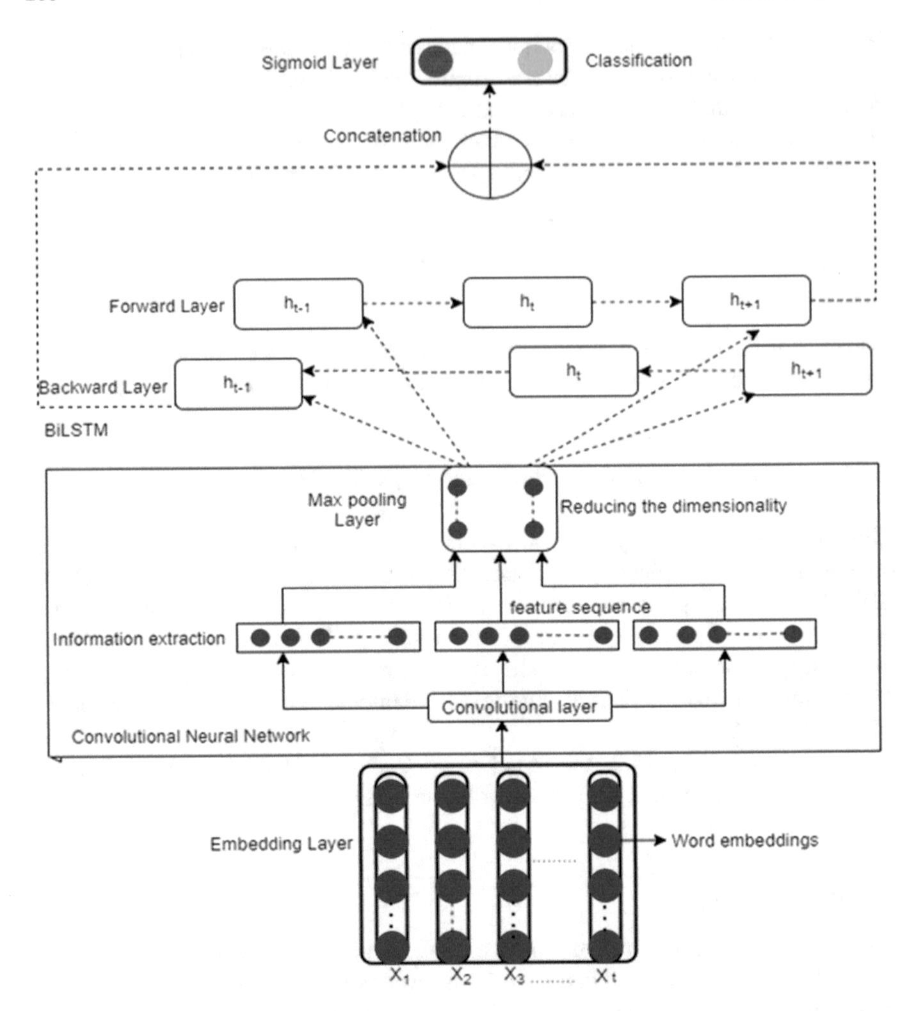

Fig. 14.2 Construction of CNN-BiLSTM model

where $E(w)$ is output embedding matrix, R indicates real number system, V denotes the vocabulary size (maximum features), and D points to the dimension of word embeddings vector.

- Convolution Layer

Convolution layer is a main layer of Convolutional Neural Network technique, conducting mathematical operation on input embeddings matrix delivered by the preceding layer. In order to obtain a sequence information and decrease the dimensions of the input sequence, the convolutional layer passes over input matrix using filters. The convolution process in this layer is performed in one dimension. We set 64 filters having windows size of 3 to operate on word-based representations to extract set of local features through sequences of words. A general equation for convolutional operation can be introduced below.

$$t_{i,j} = \sum_{l=1}^{n}\sum_{w=1}^{m} f_{l,w} \otimes P_i + l - 1, \quad j + w - 1 \tag{14.10}$$

where $\otimes$ symbolizes element-wise cross production, $t_{i,j} \in R^{l \times d}$ is demonstrating t elements of output matrix, $f_{l,w} \in R^{n \times m}$ represents the elements of the weight matrix, and $P_i + l - 1, \ j + w - 1 \in R^{lxw}$ is exemplified pth elements of the input matrix.

- Max Pooling Layer

This layer receives an input sequence from convolutional layer and executes down sampling process as well as spatial dimensionality reduction for the given sequences. It is utilized to choose a maximum value of features sequences from the pool of each filter kernel.

- Bidirectional LSTM Layers

Bidirectional LSTM networks join two hidden layers of dissimilar directions to the same output. Through this practice of reproductive deep learning, the production layer in a network is capable of obtaining sequences knowledge from historical and upcoming states instantaneously. Memory cells in the LSTM layer can eventually transfer results from historical data features into the output. Moreover, the process of features learning place in one direction, that means in a forward direction only; this ignores the backward construction and thus reduces the performance of machine learning system. For solving this weakness, processing of data in bidirectional recurrent network technique is accomplished in two directions: forward and backward. Each LSTM memory has four gates which are input i_t, forget f_t, cell state c_t, and output gate o_t. The equations of these gates are presented as follows [14].

$$i_t = \sigma(W_{ix}x_t + W_{ih}h_{t-1} + b_i) \tag{14.11}$$

$$f_t = \sigma\left(W_{fx}x_t + W_{fh}h_{t-1} + b_f\right) \tag{14.12}$$

$$o_t = \sigma(W_{ox}x_t + W_{oh}h_{t-1} + b_o) \tag{14.13}$$

$$c_t = f_t c_{t-1} + i_t * \tanh(W_{cx}x_t + W_{ch}h_{t-1} + b_c) \tag{14.14}$$

$$\overrightarrow{h_t} = o_t * \tanh(c_t) \tag{14.15}$$

$$\overleftarrow{h_t} = o_t * \tanh(c_t) \tag{14.16}$$

$$\tanh(x) = \frac{1 - e^{2x}}{1 - e^{2x}} \tag{14.17}$$

$$H_t = \left(\overrightarrow{h_t} : \overleftarrow{h_t}\right) \tag{14.18}$$

where tanh and sig are tangent and sigmoid activation functions, respectively. x is the input sequences. W and b indicate weight and bias factors. C_t is cell state, h_t refers to the output of the LSTM cell, and H_t is the output of bidirectional concatenation of $\overrightarrow{h_t}$ forward and $\overleftarrow{h_t}$ backward LSTM layers at the current time t.

- Sigmoid activation layer: This is an activation function utilized for binary classification [13]. In this proposed CNN-BiLSTM model, the sigmoid layer, known as output layer, has only one neuron or node for carrying out the classification of an input data into fake or truthful. The equation of this function is given below.

$$\sigma = \frac{1}{1 - e^{2x}} \tag{14.19}$$

14.3.7 Evaluation Metrics

To measure the performance of RF and CNN-BiLSTM models applied for discriminating between fake and truthful reviews, we have employed diverse assessment metrics for estimating the proposed models with a number of false-positive as well as false-negative samples. Evaluation measurements such as precision, recall, specificity, accuracy, and $F1$-score can be calculated from

$$\text{Accuracy} = \frac{\text{TP} + \text{TN}}{\text{FP} + \text{FN} + \text{TP} + \text{TN}} \times 100 \tag{14.20}$$

$$\text{Precision} = \frac{\text{TP}}{\text{TP} + \text{FP}} \times 100 \tag{14.21}$$

$$\text{Sensitivity} = \frac{\text{TP}}{\text{TP} + \text{FN}} \times 100 \tag{14.22}$$

$$\text{specificity} = \frac{\text{TN}}{\text{TN} + \text{FP}} \times 100 \tag{14.23}$$

$$F1\text{-score} = 2 * \frac{\text{precision} \times \text{Sensitivity}}{\text{precision} + \text{Sensitivity}} \times 100 \tag{14.24}$$

two confusion metrics depicted in Figs. 14.3 and 14.4. These measurements metrics have the following equations.

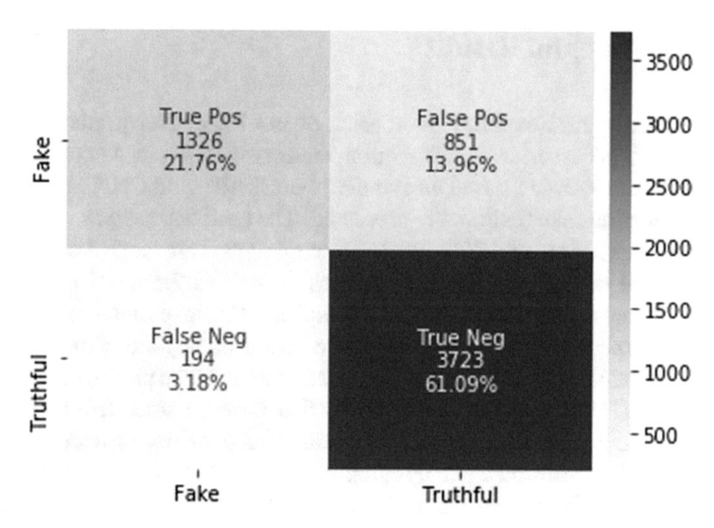

Fig. 14.3 Confusion matrix for RF

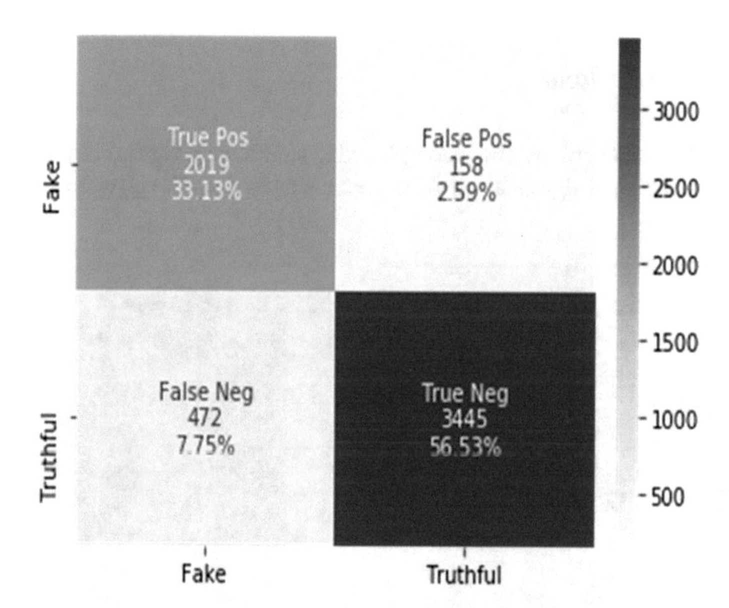

Fig. 14.4 Confusion matrix for CNN-BiLSTM

Where, true positive (TP) denotes the total numbers of the reviews texts that are effectively classified as fake reviews. FP is the total number of the reviews texts, which are imperfectly classified as truthful reviews. TN denotes the total number of reviews texts that are correctly classified as truthful reviews. FN is the total number of samples that are incorrectly classified as Fake reviews.

14.4 Experimental Results

The dataset used in these experiments consists of 30,470 Amazon product reviews for phones and their accessories. In this section, the classification results of the machine learning and deep learning based on Random Forest (RF) and CNN-BiLSTM algorithms for detecting fake reviews are presented. The used dataset was split into 70% training, 10% validation, and 20% testing. Experiments were carried out to evaluate the performance of RF with TF-IDF features as well as the CNN-BiLSTM with Word2Vec features. Dependent on these samples rates, assessment metrics which are sensitivity, specificity, precision, F1-score, and accuracy are computed to evaluate the RF and CNN-BiLSTM models for predicting the fake product reviews using a testing dataset. With comparing the results that were obtained from these experiments, it was observed that the deep learning-based model provided promising results and outperformed machine learning-based models. Figure 14.5 displays and visualizes the classification results for RF and CNN-BiLSTM models. The accuracy performance and loss of CNN-BiLSTM model are presented in Fig. 14.6.

14.4.1 Word Cloud

Word cloud is a technology always employed in natural language processing domain to visualize the most significant and frequent used words in the given text. Here, we

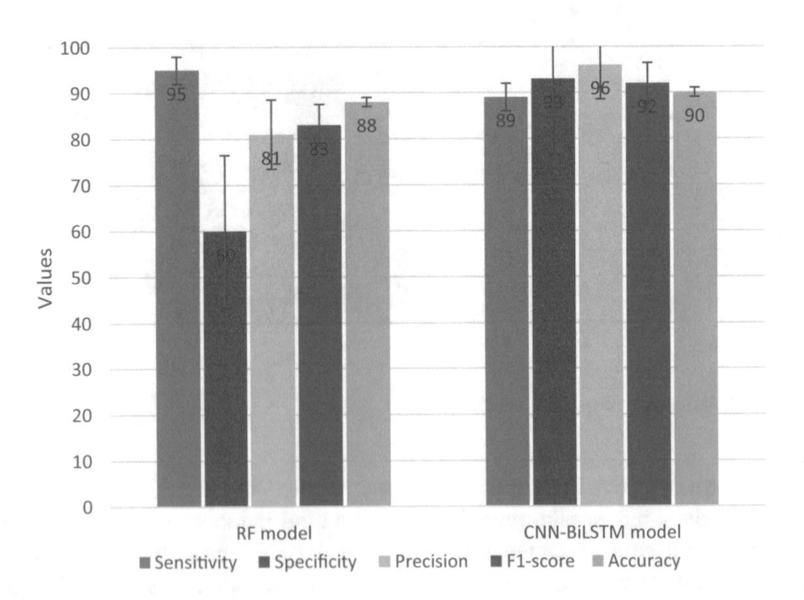

Fig. 14.5 Visualization of the classification results for RF and CNN-BiLSTM models

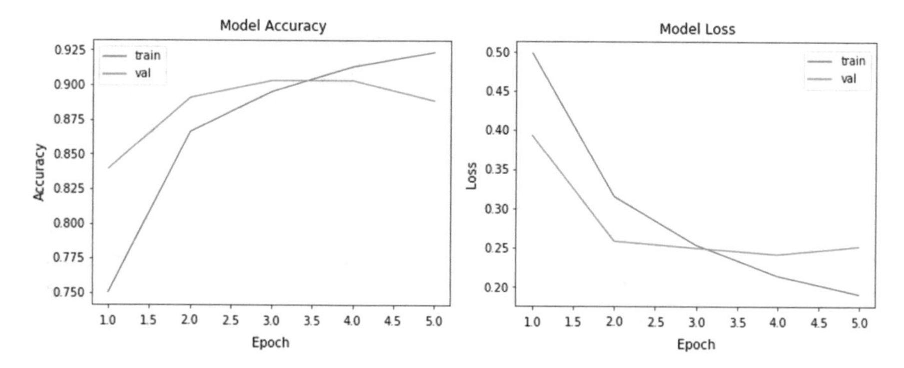

Fig. 14.6 Performance of the CNN-BiLSTM model

Fig. 14.7 Wordcloud for product reviews of the dataset

have used the wordcloud for visualizing the repeated words in products reviews of the used dataset. Figure 14.7 shows wordcloud.

14.5 Conclusions and Future Research

Recently, the problem of fake reviews has not been tackled because there are no really clear hints, indications which assist in detecting such reviews in online e-commerce websites. These reviews have an influence on both customer's decision purchasing and e-businesses revenues. However, an identification of fake product reviews has attracted the attention of academic research, and companies selling products online, and encouraged them to develop applications that are capable of identifying fake reviews. The present research work attempts to construct distinctive methodology for finding solution for fake opinions contained in the Amazon product reviews. We model semantic exemplification of reviews by combining words embeddings into sentences level representation learning. This has been achieved using hybrid neural networks incorporated with Word2Vec word embedding technique that was used to map each word of review text into real-valued vector. The findings of this research work can be summarized in four points. Due to the sparsity of words in the review text, an outstanding way of analyzing and extracting the features for fake reviews is a Linguistic Inquiry and Word Count (LIWC) dictionary. Furthermore, the best choice for labeling the reviews as fake and truthful is by using rule-based method. A Word2Vec method can find the semantic between vectors representations of each word of review, and thus, it outperforms TF-IDF method, which can only convert each word to one-dimensional vector representations, and the similarity between words is lost. Deep learning-based model provides better performance than the machine learning-based model for fake reviews detection, particularly when a dataset is large. This research has limitations such that the dataset used in the experiments is limited to English language reviews and to Amazon based e-commerce company. For possible future work, we will attempt to extract new features for fake reviews from online product reviews to know from where the reviews have been posted. Another possible area of research is to developing multi-language deep learning model for detecting Arabic and English spam reviews.

References

1. Appseconnect: https://www.appseconnect.com/importance-of-product-reviews-in-ecommerce (2017). Accessed date 8 Sept 2021
2. Ahmed, H., Traore, I., Saad, S.: Detecting opinion spams and fake news using text classification. Secur. Priv. **1**(1), 9 (2018)
3. Algur, S.P., Ayachit, N.H., Biradar, J.G.: Exponential distribution model for review spam detection. Int. J. Adv. Res. Comput. Sci. 938–947
4. Alsubari, S.N., Shelke, M.B., Deshmukh, S.N.: Fake reviews identification based on deep computational linguistic. Int. J. Adv. Sci. Technol. **29**, 3846–3856 (2020)
5. Alsubari, S.N., Deshmukh, S.N., Al-Adhaileh, M.H., Alsaade, F.W., Aldhyani, T.H.: Development of integrated neural network model for identification of fake reviews in E-commerce using multidomain datasets. Appl. Bionics Biomech. (2021)
6. Alsubari, S. N., Deshmukh, S. N., Alqarni, A.A, Alsharif, N.H.T, Aldhyani T.H, Alsaade, F.W, Khalaf, O.I: Data analytics for the identification of fake reviews using supervised learning. Comput, Mater. Contin. **70** (2), 3189–3204 (2022)
7. Anderson, E.W.: Customer satisfaction and word of mouth. J. Serv. Res. **1**(1), 5–17 (1998)
8. Banerjee, S., Chua, A.Y.: A study of manipulative and authentic negative reviews. In: Proceedings of the 8th International Conference on Ubiquitous Information Management and Communication, p. 76 (2014)
9. Chen, L., Jiang, T., Li, W., Geng, S., Hussain, S.: Who should pay for online reviews? Design of an online user feedback mechanism. Electron. Commer. Res. Appl. **23**, 38–44 (2017)
10. Chen, Y., Xie, J.: Online consumer review: word-of-mouth as a new element of marketing communication mix. Manag. Sci. **54**, 477–491 (2008)
11. Chen, Y.R., Chen, H.H.: Opinion spam detection in web forum: a real case study. In: Proceedings of the 24th International Conference on World Wide Web, International World Wide Web Conferences Steering Committee (2014)
12. Floyd, K., Freling, R., Alhoqail, S., Cho, H.Y., Freling, T.: How online product reviews affect retail sales: a meta-analysis. J. Retail. **90**(2), 217–232 (2014)

13. Hajek, P., Barushka, A., Munk, M.: Fake consumer review detection using deep neural networks integrating word embeddings and emotion mining. Neural Comput. Appl. **32**(23), 17259–17274 (2020)
14. https://msdn.microsoft.com/en-us/magazine/mt846470.aspx (2018). Accessed
15. Hu, N., Bose, I., Gao, Y., Liu, L.: Manipulation in digital word-of-mouth: a reality check for book reviews. Decis. Support Syst. **50**(3), 627–635 (2011)
16. Jindal, N., Liu, B.: Review spam detection. In: Proceedings of the 16th International Conference on World Wide Web, pp. 1189–1190 (2007)
17. Kamerer, D.: Understanding the Yelp review filter: an exploratory study. First Monday (2014)
18. Kim, S., Chang, H., Lee, S., Yu, M., Kang, J.: Deep semantic frame-based deceptive opinion spam analysis. In: Proceedings of the 24th ACM International on Conference on Information and Knowledge Management, pp. 1131–1140 (2015)
19. Kouzis-Loukas, D.: Learning scappy. Packet Publishing Ltd. England (2016)
20. Lai, S., Liu, K., He, S., Zhao, J.: How to generate a good word embedding. IEEE Intell. Syst. **31**(6), 5–14 (2016)
21. Lappas, T.: Fake reviews: the malicious perspective. In: Proceedings of the International Conference on Applications of Natural Language Processing and Information Systems, pp. 23–34 (2012)
22. Liu, B.: Sentiment analysis: mining opinions, sentiments, and emotions. Cambridge University Press, Cambridge, England (2015)
23. Louppe, G., Wehenkel, L., Sutera, A., Geurts, P.: Understanding variable importances in forests of randomized trees. Adv. Neural. Inf. Process. Syst. **26**, 431–439 (2013)
24. Luca, M., Zervas, G.: Fake it till you make it: reputation, competition, and Yelp review fraud. Manage. Sci. **62**(12), 3412–3427 (2016)
25. Mayzlin, D.: Promotional chat on the Internet. Mark. Sci. **25**(2), 155–163 (2006)
26. Mcauley, J., Pandey, R., Leskovec, J.: Inferring networks of substitutable and complementary products. In: Proceedings of the 21th ACM SIGKDD International Conference on Knowledge Discovery and Data Mining, pp. 785–794 (2015)
27. Mikolov, T., Chen, K., Corrado, G. and Dean, J.: Efficient estimation of word representations in vector space. Proceedings of Workshop at ICLR. arXiv:1301.3781v1 (2013)
28. Mukherjee, A., Liu, B., Glance, N.: Spotting fake reviewer groups in consumer reviews. In: WWW 2012, pp. 191–200. ACM (2012)
29. Mukherjee, A., Kumar, A., Liu, B., Wang, J., Hsu, M., Castellanos, M., Ghosh, R.: Spotting opinion spammers using behavioral footprints. In: Proceedings of the 19th ACM SIGKDD International Conference on Knowledge Discovery and Data Mining, pp. 632–640 (2013)
30. Mukherjee, A., Venkataraman, V., Liu, B., Glance, N.: What Yelp fake review filter might be doing? In: Proceedings of the Seventh International AAAI Conference on Weblogs and Social Media (2013)
31. Nam, S., Manchanda, P., Chintagunta, P.K.: The effect of signal quality and contiguous word of mouth on customer acquisition for a video-on-demand service. Mark. Sci. **29**(4), 690–700 (2010)
32. Newman, M. L., Pennebaker, J. W., Berry, D. S., Richards, J. M.: Lying words: Predicting deception from linguistic styles. Pers Soc Psychol Bull. **29**(5), 665–675 (2003)
33. Ott, M., Choi, Y., Cardie, C., Hancock, J.T.: Finding deceptive opinion spam by any stretch of the imagination. In: Proceedings of the 49th Annual Meeting of the Association for Computational Linguistics: Human Language Technologies, vol. 1, pp. 309–319 (2011)
34. Öğüt, H., Taş, B.K.O.: The influence of internet customer reviews on the online sales and prices in hotel industry. Serv. Ind. J. **32**(2), 197–214 (2012)
35. Shasha, D., Pengzhu, Z., Xiaoyan, Z., Xinmiao, L.: Deception detection based on fake linguistic cues. J. Syst. Manag. **23**, 263–270 (2014)
36. Smith, D.A.: Amazon Reviewers Brought to Book (2004). Available online: https://www.theguardian.com/technology/2004/feb/15/books.booksnews
37. Wang, Y., Lu, X., Tan, Y.: Impact of product attributes on customer satisfaction: an analysis of online reviews for washing machines. Electron. Commer. Res. Appl. **29**, 1–11 (2018)

38. Xu, C.: Detecting collusive spammers in online review communities. In: Proceedings of the Sixth Workshop on Ph.D. Students in Information and Knowledge Management, pp. 33–40 (2013)
39. Xu, H., Liu, D., Wang, H., Stavrou, A.: E-commerce reputation manipulation: the emergence of reputation-escalation-as-a-service. In: Proceedings of the 24th International Conference on World Wide Web, pp. 1296–1306 (2015)
40. Zhang, D., Zhou, L., Kehoe, J.L., Kilic, I.Y.: What online reviewer behaviors really matter? Effects of verbal and nonverbal behaviors on detection of fake online reviews. J. Manag. Inf. Syst. **33**(2), 45 (2016)

Printed in the United States
by Baker & Taylor Publisher Services